Hydrocarbon Processing and Refining

This book covers petroleum refining and gas purification processes, including refinery configurations comprising of relevant units *with special emphasis on processing of heavy crudes with high acid number.* It includes a short review of distillation principles, distillation column auxiliaries, critical column pressure control strategies, critical issues of crude and vacuum distillation units particularly for heavy crude processing. Different corrosion mechanisms and their prevention with regards to heavy high TAN crude processing are also included. Fundamentals are explained with support of steady-state simulation and presented with simulation flowsheets and output, supported by examples of calculations and troubleshooting case studies.

Features:

- Deals with principles and practices in the hydrocarbon industry and petroleum refinery with *emphasis on heavy crude processing*
- Focuses on operation and practices of the major process units *with simulation examples* and aimed at the professional engineer
- Covers acid gas treatment in view of increased emphasis on carbon capture and storage, and introduction of residue gasification processes
- Elucidates methodologies for safety relief load computation for distillation columns
- Explains real-life problems in reboilers, column internals, column pressure controls and corrosion in crude, and vacuum distillation and secondary units with several case studies

This book is aimed at professionals in petroleum engineering and graduate students in chemical engineering.

Hydrocarbon Processing and Refining

Principles and Practices

Ashis Nag

CRC Press
Taylor & Francis Group
Boca Raton London New York

CRC Press is an imprint of the
Taylor & Francis Group, an **informa** business

First edition published 2023
by CRC Press
6000 Broken Sound Parkway NW, Suite 300, Boca Raton, FL 33487-2742

and by CRC Press
4 Park Square, Milton Park, Abingdon, Oxon, OX14 4RN

CRC Press is an imprint of Taylor & Francis Group, LLC

ISBN: 9781032214030 (hbk)
ISBN: 9781032214054 (pbk)
ISBN: 9781003268246 (ebk)

DOI: 10.1201/9781003268246

Typeset in Times New Roman
by Deanta Global Publishing Services, Chennai, India

Contents

Preface..xv
About the Author xvii

Chapter 1   Overview of Refining and Distillation 1

    1.1   Refining: An Overview...1
    1.2   Distillation: A Review...6
        1.2.1   Review of Column Internals 8
        1.2.2   Typical Dimensions and other parameters
              of Internals...8
        1.2.3   Dumping...11
    1.3   Column Pressure Control and Reboilers 13
        1.3.1   Pressure Control by Submergence Variation.............. 14
              1.3.1.1   Pressure Balance around Column
                      Overhead, Condenser, and Reflux Drum.... 17
        1.3.2   Pressure Control by Hot Vapor Bypass...................... 18
        1.3.3   Pressure Control with Vapor Line Restrictions..........20
        1.3.4   Case Study on Pressure Control............................20
        1.3.5   Reboilers...22
        1.3.6   Once-Through Reboiler.....................................22
        1.3.7   Recirculation-Type Reboilers 23
        1.3.8   Critical Flux 25
        1.3.9   Problems in Reboilers 25
              1.3.9.1   Case Study 1 25
              1.3.9.2   Case Study 2 25
              1.3.9.3   Case Study 3: Problem of Film Boiling
                      in a Naphtha Vaporizer..........................27
              1.3.9.4   Case Study 4: Waterfall Pool Effect in
                      a Reboiler of a Sour Water Stripper...........27
    1.4   Processing of Heavy Crudes...28
        1.4.1   Introduction 28
        1.4.2   Crude Oil Upgraders 29
        1.4.3   General Approach for Induction of Heavy Crudes.......31
        1.4.4   Processing Issues in Crude Vacuum Units.................32
        1.4.5   Design/Rating of an AVU 32
              1.4.5.1   Rating of an Existing Unit.........................32
              1.4.5.2   Considerations for New Design of a
                      Crude–Vacuum Unit 38
              1.4.5.3   Features Needed in Crude Units for
                      Processing of Heavy Crudes......................40
              1.4.5.4   Features Needed in Vacuum
                      Distillation Units for Heavy Crudes 42

1.4.6   Processing Issues in Coker Units for Residues
        Derived from Heavy Crudes 50
1.4.7   Processing Issues in VGO Hydrotreater/
        Hydrocracker Units for Feed Derived from
        Heavy Crudes 50
1.4.8   Processing Issues in FCCU for Feed Derived
        from Heavy Crudes 51
1.4.9   Effect on Other Units 52
1.5   Corrosion in Crude Vacuum Units with Heavy Crudes 52
1.5.1   Acid Corrosion 52
1.5.2   Caustic Injection...55
1.5.3   Chemistry of Caustic Reaction................................56
1.5.4   Tramp Amines...59
1.5.5   Use of Filmer...62
1.6   Processing of High-TAN Crudes...62
1.6.1   Introduction 62
1.6.2   Methodologies Followed 63
1.6.3   Concerns for Use of Corrosion Inhibitors for NA......63
1.6.4   Mechanism of Naphthenic Acid Corrosion..............64
1.6.5   Influence of Temperature on Naphthenic Acid
        Corrosion..65
1.6.6   Factors Influencing High-Temperature
        Naphthenic Acid Corrosion (Flow Velocity and
        Flow Regime) 65
1.7   Processing of Crude Oils with High Calcium 65
1.8   Case Studies in Distillation Units..66
1.8.1   Case Study 1...66
1.8.2   Case Study 2...67
1.8.3   Case Study 3...69
1.8.4   Case Study of Pumparound Problem of a Crude Unit ...70
1.8.5   Case Study of a Vacuum Unit 71
1.8.6   Case Study of Vacuum System Bottom Coking.........72
Bibliography...72
Abbreviations 73

Chapter 2   Fluid Catalytic Cracking Unit (FCCU)..75
2.1   Introduction 75
2.1.1   Process Description...75
2.2   Reactor–Regenerator Heat Balance..80
2.2.1   Description of R-R Section 80
2.2.2   Overall Heat Balance 81
2.2.3   Regenerator Heat Removal.......................................82
2.2.4   Review of Requirement of Catalyst Cooler..............84
2.3   Combustion Modes of Regenerators..88

2.4   Definition of Delta Coke..88
        2.4.1   Types of Delta Coke 88
2.5   Power Recovery from Regenerator Hot Flue Gas 89
2.6   Operating Variables..91
        2.6.1   Riser Temperature 91
        2.6.2   Recycle Rate 91
        2.6.3   Feed Preheat Temperature..92
        2.6.4   Fresh Feed Rate 92
        2.6.5   Catalyst Makeup Rate 93
        2.6.6   Gasoline End Point...93
2.7   Fluidization and Related Equipment in FCCU 93
        2.7.1   Fluidization 93
        2.7.2   Typical FCCU Major Equipment/Components.........93
        2.7.3   Reactor Arrangements...94
        2.7.4   Different Regenerator Configurations
                and Internals 95
2.8   Physical Properties of Catalysts Related to Fluidization.........96
2.9   Guidelines for Design of Cyclones...101
        2.9.1   Basics..101
        2.9.2   Accepted Design Limits for Cyclones......................102
2.10   Problems Related to Cyclones 103
        2.10.1 High Catalyst Losses...103
        2.10.2 Fine Generation 106
        2.10.3 Reactor Cyclone Coking..106
        2.10.4 Poor Catalyst Circulation 106
        2.10.5 Poor Product Yields...107
        2.10.6 Conclusion on Cyclones: Reasons for Failures........107
        2.10.7 Remedies 107
2.11   Riser Separation System and Positive/Negative Pressure
        Cyclones 109
        2.11.1 Description of Riser Separation Systems.................109
2.12   Estimation of Transport Disengaging Height in
        Regenerators (source publications) 113
        2.12.1 Estimation of Bed Level in Regenerator 113
2.13   Design Guidelines for Cyclone Diplegs...................................115
        2.13.1 Dipleg Diameter 115
        2.13.2 Dipleg Termination...115
2.14   Design and Operation Guidelines for FCC Catalyst
        Standpipes..117
        2.14.1 Pressure Profiles in the FCC Standpipe 117
        2.14.2 Catalyst Behavior and Operating Range..................118
        2.14.3 Standpipe Design Guidelines 119
        2.14.4 Catalyst Flux 120
        2.14.5 Standpipe Aeration Practice......................................120
        2.14.6 Choice of Aeration Gas 122

2.14.7   Practice of Aeration in Regenerated Catalyst
Standpipe (RCSP)..122
2.14.8   Practice of Aeration in Spent Catalyst Standpipe
(SCSP) 122
2.15   Reactions in FCCU and Catalysts Features and Additives 122
2.15.1   Reaction Mechanism...122
2.15.2   Additives in Catalyst 124
2.16   Corrosion in FCCU and Its Prevention/Mitigation 125
2.16.1   Prevention of Ammonia Chloride Deposition..........126
2.16.2   Calculation of NH4Cl Deposition Temperature
in the Main Fractionator..127
2.16.3   Other Types of Salt Deposition 128
2.16.4   Use of Water Wash 128
2.16.5   Use of Polysulfides 129
2.16.6   Use of Corrosion Inhibitors.....................................129
2.17   Simulation Guidelines 129
Bibliography..131
Abbreviations 132

Chapter 3   Coker 133

3.1   Introduction 133
3.2   Process Description...133
3.2.1   General Process Scheme 133
3.2.2   Feed Device..139
3.2.3   Wash Zone..140
3.3   Operating Variables..142
3.3.1   Coil Outlet Temperature..142
3.3.2   Recycle Rate 142
3.3.3   Drum Pressure 143
3.4   Activities in Decoking Operation....................................143
3.5   Effect of Feedstock...147
3.6   Coker Furnace and Factors Affecting Fast Furnace
Fouling and frequent Decoking Operation...........................148
3.6.1   Brief Description 148
3.6.2   The Coil Outlet Temperature 150
3.6.3   Metallurgy of Tubes 150
3.6.4   Reasons for Rapid Fouling and Additional Coke
Buildup in Furnace Tubes 151
3.6.4.1   Low Mass Velocity 151
3.6.4.2   Effect of Feed Quality on Rapid Fouling.....151
3.6.5   Factors That Can Act As Catalyst for Rapid
Fouling of Furnace 152
3.6.5.1   Feed Interruptions.....................................152
3.6.5.2   Feed Preheating 152

		3.6.5.3	Burners	153
		3.6.5.4	Means to Monitor Tube Skin Temperatures	153
	3.6.6		Gist of Actions to Improve Run Length (Suggested by Experts)	153
		3.6.6.1	Transfer Line Configuration	154
		3.6.6.2	Locations in a Transfer Line that can Cause Problems (According to Experts)	154
		3.6.6.3	Ways to Fix Problems with a Problematic Transfer Line	154
		3.6.6.4	Steam-Air Decoking	154
		3.6.6.5	Pigging or Mechanical Coke Removal	154
		3.6.6.6	Online Spalling	155
		3.6.6.7	Design and Operating Parameters: Firebox	155
	3.6.7	Decoking Operation		155
3.7	Coke Drum			158
	3.7.1	Effect of Cycle Time		158
	3.7.2	Some Salient Features in Coke Drum Design		158
	3.7.3	Coke Drum and Drum Deformities		159
	3.7.4	Some Key Points of the Coking Cycle and Its Effect on Coke Drums (Expert Opinion)		159
	3.7.5	Fast Quench Issues (As Explained by Experts)		159
	3.7.6	Gist of Findings of the Reasons for Drum Deformities (from the Experts)		160
		3.7.6.1	Summary by Experts	163
	3.7.7	The Coke Drum Banana Effect Syndrome		165
3.8	Coke Morphology and Effect on Operation			166
	3.8.1	Factors Influencing Coke Morphology		166
	3.8.2	Fouling in Coker		169
3.9	Coke Storage and Handling Safety			169
3.10	Case Studies			170
	3.10.1	Case Study 1		170
	3.10.2	Case Study 2		170
	3.10.3	Case Study 3		171
	3.10.4	Case Study 4		171
	3.10.5	Case Study 5		172
Bibliography				172

Chapter 4   Hydrotreating/Hydrocracking 175

4.1	Introduction of the Process and Basic Functions		175
	4.1.1	Hydrodesulfurization (HDS)	175
	4.1.2	Hydrodenitrogenation/-denitrification (HDN)	178
	4.1.3	Hydrodemetallization (HDM)	180

4.1.4   Aromatic Saturation (HDA) 180
4.1.5   Hydrocracking 180
4.2   Catalyst Functions and Mechanism of Deactivation 181
4.2.1   Catalyst Sulfiding 181
4.2.2   Deactivation of Hydroprocessing Catalyst 182
4.2.2.1   Coking or Fouling 182
4.2.2.2   Sintering 183
4.2.2.3   Mechanical Deactivation 183
4.2.2.4   Poisoning 183
4.2.3   Contaminants in Hydrotreater Feed 185
4.2.3.1   Silicon 185
4.2.3.2   Arsenic 185
4.2.3.3   Sodium and Calcium 185
4.2.3.4   Phosphorous 186
4.2.3.5   Iron 186
4.2.3.6   Nickel and Vanadium 186
4.3   Process Variables 186
4.3.1   Pressure 187
4.3.2   Temperature 188
4.3.3   Gas Recycle and H2/Oil Ratio 188
4.3.4   Space Velocity 189
4.4   Operating Concerns in Naphtha Hydrotreater Units 189
4.4.1   Increase in Pressure Drop across Reactors 190
4.4.2   Increase in Pressure Drop in the Reactor
        Effluent Circuit 192
4.5   Diesel Hydrotreaters and Common Operating Problems 192
4.5.1   Common Problems in Diesel Hydrotreaters 193
4.5.2   Color in Treated Diesel 194
4.6   Heavy Distillate Hydrotreaters and Common Operating
      Problems 194
4.6.1   Common Problems in Heavy Oil Hydrotreaters 195
4.6.1.1   Higher Filter Backwash Frequency 195
4.6.1.2   Higher Differential Pressure Drops
          across Reactors 195
4.6.1.3   Higher Differential Pressure in Reactor
          Effluent Circuit 195
4.7   Overview of Residue Hydrotreaters 198
4.8   Hydrocrackers and Various Configurations 199
4.9   Metallurgy of Hydrotreaters and Hydrocrackers 199
4.10   Case Studies 202
4.10.1   Effluent Circuit High-Pressure Drop 202
4.10.2   Corrosion in Diesel Hydrotreaters 204
4.10.3   Color in Ultra-Low High Speed Diesel (ULHSD) 206
4.10.4   Capacity Augmentation of a NHT 208
4.10.5   Capacity Augmentation of a Hydrocracker 209
4.10.6   Hydrotreated Wild Naphtha Processing 210

4.11   Improvement of Recycle Gas (RG) Purity 211
4.12   Essentials of Simulation 214
Bibliography 216
Abbreviations 216

Chapter 5   Acid Gas Treatment.. 217

5.1   Introduction 217
5.2   Options for Acid Gas Removal...................................... 218
      5.2.1   Processes 218
      5.2.2   Choice of Process 218
5.3   Common Physical Solvents for Acid Gas Removal.............. 221
      5.3.1   MeOH (Methanol).. 222
      5.3.2   DEPG (Dimethyl Ether of Polyethylene Glycol)...... 223
      5.3.3   NMP (N-Methyl-2-Pyrrolidone) 224
      5.3.4   PC (Propylene Carbonate)................................... 225
      5.3.5   Comparison of Physical Properties of Solvent
              and Gas Solubilities..................................... 226
      5.3.6   Physical Solvent Regeneration 228
      5.3.7   Simplified Flow scheme of few Physical solvent
              processes.. 228
5.4   Physical-Chemical Solvents and Processes..................... 228
5.5   Chemical Solvents and Processes................................ 231
      5.5.1   Conventional Amine-Based Solvents Used
              for acid gas removal and Carbon Capture and
              Storage (CCS).. 231
      5.5.2   Sterically Hindered Amine Solvents....................... 233
      5.5.3   Non-Amine-Based Solvents 234
      5.5.4   Chemical Reactions in the Absorber and
              Regenerator.. 235
5.6   Recent Trends and Advances....................................... 237
      5.6.1   Solvent Blends 237
      5.6.2   Chemistry of Reaction of Piperazine 238
      5.6.3   Introduction to Ionic Liquids............................ 239
5.7   Different Types of Membrane Processes........................... 240
      5.7.1   Spiral Wound Membrane 240
      5.7.2   Hollow Fiber Membrane 241
      5.7.3   One-Stage Membrane Process 241
      5.7.4   Two-Stage Membrane Process 242
      5.7.5   Membrane Pretreatment.................................... 242
      5.7.6   Membrane Life 244
5.8   Experiences of Amine Unit in Industries........................ 244
      5.8.1   Effect of Amine Strength on Reboiler Condenser
              Duties of Amine Regenerator.............................. 244
      5.8.2   Common Problems in Amine Units........................... 246
              5.8.2.1   Amine Degradation 246

5.8.2.2   Foaming 246
5.8.2.3   Corrosion 247
5.8.2.4   Fouling...248
5.8.2.5   Poor Sweet Product Quality 249
5.8.2.6   Carry Over of Amine to Reflux Drum
           of the Regenerator...............................249
5.8.3   Sources for Contamination....................................249
5.8.4   Actions to Arrest Contamination of Amine
         Systems..250
5.8.5   Actions for Lowering Contaminations....................251
5.9   Case Studies...253
5.9.1   Higher Hydrocarbon Content in Acid Gas from
         Amine Regenerator 253
5.9.2   Sweet Gas Specification Not Achievable 255
5.9.3   Lower H2S Recovery in Tail Gas Treatment Unit
         Amine Absorber...259
5.10   Carbon Capture 259
Bibliography...262

Chapter 6   Hydrogen Generation Units..265

6.1   Introduction 265
6.2   Description of a Steam Reformer 265
6.2.1   Process Description..266
6.3   Common Process Problems...270
6.3.1   Sulfur Slippage to Pre-Reformer/Reformer 270
6.3.2   Slippage of Higher Hydrocarbons from
         Pre-Reformer to Tubular Reformer 271
6.3.3   Operation with Lower Steam-to-Carbon Ratio........271
6.3.4   Overheating of Tubes of Tubular Reformer 272
6.3.5   Quality of Steam for Reaction.............................272
6.3.6   Slippage of Chlorides 272
6.4   Mechanism for Deactivation of Catalysts 272
6.4.1   Deactivation of HDS Catalyst.............................272
6.4.2   Deactivation of Zinc Oxide (ZnO) Catalyst...........273
6.4.3   Deactivation of Pre-Reformer Catalyst..................273
6.4.4   Deactivation of Reformer Catalyst.......................273
6.4.5   Deactivation of Shift Catalyst 274
6.5   Case Studies...275
6.5.1   Case Study 1...275
6.5.2   Case Study 2...277
6.5.3   Case Study 3...277
6.6   Hydrogen Recovery 282
6.7   Capacity Augmentation 282
Bibliography...282

Chapter 7   Sulfur Recovery Unit 283

    7.1   Introduction 283
    7.2   Process Description and Reaction Mechanism 283
          7.2.1   Claus Thermal Stage 285
          7.2.2   Claus Catalytic Stage 286
                  7.2.2.1   Reheat Methods 287
    7.3   Enhanced Sulfur Recovery 288
    7.4   Common Problems in Sulfur Units 291
    7.5   Capacity Augmentation 292
    Bibliography 293

Chapter 8   Plant Safety 295

    8.1   Introduction 295
    8.2   Methodologies of Relief Load Computation of Columns 295
          8.2.1   Unbalanced Heat Method (UBH Method) 296
          8.2.2   Steady-State Simulation Method 297
          8.2.3   Illustration/Examples for Computation by the
                  Methods 298
                  8.2.3.1   Computation by UBH Method 299
                  8.2.3.2   Computation by Steady-State Method 302
    8.3   Relief Load Calculation of Crude Distillation Unit 306
          8.3.1   Top Reflux Failure 306
          8.3.2   Pumparound Failure 306
          8.3.3   Site Power Failure 308
    8.4   Relief Load Calculation of Vacuum Column 310
          8.4.1   Ejector Steam Failure 311
          8.4.2   Cooling Water Failure 312
          8.4.3   Pumparound Failure 312
          8.4.4   Site Power Failure 313
    8.5   Summary of Contingencies and Accumulation of Pressure 315
          8.5.1   Overpressure 315
          8.5.2   Water into Hot Oil 315
          8.5.3   Accumulation of Pressure in Various Cases 315
    8.6   Safety Measures in High-Pressure and Low-Pressure
          Interconnections 316
    8.7   Location of Safety Valves and Special Features 317
          8.7.1   Location of Safety Valves in High-Pressure
                  Systems 317
          8.7.2   Location of Safety Valves in Columns 318
          8.7.3   Safety Valve Assembly in Refinery Columns 318
    8.8   Relation between Design Pressure and Hydrostatic Test
          Pressure 319
    Bibliography 319
    Abbreviations 320

Chapter 9   Challenges in Refining 321

   9.1 Lower Crude and Product Differential 321
     9.1.1 Actions Required by Refiners 321
   9.2 Stricter Product Specifications 322
     9.2.1 Revised Specifications 322
        9.2.1.1 Gasoline or Motor Spirit 323
        9.2.1.2 Diesel 326
     9.2.2 Actions Initiated by Refiners 326
     9.2.3 Lube Oil Base Stock 327
   9.3 Stricter Emission Norms 328
     9.3.1 Emission Guidelines 328
     9.3.2 Actions Necessary for Reduction of Air
        Emissions 329
     9.3.3 Actions to Reduce Generation of Liquid Effluent
        (Recycling and Reuse) 330
     9.3.4 Improvement in Facilities 331
   9.4 Limited Sales of Heavy Ends and Coke 331
     9.4.1 Actions Taken by Refiners 331
     9.4.2 Downstream Integration with Petrochemicals 332

Chapter 10 Renewable Energy 335

   10.1 Introduction 335
   10.2 Types of Renewable Energy 335
   10.3 Types of Biomass 336
   10.4 Conversion of Biomass to Energy and Byproducts 337
     10.4.1 Conversion Processes 338
   10.5 Forms of Biofuels and Benefits of Biofuels 338
     10.5.1 Biogas 338
     10.5.2 Gasohol (A Blend of Gasoline and Ethanol) 339
     10.5.3 Biodiesel 340
     10.5.4 Benefits of Biomass Fuel 341
   10.6 Biorefinery 341
   10.7 Hydrogen Fuel 343
     10.7.1 Introduction 343
     10.7.2 Renewable Hydrogen Production Pathways 343
     10.7.3 Relative Merits of the Technologies 344
     10.7.4 Hydrogen Storage and Transport 344
     10.7.5 Use of Hydrogen and Fuel Cells in Transport
        Sector 345
   Bibliography 346
   Abbreviations 346

Index 347

Preface

This book is primarily intended for students of chemical engineering and petroleum engineering, and for practicing engineers engaged in hydrocarbon processing and refining. I have found that the plant design and associated principles for efficient operation are not covered very explicitly anywhere. They are mostly described in operating manuals of the units prepared by process licensors and designers.

As a person engaged in both plant operation, process monitoring, and design of process plants, I realise that there is a strong need to compile the principles relating to plant design and efficient operation together in the form of a book.

This work is my humble effort to compile and demonstrate the principles and practices in a concise form. The book started with an overview of refineries with high emphasis on heavy and high-acid number crude processing and processing concerns at length. A short review of distillation fundamentals is included with examples of simulation and emphasis on distillation column auxiliaries that largely contribute to operation success.

The fundamentals associated with secondary process units like the fluid catalytic cracking unit (FCCU) and delayed coker unit for upgradation of the bottom of the barrel are covered at length. Hydrotreating and hydrocracking, essential for fuel quality improvement and bottom of barrel upgradation, are extensively dealt with in the book, including the principles associated with operation. The hydrogen generation unit and hydrogen management is given due emphasis. A fresh review of the acid gas treatment is included in view of increased emphasis on carbon capture and storage and reduction in greenhouse gas emissions. A concise review of safety in design, including the most sought after subject of safety relief load computation for distillation columns, is covered for many industrial columns that may be useful for process engineers. A good number of case study illustrations are also included in the chapters elaborating the principles, with help of simulations, and they are based on my own experiences of plant operation, design, and troubleshooting.

The present challenges faced by refiners are also sufficiently elaborated. Finally, the different forms of renewable energy, biomass-derived green fuels, methodologies for production of biofuels, and benefits are highlighted, including the emerging green fuels of the future.

I earnestly hope that this work will be useful and add value to your academic and professional pursuits.

About the Author

Ashis Nag is a professional chemical engineer with 40-plus years of rich experience in various fields of petroleum refining and related downstream process technologies. He has worked in various refineries of Indian oil corporations and at the Essar Oil Refinery, Jamnagar. He served as head of special central process engineering and design cell of Indian Oil Corporation Limited (IOCL). Presently, he is working as a process consultant. Internationally traveled and the author of several papers for international journals and forums, he has also authored the book *Distillation and Hydrocarbon Processing Practices* (published by PennWell). He has received awards for innovative designs and earned a BCHE degree in chemical engineering from Jadavpur University, Kolkata, India, in 1976.

1 Overview of Refining and Distillation

1.1 REFINING: AN OVERVIEW

A refinery converts crude to different hydrocarbon fractions or products. A refinery is essentially deployed for refining of crude oil for production of fuels, lubricant base oils, and preparation of starting stock for petrochemicals. The typical finished products produced by a fuel-oriented refinery include:

- Liquefied petroleum gases (LPG)
- Naphtha
- Motor spirit (MS) or gasoline
- Aircraft turbine fuel (ATF)
- Kerosene (SK)
- Diesel (HSD)
- Fuel oil (FO)
- Heavy petroleum stock
- Fuel oil/ Furnace oil
- Coke
- Bitumen
- Sulfur
- Lube base oils

A refinery comprises many process units, as shown in Figure 1.1 and Figure 1.2.

The crude refining process begins with the atmospheric crude distillation unit followed by the vacuum distillation unit. Other secondary units deployed are primarily for heavy oil upgradation, quality improvement, and emission control units, and can be divided into different groups based on the objectives as described next.

Heavy oil upgradation units are as follows:

- Delayed coker unit (DCU)
- Hydrocracker unit (HCU), residue hydrodesulfurization (RDS) unit, slurry hydrocrackers
- Fluid catalytic cracking unit (FCCU), residue FCCU unit (RFCCU)
- Residue deasphalting unit for upgradation of deasphalted oil (DAO)

These are the major secondary units of a refinery for upgradation of heavy oils (bottoms of barrel) and production of higher proportion of distillates.

DOI: 10.1201/9781003268246-1

1

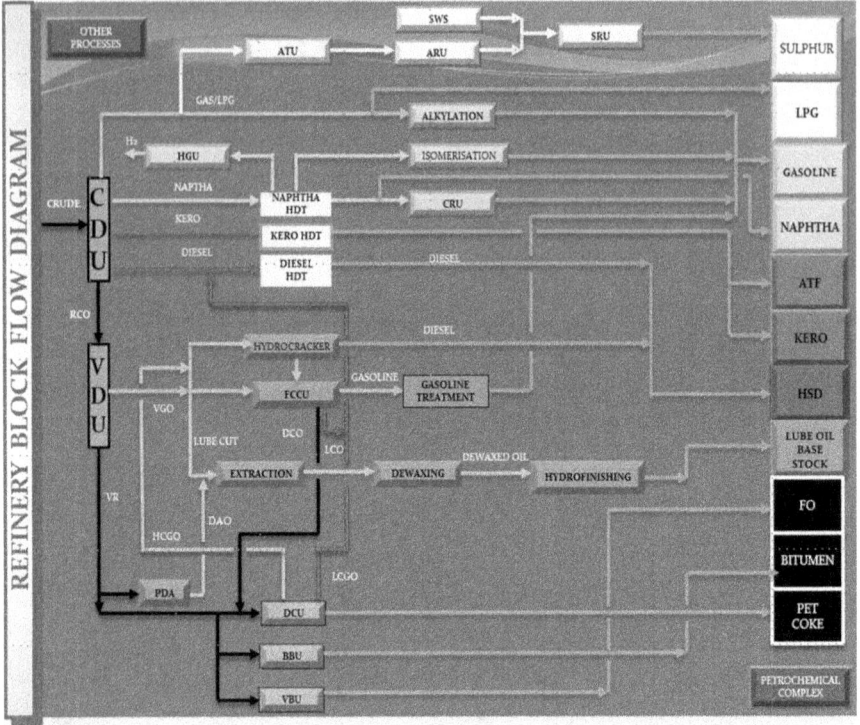

FIGURE 1.1 Typical block diagram of a refinery for production of fuels and lube base oils.

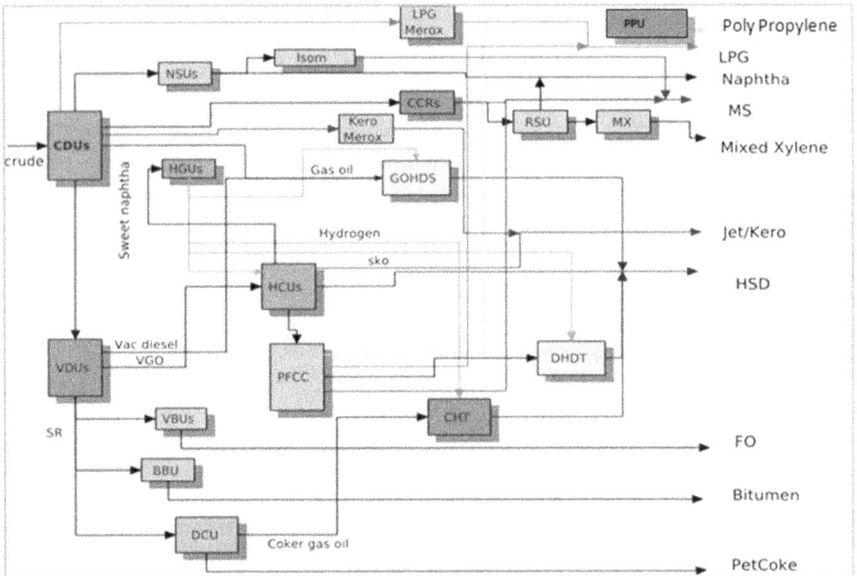

FIGURE 1.2 Typical block diagram of a fuel-oriented refinery.

Product quality improvement units are as follows:

- Kerosene hydrotreatment is required to lower sulfur of finished kerosene and raise its smoke point.
- A diesel hydrotreater (DHDS/DHDT) is required to lower the sulfur content of finished diesel fuel and increase its cetane number.
- A naphtha hydrotreater and catalytic reformer unit (CRU) are required to produce a high research octane number (RON) reformate (typical RON 98–103) as a component for blending in the gasoline pool.
- An isomerization unit (ISOM) is required to enhance the octane number of hydrotreated light straight run naphtha steams, primarily the C5 and C6 streams (typically to RON 88–91), for finally blending to the gasoline pool.
- An alkylation unit: is deployed to produce alkylate of a high octane number (typical RON around 95) by the reaction of C_4 and C_5 olefins produced from the FCCU and isobutane.
- Units to produce oxygenate like MTBE, ETBE, and TAME are required for production of high octane oxygenate streams (RON above 100) for blending into the motor spirit/gasoline pool. MTBE and ETBE are produced from isobutylenes (mostly sourced from FCCU LPG or steam crackers) by the reaction with methanol and ethanol, respectively. TAME is produced by reaction of isoamylenes with methanol. Oxygenates with high octane components (RON >100) are blended into reformulated gasoline.
- Sweetening units are deployed to extract light mercaptans from LPG and light naphtha to improve quality of LPG and gasoline. ATF sweetening units are deployed to convert mercaptans (thiols) to alkyl disulfides to make the ATF pass tests like silver corrosion.

Reformulated gasoline is typically constituted by blending of the necessary components to meet the revised specifications of MS (EURO VI):

- Reformate
- FCCU gasoline
- Isomerate
- Alkylate
- Oxygenates like MTBE, ETBE, TAME, ETOH

Acid gas treatment units, sour water strippers, and *sulfur units* are also very important facilities of a refinery for control of emissions from acid gases and meeting the overall environmental objectives.

A *hydrogen generation unit* (HGU) is required for production of necessary hydrogen for hydrotreatment of different fractions.

Other units can be a bitumen blowing unit (BBU) for production of Bitumen and Visbreaker unit (VBU) for lowering the viscosity of heavy residues for production of Fuel Oil.

Units for production of lube oil-based stocks are aromatic extraction units like a furfural extraction unit (FEU), a propane deasphalting unit (PDA) for rejection of asphalts from vacuum residue (VR), a solvent dewaxing unit (SDU), and a catalytic dewaxing unit for lowering the pour point of lube oil-based stocks and hydrofinishing unit (HFU) for improvement of color stability and other properties of lube base oils.

Earlier, production of lube oil needed special lube-producing crude oils that were processed in crude and vacuum units (having side strippers) to produce fractions having the desired viscosities to convert the fractions into different lube base oils. The fractions, require further processing in different process units as follows for each grade:

- Spindle oil (SO) → FEU → SDU → Hydrofinishing (HFU) → 70 N and 100 N (Neutral) base oil
- Light neutral (LO) → FEU → SDU → Hydrofinishing (HFU) → 150 N base oil
- Intermediate oil (IO) → FEU → SDU → Hydrofinishing (HFU) → 500 N base oil
- Heavy oil (HO) → FEU → SDU → Hydrofinishing (HFU) → 850 N base oil
- VR → PDA → FEU → SDU → Hydrofinishing (HFU) → Bright stock (BS)

The furfural extraction unit (FEU) is required to lower the aromatic content of lube oil cuts to improve its viscosity index (VI). The aromatic rich streams are rejected in the extraction process and the aromatic stock is often used for production of carbon black.

A solvent dewaxing unit (SDU) is required to improve the pour point of lube oil fractions and reject wax as a byproduct. Catalytic dewaxing units (CDW) or ISO-dewaxing catalytically convert the waxes to lubes and increase the yields of lubes while meeting the desired pour point of lube base stocks.

A propane deasphalting unit (PDA) is deployed to remove the asphalts from the vacuum residue (VR) stream and produce deasphalted oil (DAO) for production of high-viscosity lube base oils. Sometimes the DAO is fed to residue hydrocrackers or RFCCUs for upgradation to distillates. The rejected asphalt is traditionally used for blending with blown bitumen to produce lower penetration/industrial grades of bitumen or can be gasified.

A hydrofinishing unit (HFU) is used to improve the color of lube base stock by removal of nitrogen compounds and to enhance the color stability by saturation of olefins and di-olefins, if any.

With the advent of hydrocrackers, special lube-producing crude oils are no longer required for production of the majority of lube oil grades (e.g., 70 N, 150 N, 500 N). The unconverted stream from the hydrocracker unit can be used as starting stock for production of lube oil fractions by splitting the cuts in a vacuum unit, which are subsequently treated and converted to different lube oil base stocks. The group II/III lube oils are of higher demand having lower sulfur (S <0.03% wt.) and higher percentage of saturates >90% wt. and higher viscosity index.

1. *Coke and heavy oil gasification* were recently included in refinery configuration. Acid gas treatment finds extensive use in the purification section in gasification complexes and other areas.

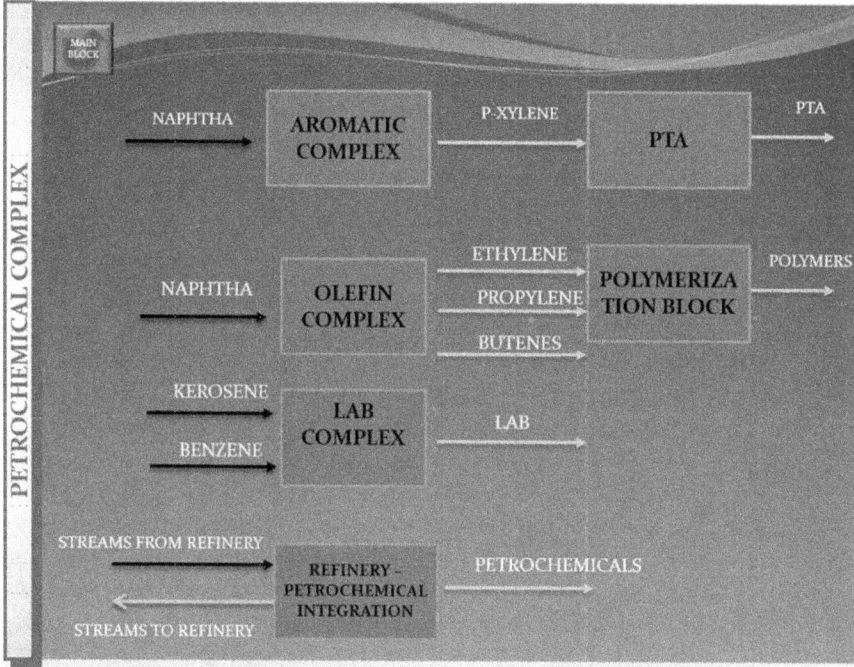

FIGURE 1.3 Block diagram of a typical refinery and petrochemical integration.

2. Modern refineries also have introduced *downstream integration and installed petrochemical units* like standalone steam crackers based on naphtha or natural gas as feed to produce olefins like ethylene, propylene, butylene, butadiene (Figure 1.3). These olefins are also produced by secondary conversion units like the fluid catalytic cracking unit, often called the petrochemical FCCU, and operate at more severe conditions. Propylene can be also produced from FCCU LPG, and ethylene can be produced from FCC sponge gases. Often the propylene produced from the FCCU and coker is integrated with the steam cracker distillation section for further purification and recovery. Refineries have further gone for production-of finished petrochemicals and started producing petrochemicals like polyethylene, MEG, polypropylene, butadiene, and styrene butadiene rubbers (SBR).

3. *Facilities for production of aromatics* like benzene, toluene, and paraxylene are also deployed in some refineries. Some refiners are producing finished PTA from paraxylenes. The starting unit for production of aromatics is a continuous catalytic reforming unit (CCRU), typically operating at a higher severity than that used for production of reformate for blending in MS.

4. Some refineries have *complexes to produce linear alkyl benzene* (LAB) from heart cuts of kerosene, used in manufacture of detergents. Some refineries are known to have gone for *upstream integration* as well.

Refineries usually have a captive power plant (CPP) and steam generation and distribution facilities to generate and supply power and steam to meet the requirement of a refinery. Compressed air system, Instrument air facilities, nitrogen generation units, demineralized water plants are also provided as utilities.

1.2   DISTILLATION: A REVIEW

Distillation is a most practiced unit operation deployed in refineries and in the majority of chemical industries for separation of components based on their boiling points or relative volatility. The theories of distillation are well known and documented in many books, journals, and research papers. But only a few experiences were found worth including in this volume along with a brief description of the fundamentals. Only a review of important issues are presented in the following.

A distillation column is provided with a reboiler at the bottom and condenser at the top. The reboiler provides the heat required to vaporize part of the feed and generate vapor (boil up). The condenser condenses the overhead vapors and provides the reflux. The low boiling overhead liquid product is recovered from the top and the product with a higher boiling point is drawn from the bottom. The column is provided with internals (trays or packings) for intimate contact of vapor and liquid to achieve enrichment of components and constitute the stages. At a specified reflux ratio, a minimum number of theoretical trays (stages) is required for a desired separation. A minimum reflux is always required to attain the specified separation. The requirements for stages and reflux can be computed by Fenske–Underwood–Gilliland (FUG) or by Fenske–Underwood and Erbar–Maddox (FUEM) correlations. The minimum number of stages are calculated by Fenske's equation at total reflux, and minimum reflux is calculated by the Underwood equation. The actual reflux ratio is typically taken as 1.2–1.5 times the minimum reflux, evaluated by the Underwood equation. Later, the required number of stages are computed, at the actual reflux by either Gilliland or Erbar–Maddox correlations. Similar to the required reflux ratio, the minimum boil up ratio is also fixed for a desired separation, with a fixed number of theoretical stages available primarily in the column stripping section.

The column pressure is a very important variable and needs to be defined for a distillation calculation, along with the specified degree of separation desired. The use of steady-state simulators is common for design and rating of distillation columns. However, the McCabe–Thiele method is always considered very valuable and essential to interpret many distillation features and problems. A simulation of distillation for benzene and toluene separation is presented as an illustration in the following example.

A feed of 50,000 Kg/Hr having 50:50 wt.% of benzene:toluene mixture at a temperature of 80°C needs to be separated into 95% wt. of benzene in the top product and 5% wt. of benzene in the bottom product.

The column is operated at top pressure of 1.6 Kg/cm2 (a) having 15 stages, including a condenser and reboiler. A simulation of the problem carried out and

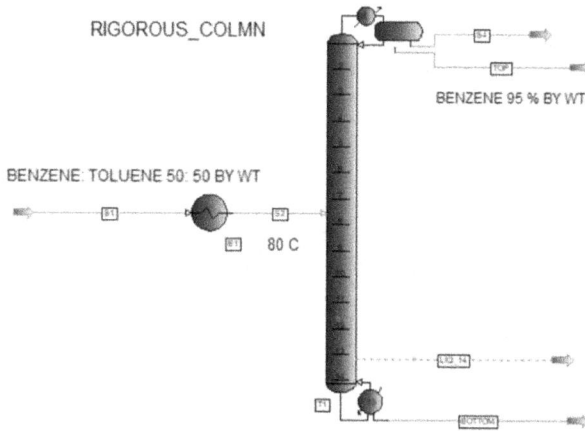

RIGOROUS_COLMN

BENZENE 95 % BY WT

BENZENE: TOLUENE 50: 50 BY WT

80 C

Stream Name		TOP	BOTTOM
Total Mass Rate	KG/HR	25000.085	24999.915
Temperature	C	95.928	128.859
Pressure	KG/CM2	1.600	1.800
Total Weight Comp. Percents			
BENZENE		94.9996	5.0001
TOLUENE		5.0004	94.9999
H2O		0.0000	0.0000

Column Name		T1
Column Condenser Duty	M*KCAL/HR	-6.2498
Column Reboiler Duty	M*KCAL/HR	6.9630
Column Reflux Ratio		1.7539

FIGURE 1.4 Simulation of the example problem of distillation column.

the simulation flow sheet and results of simulation are presented in Figure 1.4 and Table 1.1, respectively.

Thus, the simulation delivers all the essential parameters like condenser duty (Q_C) and reboiler duty (Q_R) along with top and bottom temperatures and temperature through stages. The simulation also generates flow of liquid and vapor through stages of the column along with the relevant properties like density, molecular weight, viscosity, and surface tension. Based on flows of vapor and liquid through the column and the required properties, the diameter of the column can be computed by the simulator including the design of internals. However,

TABLE 1.1
Results of Simulation

Column Name	Units	T1
Condenser duty, Q_C	MMKCAL/HR	−6.25
Reboiler duty, Q_R	MMKCAL/HR	6.96
Reflux ratio	—	1.754

design of column internals by experts is always advisable. But a column can be conveniently rated by a simulator with the input of column diameter and type of internal used, along with other details like type and number of dispersers, downcomer detail, number of liquid passes, and foaming behavior (system factor). During rating of a column, a diameter is considered acceptable if the flooding denoted by the simulator is found within 80%.

The problems normally experienced in distillation columns are not due to theories adopted in design but mostly due to problems with internals, pressure control strategies, and the arrangements provided in column bottom spout and in reboilers. A review of the these is presented in the following section.

1.2.1 Review of Column Internals

The column internals are devices deployed for intimate contact between vapor and liquid in a column to achieve mass transfer and heat transfer. Two types of contacting devices are generally in use: trays and packings. Trays can be cross flow and counter flow. Packings are inherently counter-flow devices. Actual product qualities compared with design is the best indicator of performance of internals. Pressure drop in a column across trays/packings can act as a guide for evaluating the performance of column internals. Typical pressure drop in a tray is 0.07 psi to 0.12 psi per tray. A higher value may indicate flooding. A very low value will indicate dumping or dislodgement of tray sections. The subject of column internals has been sufficiently dealt with in various books, journals, and publications. A few important aspects are covered in the subsequent sections. Typical dimensions and other few parameters of trays may be useful for process engineers and are discussed next.

1.2.2 Typical Dimensions and other parameters of Internals

1. Typical weir height: 50 mm. (For vacuum applications, a lower weir height of 6–12 mm is practiced to reduce pressure drop.)
2. Hole diameter: 5 mm for sieve trays. Valve diameter: 47.6 mm for valve trays (with floating valves).
3. Height under downcomer: 40 mm.
4. Downcomer liquid seal: 8–10 mm. The downcomer should always have a liquid seal to avoid vapor short circuiting through the downcomer. Due to installation errors, the downcomer seal is often found absent in columns. This is a serious error in fitting. Sometimes dynamic seals are provided in downcomers in services with high liquid rates like in strippers (below the feed tray) and in pumparound zones. However, when a product withdrawal is involved, it should necessarily have a static seal.
5. Weir length: Normally weir length is around 60%–85% of the column diameter. A good initial value recommended by experts is 76%, and the equivalent downcomer area is 10%–12% of the column cross-sectional area. For double pass trays, the typical width of the central downcomer

is 200–250 mm. Weir length should be adequate to deliver a weir loading below 8 gallons per minute (GPM)/linear inch of weir length.

6. Hole area: The hole area is approximately 10%–14% of active area for trays, whereas the open area for packings is typically 40% or higher. Thus, pressure drop with packing is far less and is deployed where the vapor volume handled is very high like in vacuum columns.

7. Pressure drop per tray: 0.07 to 0.12 pounds per square inch (psi).

8. Tray spacing (typical): 450 or 600 mm.

9. Typically for columns operating at higher pressure, trays are preferred over packings as the density difference between vapor and liquid reduces with increase in pressure and needs liquid disengagement space, available in trays.

Some fundamental descriptions are presented next before proceeding further.
Column flooding mechanisms described by experts are

- Spray entrainment flooding
- Froth entrainment flooding
- Downcomer backup flooding
- Downcomer choke flooding

Spray entrainment flooding refers to carryover of small liquid droplets from the tray below to the tray above by the velocity of vapor. The limiting velocity of vapor is known as flooding velocity and computed traditionally with the Souders–Brown equation:

$$U_{sflood} = C_{SB} \sqrt{\left((\rho_L - \rho_V)/\rho_V\right)} \text{ and } C_{SB} = U_{sflood} \sqrt{(\rho_V/(\rho_L - \rho_V))}$$

C_{SB} is a constant of the Souders–Brown equation and can be correlated with tray spacing, vapor, and liquid flows. ρ_L and ρ_V are liquid and vapor densities, respectively.

Froth entrainment flooding occurs when the liquid froth height in a tray approaches the upper tray with high vapor velocities. This happens where liquid load is very high.

Downcomer backup flooding occurs when the height of liquid with froth in the downcomer approaches the tray above. This can be as a result of higher pressure drop in the tray dispersers and also due to resistance in the flow path of liquid in the downcomer inlet or at the outlet. The downcomer backup h_{dc} is shown in Figure 1.5.

h_{dc} can be computed by the equation $h_{dc} = h_{dry} + h_w + h_{ow} + h_d$, where h_{dc} is the downcomer backup height, h_{dry} is liquid head equivalent to dry tray pressure drop, h_w is weir height, h_{ow} is the liquid crest above the weir, h_d is the head loss in the downcomer apron due to flow of liquid in the downcomer. h_{dc} should not cross 50% of tray spacing to avoid downcomer backup flooding.

Downcomer choke flooding results due to accumulation of liquid with froth at the entrance of the downcomer due to less area available at the inlet or due to the

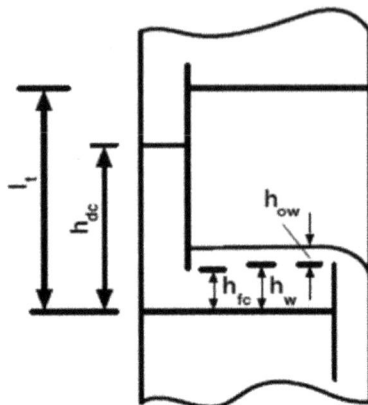

FIGURE 1.5 Downcomer backup.

recycling of liquid at the inlet. Downcomer load at the entrance of the downcomer is typically maintained as 8–10 gallons per minute (GPM)/linear inch of weir length. Beyond the value number of liquid pass is increased to contain the liquid load. The liquid flow arrangement for a multipass tray is shown in Figure 1.6 and Figure 1.7.

Anti-jump baffles are provided for on-center and off-center downcomers to arrest the liquid throw over the weirs (and direct them toward the downcomer) and are known to increase the capacity of the downcomer by around 20%. The author observed dislodgement of these baffles in some units and blocking of the downcomer liquid flow path, resulting in flooding. The anti-jump baffle is shown in Figure 1.8.

A *splash baffle* or vapor hood is often used for the spray regime design when the liquid flow rate is very low. This is both to avoid the liquid droplets being flung directly to the downcomer and to collect the droplets in order to induce the froth regime. A splash

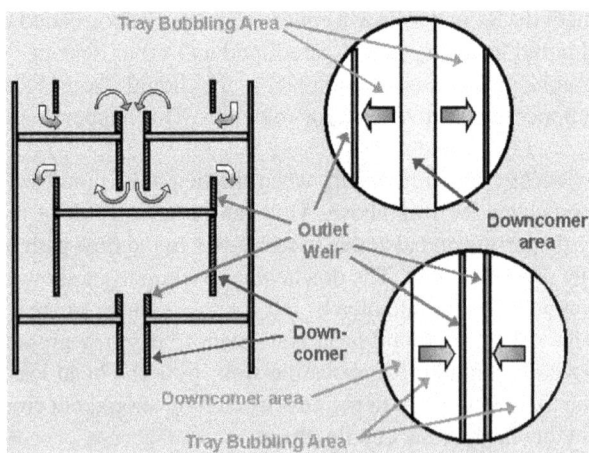

FIGURE 1.6 A two-pass tray.

(a) 2-Pass Crossflow

(b) 3-Pass Crossflow

(c) 4-Pass Crossflow

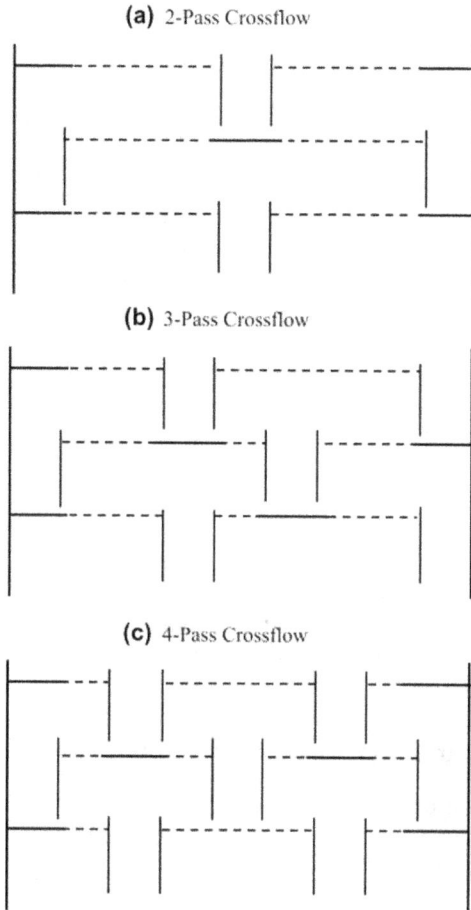

FIGURE 1.7   Multipass trays.

baffle/vapor hood should be used only with low liquid flow rates and never in high liquid flow rates as splash baffles/vapor hoods can choke the liquid flow to the downcomer and may lead to premature flooding. The arrangement is shown in Figure 1.9.

Often proprietary Z bars are used to increase the flow area under downcomers. The Z bars take support from the existing bolting bars and tray support ring (TSR). The fitting avoids cutting of old bolting bars and reweld of new bolting bars to increase area within downcomers during revamp applications (Figure 1.10).

1.2.3   DUMPING

Dumping is a phenomenon where the majority of liquid from the tray drops from the tray before reaching the outlet downcomer, due to low vapor velocity or improper fitting of tray segments. This adversely affects performance and particularly the

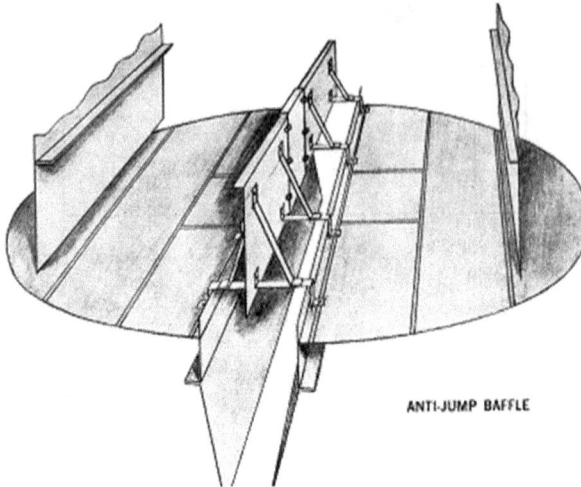

ANTI-JUMP BAFFLE

FIGURE 1.8   An anti-jump baffle in central downcomer.

separation in the column. In valve trays, the typical maximum slot velocity can be determined from the following empirical correlation: V maximum = 15/(vapor density)$^{1/2}$ in ft/sec. The minimum slot velocity in feet per second can be estimated from the following equation:

$$V \text{ minimum} = 6.5/\left(\text{vapor density}\right)^{1/2} \text{ in ft/sec}$$

where vapor density is expressed in lb/cu ft.

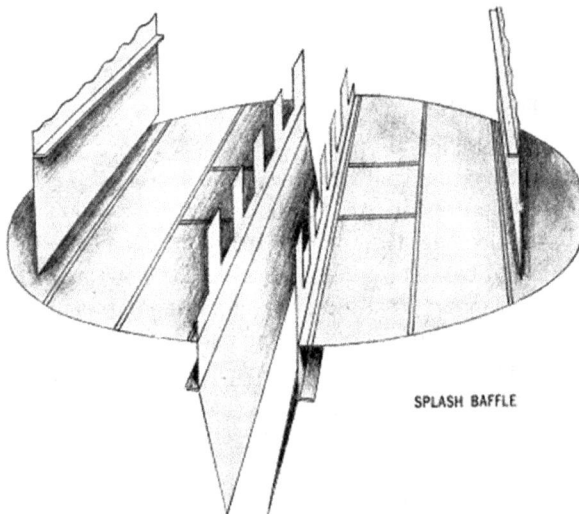

SPLASH BAFFLE

FIGURE 1.9   A splash baffle.

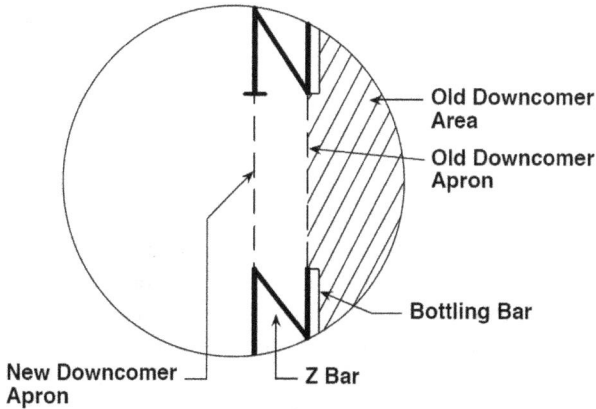

FIGURE 1.10   Tentative schematic drawing of Z bars.

The minimum slot velocity calculation is based on minimum vapor flow and the area of the opening around the valve periphery (fully opened). As a rule, the slot area for Koch and Glitch valves is equal to approximately 0.012 sq. ft. per valve. But this should be checked with the tray supplier. Often the peripheral open area (slot area) is termed the *open curtain area*, typically indicated by the tray designer. When valves with different weights are specified for a single deck, for meeting turndown, the minimum velocity should be computed for the minimum vapor flow using the slot area for the lighter valves plus the area of the fixed opening (valves having a dimple to keep it off the deck) of the heavier valves. If a velocity of vapor is lower than the minimum slot velocity, the tray is likely to dump.

1.3   COLUMN PRESSURE CONTROL AND REBOILERS

Pressure control is the most important and rudimentary control for a distillation column. Variation of column pressure can cause swing in operation and can adversely affect the product qualities. Different pressure control schemes are in use based on the application. When there are non-condensables in the overhead product, pressure of the column usually is controlled by restricting the bleed (the control valve is located on the bleed line). If a compressor is used to compress the net gas, then pressure control is achieved by controlling the spill back flow from the compressor discharge to the inlet of the overhead condensers. If an air cooler is used in the overhead as a condenser, then pressure of the column can be controlled by varying the condensate temperature, usually achieved by blade angle variation of air cooler. The aforementioned cases of pressure control are relatively straightforward.

Pressure control of a column with condensate at the bubble point presents bigger challenges for pressure control and often causes problems unless implemented correctly. The author has experienced several cases of operational problems triggered by column pressure control malfunctions in columns with bubble condensers.

Accordingly, few important pressure control schemes of distillation columns for bubble condensers are discussed in detail (where problems with pressure control were observed and likely to occur).

1.3.1 PRESSURE CONTROL BY SUBMERGENCE VARIATION

Pressure control schemes with submergence variation are used mainly for condensers where condensate is at a bubble point. In the scheme in Figure 1.11 and Figure 1.12, the condenser is located above the level of the drum and a control valve is provided at the condensate outlet line to vary the submergence of the condenser tubes and thereby alter the heat transfer area and in turn condensate temperature for control of pressure of the drum and the column. In these schemes a small equalization line from overhead to the drum is provided to control the drum pressure (1.5 to 2 inches in diameter). Alternatively, a hot vapor line with a control valve can be provided to control the drum pressure, as shown in Figure 1.12. The condensate can enter from the top (with extension of the inlet pipe below the liquid level in the drum), as in

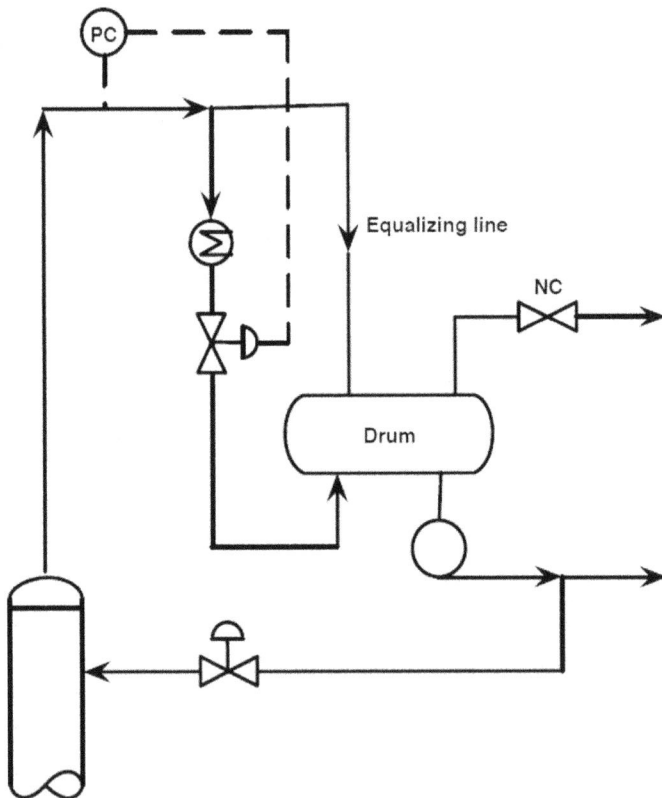

FIGURE 1.11   Pressure control with submergence variation.

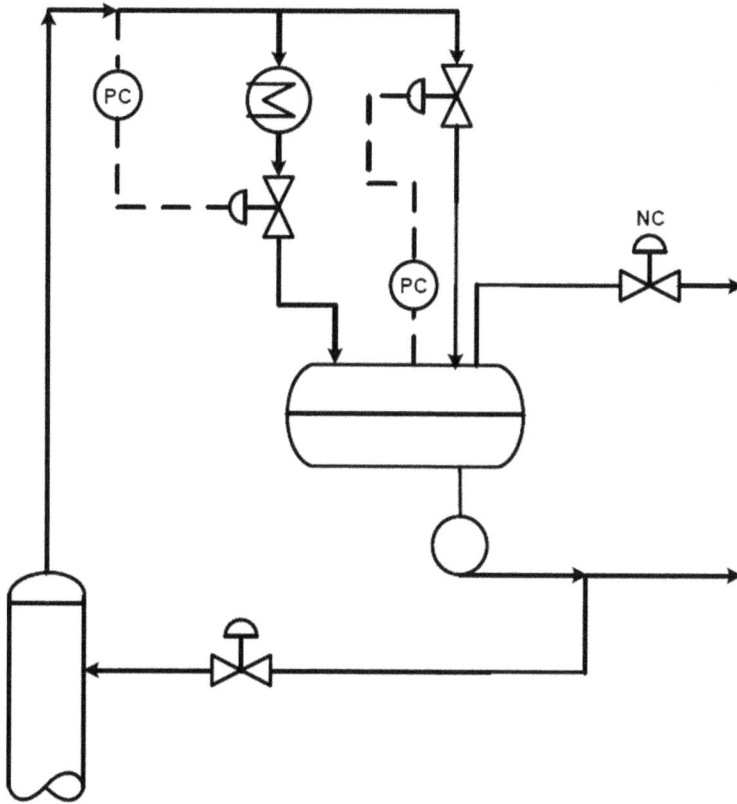

FIGURE 1.12 Existing pressure control scheme, subcooled condensate entering from top (incorrect scheme).

Figure 1.13. The condensate can be also admitted from the bottom to avoid contact of subcooled liquid with dew point vapor (thereby avoiding tower pressure fluctuations due to *vapor collapse*).

The height of the condenser above the drum is extremely critical to avoid flooding of the condenser (with condensate) that can adversely affect the heat transfer. In the scheme with an equalization line, the height of condenser should be at least equal to the height of condensate to balance the sum of pressure drop of the condenser and the pressure drop in the control valve at the condensate outlet, i.e., a minimum height of condenser above drum liquid level: $(H) = (DP_C + DP_L$ control valve$)/\rho_c$ (refer Figure 1.14), where DP_C is the pressure drop across the condenser, DP_L is the pressure drop in the condensate (liquid) line across the valve, and ρ_c is the density of condensate at condensate outlet temperature. This neglects the pressure drop due to friction in the condensate line.

Problems of flooding of the condenser may be likely during capacity revamp, unless checked critically and implemented correctly. The additional condensate flow

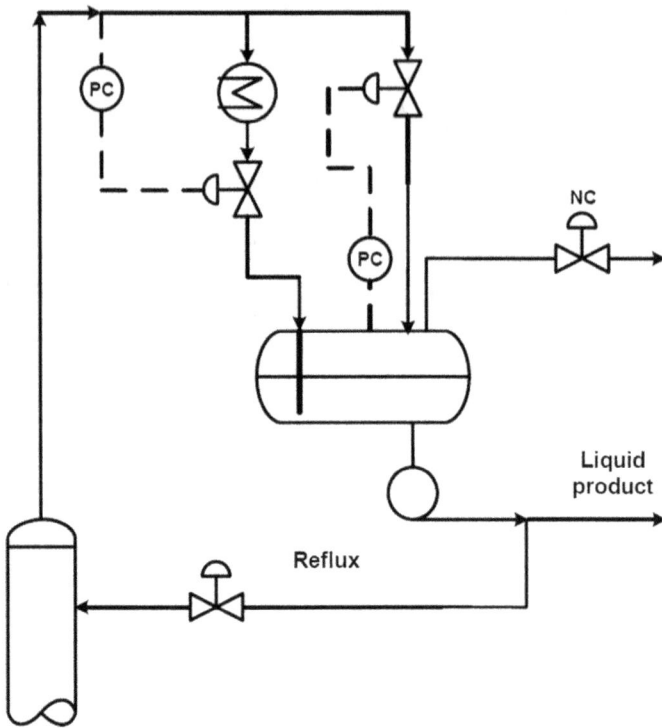

FIGURE 1.13 Condensate entering from top and extended inside liquid level (correct scheme).

may cause additional pressure drop in the condensate line and in the condensate outlet valve, and may eventually result in partial flooding of the condenser above. This flooding may reduce the heat removal capacity of the condenser. The author retained the condensate outlet valve in a revamp and later checked and found that 5%–10% of the condenser heat transfer area was likely to get flooded with higher flow as a result of higher pressure drop in the outlet. The condensate valve was later changed with a lower pressure drop valve to resolve the problem.

In quite a few column pressure controls, where the drum liquid is at the bubble point, the scheme Figure 1.11 is practiced. When the condenser is located above the drum, the condensate outlet line is admitted from the bottom or is extended inside the drum liquid level when admitted from the top to avoid the contact of subcooled liquid with dew point vapor and avoid incidences of *vapor collapse*. An equalizing line or an interconnecting line is necessary to control pressure of the reflux drum with or without a valve. The advantage in the scheme is that condensation takes place at higher pressure and also needs a smaller control valve on the condensate liquid line.

In the scheme in Figure 1.12, pressure control is achieved by varying submergence with condensate entry from the top. Drum pressure is controlled by the valve on the hot vapor bypass line. But there was problem of continuous pressure fluctuations.

Subsequently, the problem was analyzed and found that the condensate line was not extended inside the liquid level. The arrangement was corrected and the condensate line extended deep inside the liquid level in the drum to avoid contact of subcooled liquid with dew point vapor, as shown in Figure 1.13. The problem was resolved. Further, dipping of the condensate line below liquid also ensures a seal and arrest vapor blow through the condenser.

1.3.1.1 Pressure Balance around Column Overhead, Condenser, and Reflux Drum

A pressure balance around the column top and reflux drum is explained with the help of the Figure 1.14.

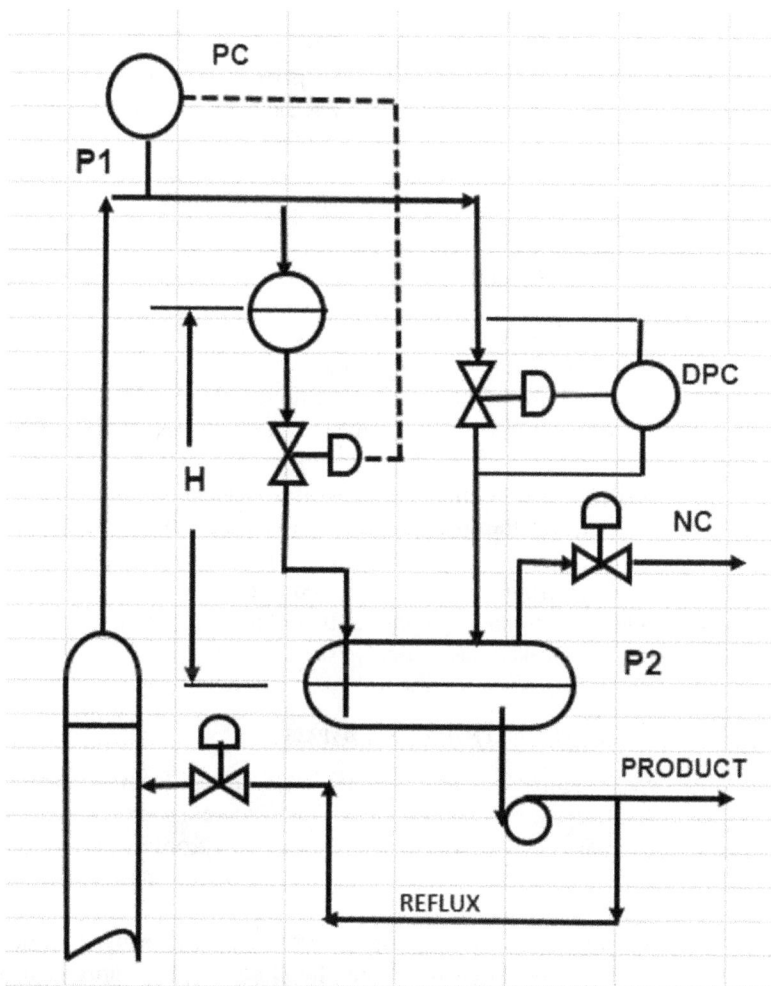

FIGURE 1.14 Generalized pressure balance around column top and reflux drum.

If the top pressure of the column is P1 and drum pressure is P2, then the correlations between P1 and P2 are $P1 = P2 + DP_V$ (equation 1) and $P1 = P2 + DP_C + DP_L - H\rho_c$ (equation 2), where DP_V, DP_C, and DP_L are the pressure drops in the hot vapor line valve, condenser, and condensate liquid line valve, respectively. H is equal to the height of the liquid leg above the liquid level in the vessel (Figure 1.14), and ρ_c is the density of the condensate in conformable units. The line frictional losses for vapor and condensate lines are considered nil (negligible). Equating equations 1 and 2 and eliminating P1 and P2 gives the correlation as stated: $DP_V = DP_C + DP_L - H\rho_c$ or $H\rho_c = DP_C + DP_L - DP_V$. Thus, a higher pressure drop in the condenser and condensate valve will increase the height of liquid leg, whereas an increase in pressure drop in the hot vapor valve will reduce the drum pressure and in turn will reduce the liquid leg above the drum liquid level. In case there is no valve in the vapor line (the column is directly connected to the drum through an equalization line where $DP_V = 0$) the equation will reduce to $H\rho_c = DP_C + DP_L$. Thus, higher pressure drops in both the condenser and condensate outlet valve will result in higher liquid leg and may partially flood the condenser.

In most cases, a condenser is provided with vertical cut baffles to allow free horizontal zigzag travel of condensate along the length of the condenser. In some condensers the baffles have horizontal cuts or slanted horizontal baffles are used. In those, a seal is automatically established. Sometimes a seal (with an inverted U) is provided in the outlet pipe leading to the drum. If the condenser has a vertical cut baffle, then the condensate pipe needs to be dipped inside the liquid level in the drum and this will provide a condensate seal and likely arrest vapor blowing through the condenser.

Normally, whenever a condensate line with a restriction type of arrangement is used or a U seal is provided, a non-condensable vent line is required from the cold end of the condenser to the top of the drum with a valve. Vapor ingress may be likely into the condenser and the vapor may accumulate and blanket the heat transfer area unless it finds a way out (to the drum or elsewhere).

There are lots of different opinions and doubts about maintaining the position (open or close) of the valve in the condenser vent line. The author had doubts about a couple of control schemes. So, one must thoroughly analyze and conclude. Normally, condensers are located at heights and are not visited frequently by operators (to open the non-condensable vents to vent any light and then close). A pressure balance equation, as shown earlier, is worked out for analysis.

1.3.2 PRESSURE CONTROL BY HOT VAPOR BYPASS

The hot vapor bypass method is a very common method used for fractionation column pressure control. Hot vapor bypass is used where column overhead is saturated condensate (bubble point condenser). The name originates from the use of part hot vapor that bypasses the overhead condenser. Several pressure control schemes are presented next.

In the scheme in Figure 1.15 of pressure control by hot vapor bypass, the condenser vent to the drum should be in a normally closed (NC) position and occasionally it can be opened to vent the light to the drum. In a case when a number of parallel condensers are deployed, a common vent manifold can be laid and a

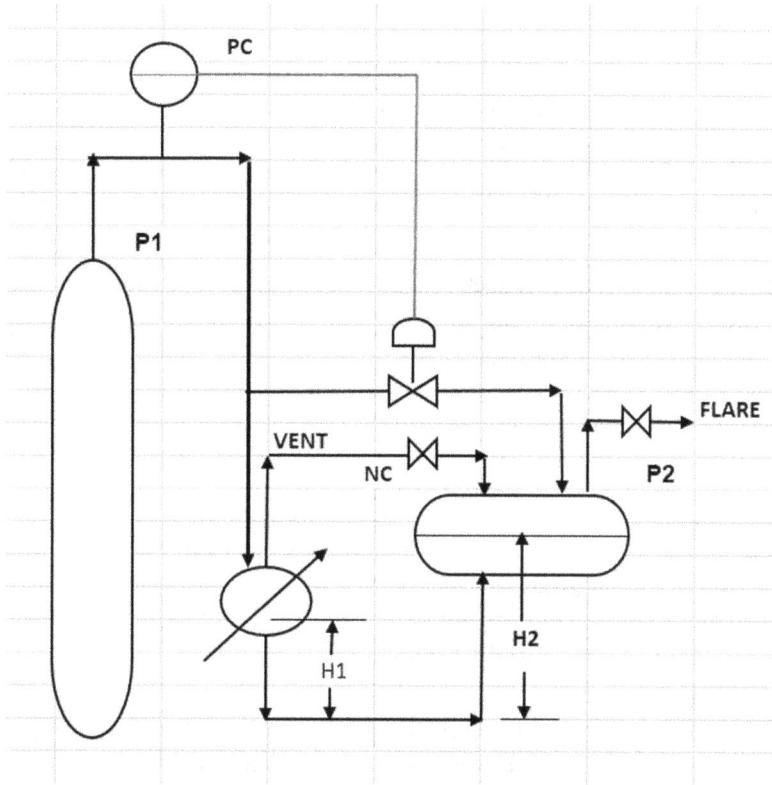

FIGURE 1.15 Hot vapor bypass with condenser located below drum.

remotely operated vent valve can be provided for occasional venting to the drum or flare where the condenser/drum is not frequently visited by operators. A drum pressure control by hot vapor makeup and bleed to the flare can be also provided for control, where frequent non-condensable ingress is expected.

Now, if the condenser is located below the drum to minimize cost for the platform structure, the column pressure is maintained with hot vapor bypass to the drum. It is usual to provide a non-condensable bleed from the drum (with a control valve to the flare). The pressure balance equation when the condenser is at lower elevation than the drum (Figure 1.15) can be written as $P1 = P2 + DP_c + H\rho_c$ and $P1 = P2 + DP_V$, where $H = H2 - H1$ or $DP_V = DP_C + H\rho_c$. The pressure drop across the vapor valve thus balances the pressure drop in the condenser and the differential head of liquid between the drum and the condensate limb.

Doubts came up in one unit for hexane production. The depentanizer and hexane column both had a condenser located on the ground with hot vapor makeup and bleed split range control to the flare, as in Figure 1.15. The unit also had 1-inch non-condensable vents from the condenser connected with valves to the drum. The unit started with the vent valves open to automatically vent any non-condensable (propane and butane in this case), and a continuous pressure fluctuation was observed. The reason for fluctuation was not understood. The phenomenon was difficult to comprehend. Initially, it was thought that when the vent valve is open from the condenser to the drum, the condenser outlet pressure will try to equalize with the drum pressure, which has lower pressure than the column, and there should be no problem in getting the non-condensable vented to the drum, and hot vapor bypass valve should marginally close to maintain the column pressure at the set value. Later, it was further analyzed in the following equation:

Pressure at the condenser outlet: $P_{CO} = P2 + H\rho_c$ and $P1 = P_{CO} + DP_C$

Scenario with the equalization line open. When the condenser outlet is connected to the drum, then P1 nearly equals to P2 (higher by only pressure drop in the vent line). Thus, the equation can be written as $P_{CO} = P_{CO} + DP_C + H\rho_c$, and consequently, H equals DP_C. Thus, the condenser is likely to get flooded. Expressed differently, the liquid buildup in the condenser counterbalances the condenser pressure drop and pressure control may fail. Having analyzed as earlier, the condenser vent valves in both columns were closed. The pressure fluctuation reduced. During normal operation they are periodically opened for some time and again closed. In general, the vent valve should remain as a normally closed (NC) valve.

1.3.3 Pressure Control with Vapor Line Restrictions

The majority of condensers at the bubble point use the scheme in Figure 1.16. Response of the scheme is learnt to be fast and big butterfly control valves are easily available for installation in vapor lines. The pressure control is achieved by restriction of valves on the vapor line, and drum pressure is manipulated by the vapor bypass line. The pressure balance works out as $P1 = DP_{OV} + DP_C + P2 - H\rho_c$ and $P1 = DP_V + P2$, where P1 is the column top pressure and P2 is the drum pressure, H is the condensate leg above liquid level in the drum, and DP_{OV} and DP_V are the pressure drop in the overhead line valve and condenser bypass valves, respectively. However, in this scheme condensation will occur at lower pressures than the column pressure and may need marginal additional heat transfer area in the condenser. A condenser vent is not required as the noncondensables will freely flow along the condensates and get into the drum. However, this scheme may not be appropriate for having a condenser the below drum as there may not be adequate differential pressure for pushing condensate to the drum. An over-pressure bleed control valve from the drum may be provided for better performance.

The control valve in the equalization line has the advantage of keeping the drum at lower pressure than the condenser inlet. Thus, height requirement for the condenser (above the liquid level of the drum) can be lower. This may help revamp applications at higher loads of condensate to avoid flooding of the condenser.

1.3.4 Case Study on Pressure Control

A crude column in a unit was provided with a fuel gas makeup and bleed to flare arrangement for pressure control. There was a seal for ensuring subcooling

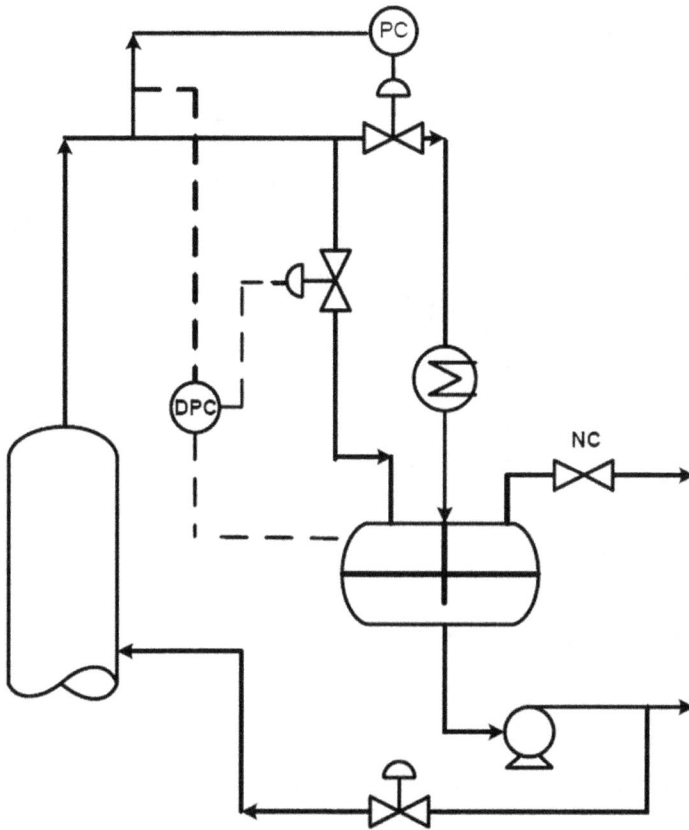

FIGURE 1.16 Pressure control with vapor line restriction.

downstream to the condenser. Non-condensable vents from the condenser and a seal loop was connected to the drum. The control valve to the flare was always found open, and LPG production from the stabilizer was found much lower than the potential. The problem was seen in detail at the site and it was found that the vent from the condenser originates from the hot end of the condenser instead of the cold end. Thus, the hot vapors through the vent line used to directly short circuit to the drum and used to bleed to the flare. This has led to loss of LPG components from the drum. The valves of this vent line were closed and a bypass across the liquid seal was made, and for column pressure control by makeup and bleed in the drum continued. LPG yield from the stabilizer increased with the change in the pressure control scheme as a result of less bleed of LPG components from the reflux drum of the crude column. Later it was understood that the error got introduced during modifications to revamp the unit. The original arrangement is shown in Figure 1.17.

PROCESSING SCHEME

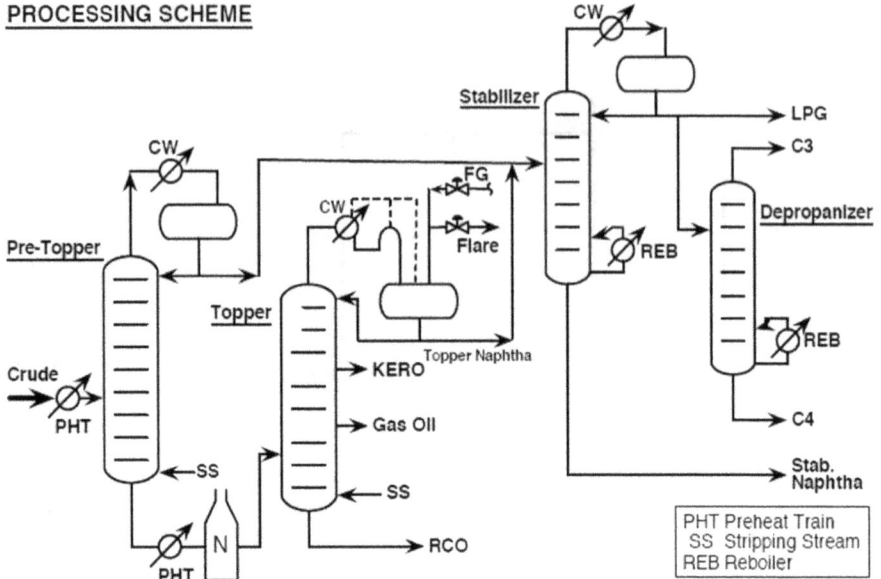

FIGURE 1.17   Original processing scheme.

1.3.5   REBOILERS

Reboilers are also extensively described in many books and articles. But some problems experienced were found worth including in this volume along with a brief description of the fundamentals. There are essentially two types of reboilers primarily used in distillation columns:

1. Once-through
2. Recirculation

Each may have two subtypes of configurations:

1. Thermosiphon
2. Forced circulation

1.3.6   ONCE-THROUGH REBOILER

In a distillation column with a once-through reboiler, the internal reflux is collected from the tray and directly fed to the reboiler. The reboiler outlet vapor and liquid are delivered to the column bottom. The internal reflux liquid is not directly routed to the column bottom without reboil. In the once-through arrangement, the reboiler can be considered as one theoretical stage. These reboilers are deployed only for low boil-up services like in strippers.

1.3.7 RECIRCULATION-TYPE REBOILERS

In a recirculation-type reboiler, liquid from the bottom spout is drawn and passed through the reboiler. The liquid from the bottom of the column to the reboiler is constituted by the internal reflux liquid from the tray above and part of the liquid from the reboiler outlet. In this type of reboiler, the liquid recirculates through the reboiler. The recirculation-type reboilers are used in columns operating with *high reflux ratios, as in splitters.* Recirculation and once-through reboiler arrangements are given in Figure 1.18. In the recirculation type of reboilers there can be arrangements with baffles and without baffles (*unbaffled arrangement*).

Often, a column bottom reboiler with a baffle is used. The partition baffle or preferential baffle separates the bottom spout into two compartments. The internal reflux is routed into one chamber and the liquid is then sent to the reboiler either by using a thermosiphon or forced circulation and the reboiler outlet is routed to the top of the same compartment, where the liquid will overflow to the product side compartment before being sent out. This ensures a constant head of liquid at the inlet of the reboiler, which is necessary for consistent good performance and is referred to as *baffles with overflow,* shown in Figure 1.19.

The other arrangement has the reboiler outlet returning to the product side compartment. Part of the liquid from the product side compartment enters the reboiler side with a large hole in the baffle (for recirculation of part of the reboiler outlet liquid to the reboiler inlet) and the arrangement is referred to as *baffles with underflow,* shown in Figure 1.20. The reboilers with baffles can deliver a fraction of a stage, and the ones without baffles cannot. It can be seen in Figure 1.20 that the liquid from product side recirculates to the reboiler side compartment through the big hole in the partition baffle, thus there is always a fixed level in the reboiler side compartment, providing a steady head at the inlet to the reboiler or to the suction of the pump, if used to feed the reboiler.

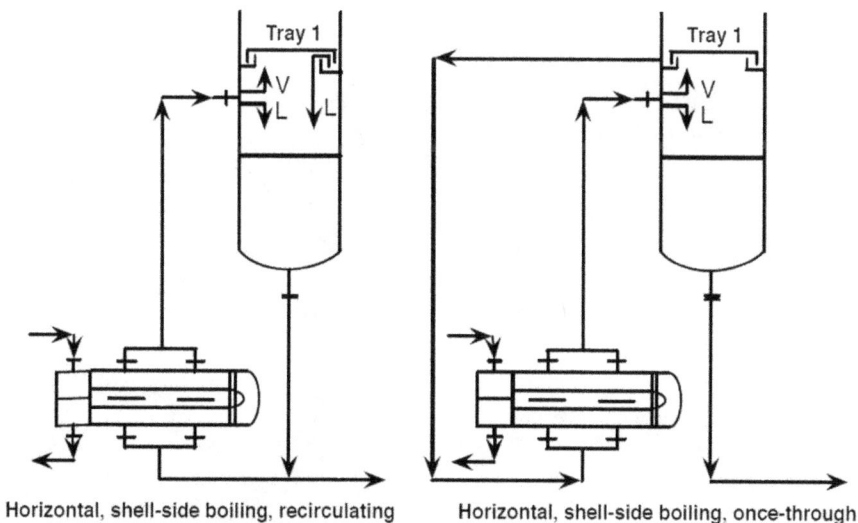

Horizontal, shell-side boiling, recirculating   Horizontal, shell-side boiling, once-through

FIGURE 1.18 Recirculating and once-through arrangement.

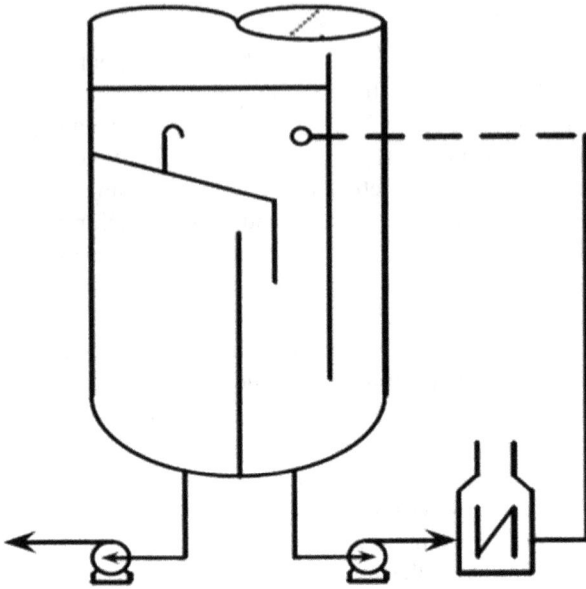

FIGURE 1.19   Baffle with overflow.

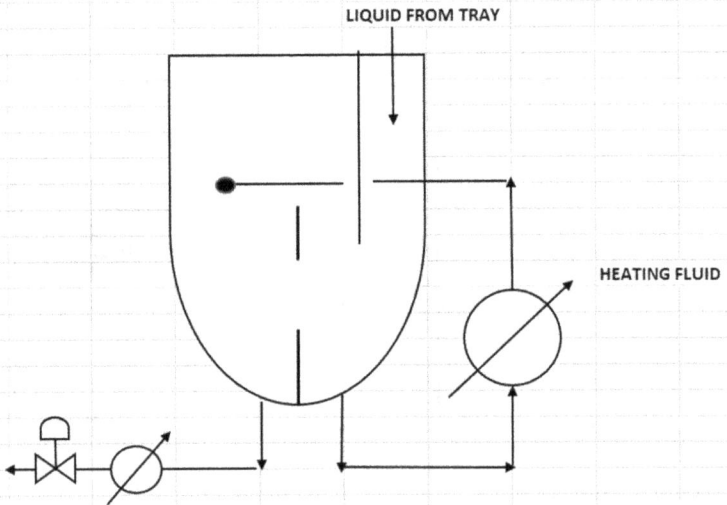

FIGURE 1.20   Baffle with underflow.

1.3.8 CRITICAL FLUX

In a reboiler, heat flux needs to be lower than critical flux to avoid film boiling and consequent drop in heat duty. Typically, when designing a thermosiphon reboiler, vaporization at the reboiler outlet is always kept below 30% to avoid critical flux and film boiling. Software for thermal design of reboilers performs rigorous calculations of critical flux when designing. Experts recommend limiting the vaporization at the outlet within 50% when a furnace is used as reboiler to avoid overheating of the tubes and consequent degradation of the fluid inside the tubes.

The reduced pressure correlation given by Mostinski is simple to use and gives values that are as reliable as those given by more complex equations. Mostinski's correlation uses reduced pressure for predicting the maximum critical heat flux:

$$q_c = 3.67 \times 10^4 P_c \left(\frac{P}{P_c} \right)^{0.35} \left[1 - \left(\frac{P}{P_c} \right) \right]^{0.9}$$

where P is the operating pressure (in bars), P_c is the liquid critical pressure (bars), and q_c is the critical heat flux (W/m2). A few other correlations are also available for estimation of critical flux.

1.3.9 PROBLEMS IN REBOILERS

1.3.9.1 Case Study 1

A reboiler like in Figure 1.21 (without the interconnection shown between two chambers) was found in two cases, one with a pump as shown (the other one was with thermosiphoning, not shown). In both cases, internal reflux liquid is collected in the reboiler side chamber, then is sent to the reboiler, and the reboiler outlet returns it to the product side chamber. The one with a pump for feeding the reboiler exhibited frequent loss of suction due to availability of lower NPSH than required for the pump. The thermosiphon one used to have frequent problems of lower duty. This was corrected by an interconnection with a valve between the chambers (product side and reboiler side), as highlighted with dark line or shown clouded (in Figure 1.21), thus converting the arrangement essentially to a baffle with underflow type.

1.3.9.2 Case Study 2

A thermosiphon reboiler of a FCCU stabilizer, heated by HP steam, was not delivering the heat duty, and the naphtha Reid vapor pressure (RVP) was found at the higher side. LPG production dropped. It operated in this condition for quite some time until the shutdown of the refinery. In the turnaround of the unit, the reboiler was opened and found totally coked up. It was even difficult to pull out the bundle from the shell. The reboiler was thermosiphon type with a baffle with overflow arrangement (Figure 1.22). The reboiler side was found completely coked, including the 20 mm weep hole in partition baffle bottom. The FCC gasoline is likely to have a considerable amount of olefin, which can oligomerize/polymerize and form coke. But surprisingly, in the same refinery the reboiler for the coker naphtha stabilizer was found without coke. The coker

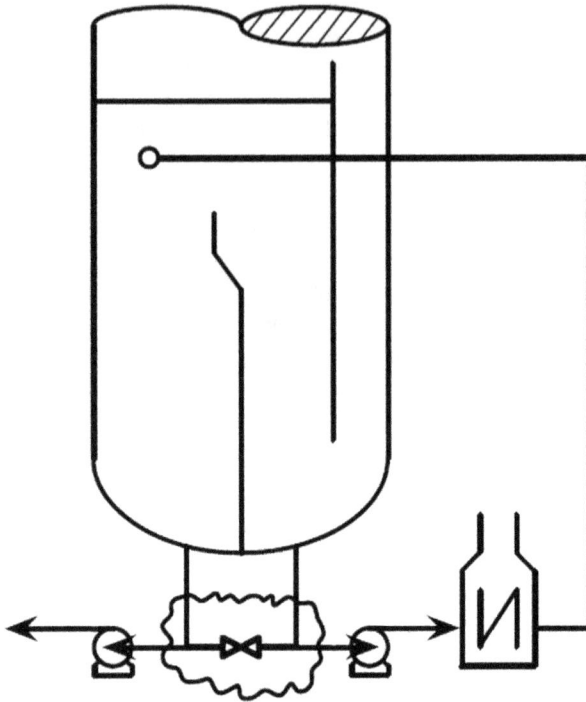

FIGURE 1.21   Resolving the problem with the shown modification.

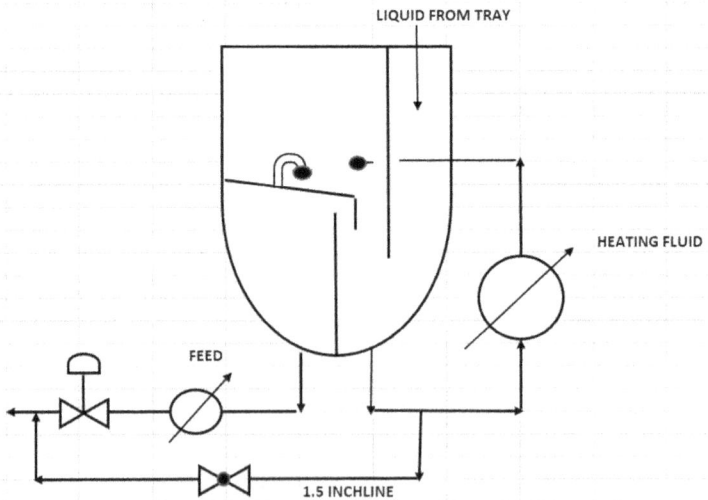

FIGURE 1.22   Modification showing a dump line with a valve.

gasoline is likely to have a higher proportion of olefins and diolefins than FCC gasoline, but still it did not coke up. The coker stabilizer bottom arrangement was examined and found that the bottom had a baffle with under flow arrangement, similar to Figure 1.20. In the arrangement the bottom liquid from the reboiler side compartment would flow through the reboiler and be routed to the product side compartment. Thus, there was no accumulation or stagnation of the bottom liquid so it would not polymerize to form coke.

In the FCCU stabilizer bottom, an arrangement to purge the liquid through a dump line of 1.5-inch diameter was made from the lowest point with a glove valve, as shown in Figure 1.22. It was apprehended that some bottom content would be forced under pressure to flow out. No lowering of the reboiler heat duty was further experienced in the continued run. In the next shutdown, taken three years later, the reboiler was again inspected and the bottom (reboiler side) was found clean and free of coke.

1.3.9.3   Case Study 3: Problem of Film Boiling in a Naphtha Vaporizer

A very strange problem was observed in a naphtha vaporizer of a steam reforming unit for H2 generation in a refinery using naphtha as feed. In the unit, naphtha was heated by HP steam (in tube side) and vaporized. The exchanger was of the TEMA and BEU type. After operation of the unit for quite some time, a shutdown of the hydrogen generation unit was taken up along with the turnaround of the total refinery. The vaporizer tube bundles were removed and considerable fouling was observed in the bundle. The maintenance team replaced the bundle with a spare and took the old bundle for cleaning/repair and testing. The unit was being started with the new bundle in the vaporizer, but very little temperature differential (heat transfer) was observed in the naphtha side, and condensation of steam in the hot side (tube side) was also marginal, corroborating very low heat transfer. Every possible likelihood was examined that could cause the problem and to identify the source. But nothing was found and the problem continued for a third day. Then the maintenance team came again, removed the new bundle from the vaporizer and fitted the old one after cleaning, repair, and testing. The reboiler started normally and problem totally disappeared. Normal heat transfer rate was observed. Later, the external surface of the tubes of the new bundle were examined was found slippery by the author and team. It was learned that some protective chemical spray was applied on the tube bundle for preservation. It was concluded that protective spray had modified the surface that induced film boiling and caused the problem.

1.3.9.4   Case Study 4: Waterfall Pool Effect in a
Reboiler of a Sour Water Stripper

In a sour water stripper, the H2S and NH3 content of stripped sour water were often used to go higher than designed (50 ppm H2S and 50 ppm NH3), for no perceptible reason. The stripper had two liquid passes in trays the section below the feed tray and had preferential baffle at the bottom dividing the bottom spout into two compartments. The liquid used to fall freely in the reboiler side compartment from the tray above and was fed to a thermosiphon reboiler heated by LP steam. The reboiler return was introduced to the product side compartment (a once-through reboiler). A

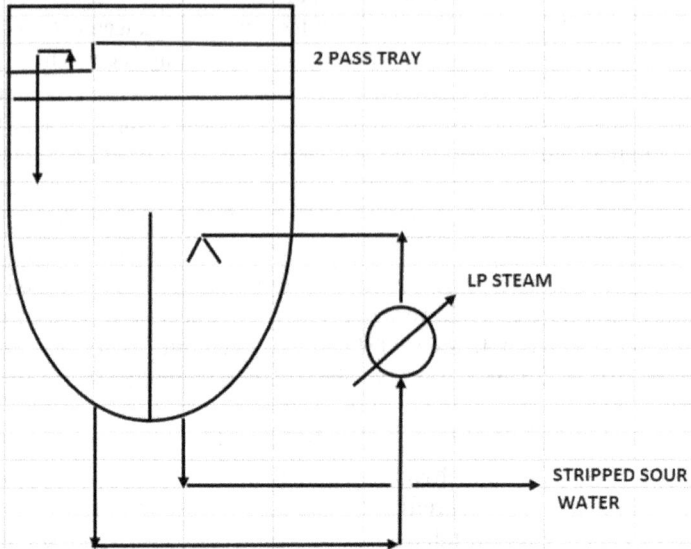

FIGURE 1.23 Sour water stripper column bottom section.

large height of free fall of internal reflux was noted (around 3.5–4 m) and suspected to be causing the foams that may get into the reboiler frequently (causing a *waterfall pool effect*) and reducing reboiler duty, and thus frequently affecting the heat input and quality of stripped sour water. The bottom spout of the column is shown in Figure 1.23.

An impingement arrangement (plate with packing) above the outlet nozzle to the reboiler was installed, resolving the issue. An improvised arrangement was made of a low height iron stool having the top surface constituted with layers of old packing held firmly in position. The entire assembly was placed above the column outlet nozzle to the reboiler to take the impact of liquid falling from height and divert relatively calm liquid to the bottom and to the reboiler. The problem was resolved.

1.4 PROCESSING OF HEAVY CRUDES

1.4.1 INTRODUCTION

Crudes are primarily classified as per API gravity. API gravity or ^{0}API is defined as API Gravity $= (141.5/\text{sp. gr.}) - 131.5$. Crudes having ^{0}API < 20 degree are called heavy crudes. Crudes having ^{0}API < 10 degree are called extra-heavy crudes. The extra-heavy crudes are mostly available from the Orinoco Belt of Venezuela and the Tar Sands Region of Canada. There is a larger reserve of heavy and extra-heavy crudes than regular crudes. The ratio of regular to heavy/extra-heavy crude reserve is known to be 1:5. Further, the heavy crudes are normally cheaper by about US$5–$15 per barrel than regular crudes and accordingly included in crude mix to reduce feed

TABLE 1.2

API Gravities of Extra-Heavy Crude Oils

Crudes (Latin American Origin)	°API (approx.)
Mexican Maya	17
Merey	8–10
Zuata	8
Crudes (Canadian Origin)	**°API (approx.)**
Cold Lake	8–10
Lloydminster	8–10
Tar Sand blends	8–10

cost. Thus, induction of cheaper heavy crudes in the crude basket provides an opportunity for the refiner to increase profitability. API gravity of some extra-heavy crudes (bituminous crudes) are listed in Table 1.2.

For the purpose of transportation and selling to customers, they are upgraded to lighter crudes. Some heavy crudes available are listed in Table 1.3.

1.4.2 CRUDE OIL UPGRADERS

Extra-heavy crudes cannot be transported through oil tankers. They are mostly processed in crude oil upgraders where traditionally carbon is rejected as coke and

TABLE 1.3

API Gravity and TAN of a Few Heavy Crudes

Crude	°API	TAN (mg KOH/g crude)
Kuwait export	30.8	0.14
Arab heavy	26.6	0.20
North Gujarat (INDIA)	22	2.06
Mexican Maya	20.2	0.16
BCF 22	21.5	0.9
Merey	16.6	2.21
Jubarte (Brazil)	17.3	3.3
Zuata (Brazil)	Around 21	2.8
Frade (Brazil)	19.8	1.0
Ras Gharib (Egypt)	22	2.2
Western Canadian select	20.7	1.45
Access western blend	20.3	1.9
Cold Lake	21.1	1.1
Lloydminster	Around 20	0.7
Canadian blend	19–21	2.3

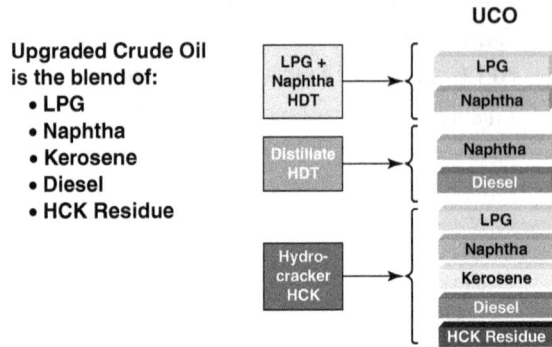

FIGURE 1.24   Constitution of upgraded crude oil (UCO).

liquid products are blended to constitute the synthetic crude. Traditional crude oil upgraders have a refinery-like configuration comprising of crude oil distillation, vacuum distillation, and the vacuum residue produced are further processed in a coking unit. Coke is rejected and liquid products are blended back to produce crudes of higher API usually 32° API or 42° API. The constitution of synthetic crude is shown in Figure 1.24.

The crude oil upgraders can also have hydrogenation facilities for vacuum gas oil (VGO) and heavy diesel fractions: extra-heavy crudes (8°–10° API) are upgraded to synthetic crude oil of around 32°/42° API. The essential features of crude oil upgraders are as follows:

- Upgraders have a typical refinery-like configuration, essentially having facilities that are carbon rejecting and/or hydrogen injecting or a combination of both.
- Typical carbon rejection facilities can be VR coker units and VR deasphalting units.
- Typical hydrogen-injecting facilities are hydrotreaters, mild hydrocrackers (MHC), and hydrocrackers.

Simplified flow diagrams of extra-heavy crude upgraders are presented in Figure 1.25.

The upgraded crude oil or synthetic crude oil (SCO) is normally blended with extra-heavy crudes to produce different grades of heavy crudes of various API gravities. Most of the heavy crude sold have an API gravity ranging from 16 to 22. The API gravities of a few heavy crudes are presented in Table 1.3.

The heavy crudes are characterized by low API, usually having a high total acid number (TAN), high sulfur, high viscosity, lower straight run distillate, higher VGO yield, and higher yields of residues. The VGO produced usually has a high TAN, higher CCR, sulfur, and Ni and V, and need special care in downstream VGO hydrotreaters and hydrocrackers. Similarly, the VR produced has high higher CCR,

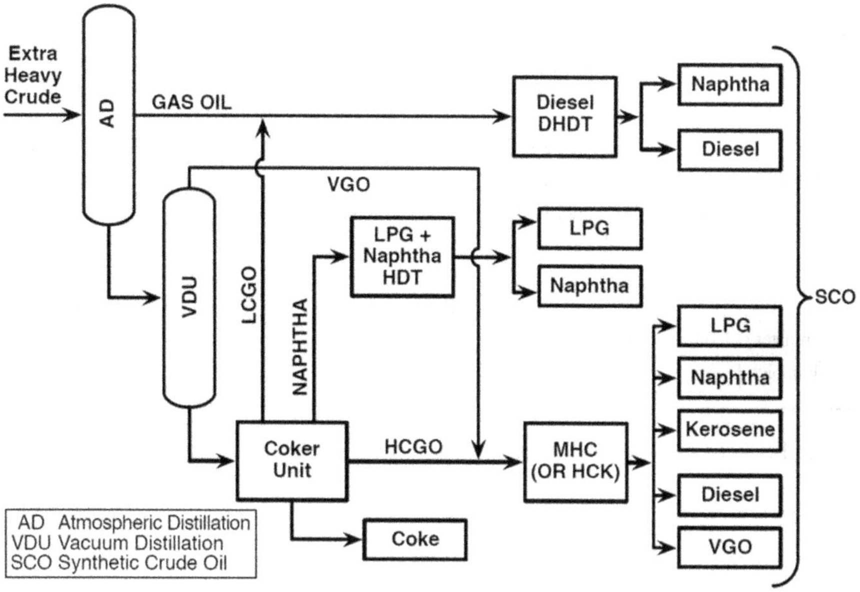

FIGURE 1.25   A simplified sketch of an upgrader using a coker unit for carbon rejection.

sulfur, and Ni and V as a feed to coker. Heavy crudes are received and normally blended before processing in refineries following these strategies:

- Compatible lighter crudes (that do not reject asphaltene upon blending)
- Hydrotreated lighter products from bitumen upgraders or extra-heavy crude oil upgraders

Usually, the heavy crudes available have high total acid number (TAN expressed in mg KOH/gm crude) as tabulated in Table 1.3. The bulk properties of some heavy crudes are given in Table 1.4.

1.4.3   GENERAL APPROACH FOR INDUCTION OF HEAVY CRUDES

Before introduction of the heavy crude in the existing crude mix, compatibility of the heavy crude with others needs to be assessed first (to ensure no asphaltene pre-cipitation). All other aspects like PONA of CRU feed and VGO yield, and properties like metal, CCR, and asphaltene content are evaluated and compared with respect to the feed specifications of the units. Similarly, the VR yield and the VR properties like CCR, asphaltenes, and metals are also checked with respect to the processing capabilities of the secondary units. Based on the findings, the proportion of heavy crude in the mix is decided. The heavy crudes produce lower atmospheric distillates but more VGO and VR. Thus, some revamp or debottlenecking of the secondary units may be necessary, particularly in the vacuum distillation unit (VDU), VGO

TABLE 1.4
Crude Bulk Properties of Heavy Crudes

Crude Type	Vanadium Content, ppm w	Viscosity at 100°F, cst	Salt Content, lb/1000 bbl of crude
Maya	290 (approx.)	95 (approx.)	6–8
Merey	295 (approx.)	460 (approx.)	45–65
Zuata	260(approx.)	330 (approx.)	55–65
Cold lake	125 (approx.)	75–80	20–25
Lloydminster (LLB)	100 (approx.)	70 (approx.)	40–45
Canadian blend	150 (approx.)	80 (approx.)	40–45

HT, and coker unit, if it is desired to increase the proportion of heavy crudes in the crude mix.

1.4.4   Processing Issues in Crude Vacuum Units

In general, the furnace duty of a crude unit reduces due to the lower proportion of straight run distillates present in the heavy crude and consequent lower vaporization in the crude unit furnace. But the vacuum unit furnace duty increases due to the higher RCO yield and the presence of higher proportions of VGO components in the RCO. Thus, vapor load in the atmospheric crude tower reduces, but the vacuum tower increases considerably. The higher vapor load in the vacuum column may result in high velocity (C_{SB}) in the wash zone and needs to be addressed. General required changes in operating parameters and conditions for processing heavy crudes in the atmospheric column and vacuum column are given in Table 1.5.

A typical configuration of a crude–vacuum unit is shown in Figure 1.26, and a typical crude unit with a pre-flash drum is shown in Figure 1.27.

1.4.5   Design/Rating of an AVU

1.4.5.1   Rating of an Existing Unit

The simulation flow sheet of a crude section of crude–vacuum unit, processing heavy crude in admixture, is presented in Figure 1.28 and material balance in Table 1.6.

TABLE 1.5
General Requirement for Processing Heavy Crudes

Variable	Atmospheric Column	Vacuum Column
Column inlet temperature	Higher	Higher
Flash zone oil partial pressure	Lower	Lower
Residue stripping efficiency	Higher	Higher

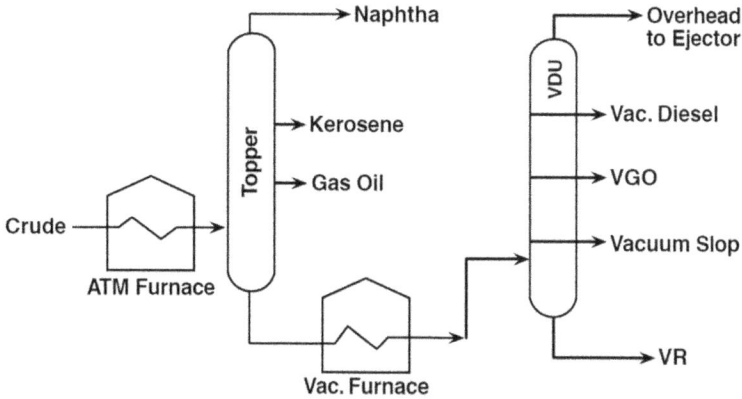

FIGURE 1.26 A combined atmospheric vacuum unit.

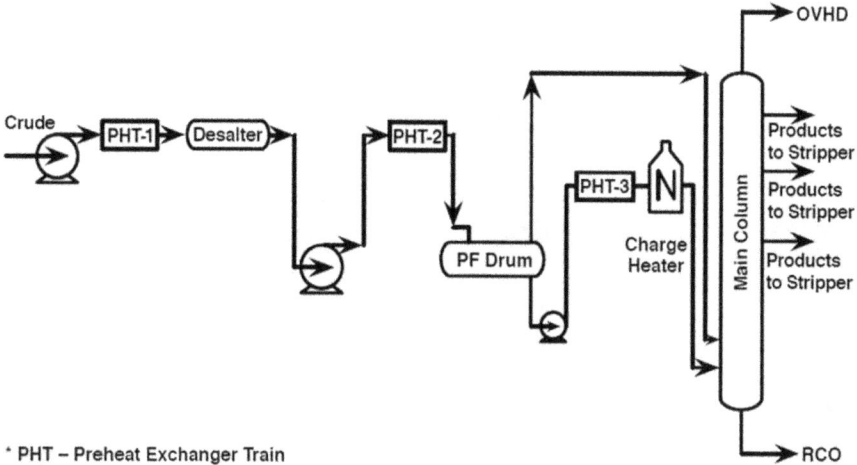

* PHT – Preheat Exchanger Train

FIGURE 1.27 Crude unit with pre-flash drum.

Theoretical stages used in crude column simulation are based on existing actual trays and efficiencies. The crude–vacuum unit simulation of a combined atmospheric vacuum unit (AVU) (shown in Figure 1.28) has a pre-flash vessel (PFV). The PFV preferred is horizontal to avoid foaming and carry over. The crude unit is provided with a top pumparound and a compressor to pick up the gases from the reflux drum to compress the vapors, and cool and recontact with naphtha in a recontactor. The hydrocarbon from the recontactor is fed to the stabilizer (not shown).

Essential inputs required for simulation are crude oil TBP and bulk density along with the flow rate and pressure and temperature of the crude. The column specifications used are pressure of reflux drum (1.5 kg/cm2 (a)), and the pressure drop across the column used in simulation is based on actual plant data. Other specifications

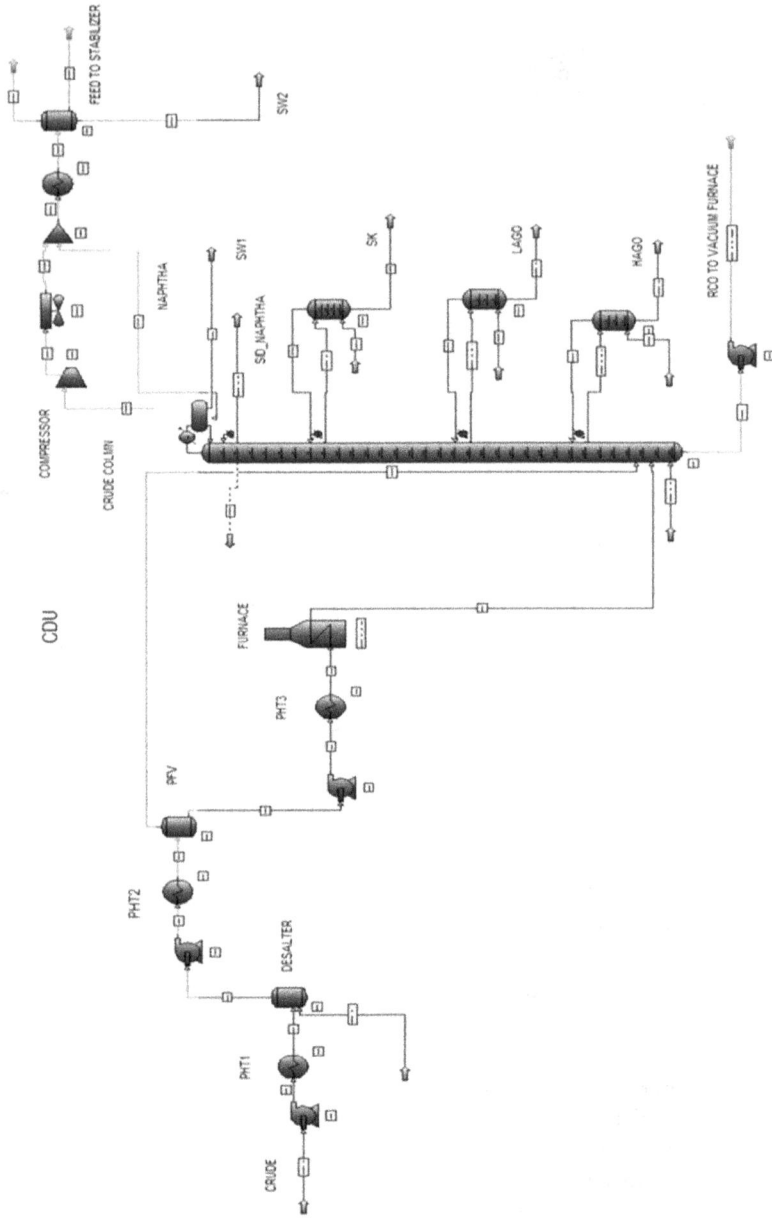

FIGURE 1.28 Crude section of crude distillation unit.

TABLE 1.6

Material Balance of Crude Section

Stream Name		CRD1	NAP	SID-NAP	SK	LAGO	HAGO	RCO
Total Mass Rate	KG/HR	1850000.000	175000.000	30000.000	281927.466	175000.000	175000.000	969801.057
Temperature	C	35.000	55.000	122.709	163.902	228.474	284.344	348.962
Pressure	KG/CM2	10.000	1.500	2.219	2.500	2.700	2.800	2.800
Total Std. Liq. Density	KG/M3	859.887	719.532	747.113	789.007	826.945	861.493	943.784
ASTM D86 at 760 MM HG (LV)	C							
IBP		-11.744	13.918	39.332	134.136	169.487	215.502	242.759
5%		31.124	37.044	82.805	152.997	205.986	261.471	336.472
10%		104.773	46.249	100.033	161.515	224.510	284.606	386.847
30%		208.579	70.689	114.836	173.280	245.484	308.169	451.688
50%		325.350	89.443	123.651	185.246	257.458	321.372	507.362
70%		492.029	105.612	131.840	200.287	269.060	336.636	566.763
90%		682.000	124.662	142.504	222.289	286.267	362.672	653.063
95%		701.648	131.249	148.372	230.000	293.104	371.134	665.474
EBP		714.302	137.969	154.424	236.993	299.616	379.987	675.362
ASTM D1160 at 760 MM HG (LV)	C							
IBP		-0.333	6.868	41.927	129.906	173.297	221.501	254.435
5%		36.103	23.893	75.209	145.456	201.905	259.819	339.121
10%		98.691	30.692	88.436	152.499	216.467	279.141	384.460
30%		213.327	65.903	112.181	172.397	247.869	313.716	464.400
50%		335.303	89.291	124.463	188.326	263.835	331.104	528.559
70%		486.260	111.576	137.220	209.327	280.809	352.361	593.336
90%		624.926	135.947	152.947	236.499	303.379	383.215	669.583
95%		667.590	144.747	160.245	247.334	312.450	396.709	692.158
EBP		695.932	153.722	167.773	257.160	321.089	410.825	710.142

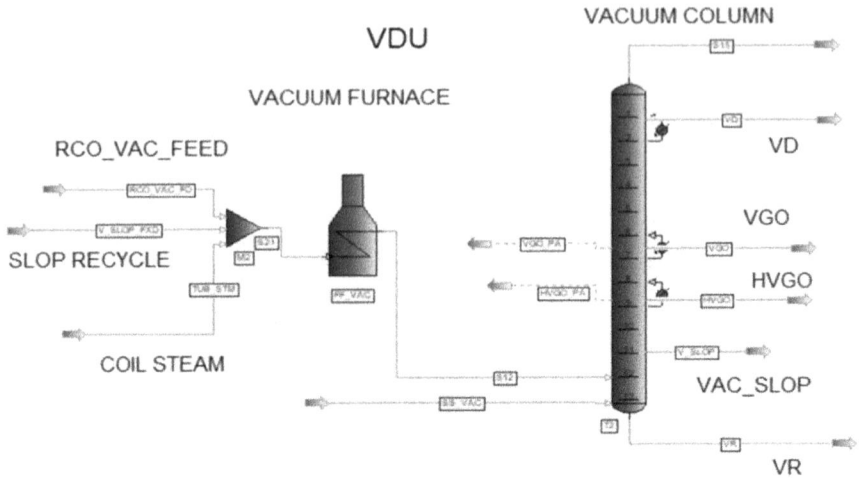

FIGURE 1.29 Vacuum section of the crude distillation unit.

like stripped product flow rates, pumparound duties and differential temperatures, top pumparound return temperature (90°C), and reflux drum temperature (55°C) are supplied based on actual plant data. The crude furnace coil outlet temperature (COT) is fixed to provide the required overflash of around 3–4 LV% on crude flow.

The hot RCO from the crude column is directly fed to the vacuum section. The simulation flow sheet of the vacuum section is presented in Figure 1.29, and the material balance of the vacuum section is given in Table 1.7. The vacuum column stages considered is based on the HETP of the existing packed beds, indicated by the packings supplier. Other vacuum column specifications like vacuum column top pressure, column pressure drop between the top and flash zone, pumparound duties, pumparound differential temperatures, and product draw rates used in the simulation are based on actual plant data. The vacuum furnace COT is fixed to provide required wash oil flow on the wash bed. The purpose of the simulation was to rate the columns and furnaces, and identify bottlenecks, if any. The column internals were also rated by supplying necessary details like column diameter, type of valves on tray, number of valves, and downcomer and packing detail. In this particular example, the crude column top pumparound section was found limiting with the actual flow of pumparound. It was advised to lower the return temperature of the pumparound by 10°C–15°C to lower the flow of top pumparound maintaining the same duty (PA return temperature must be above the water dew point).

The rating revealed that the stripping section of the vacuum column was approaching incipient flooding. The downcomer area was found limiting in the bottom stripping trays. Accordingly, the refinery was advised to modify the existing cross-flow stripping trays with the support of column internal experts. The column vapor–liquid flow profile was generated along with properties and given to the internal supplier as an input for modification of the trays.

TABLE 1.7

Material Balance of Vacuum Section

Stream Name		RCO-AC_FD	VD	VGO	HVGO	V_SLOP	VR
Total mass rate	KG/HR	969801.057	44006.345	357000.000	110000.000	60000.000	452390.778
Temperature	C	349.506	116.166	241.436	336.559	369.288	364.935
Pressure	KG/CM2	12.000	0.029	0.041	0.045	0 050	
Total Std. Liq. Density	KG/M3	943.784	860.058	912.571	943.809	962.040	981.271
ASTM D86 at 760 MM HG (LV)	C						
IBP		242.759	255.430	338.026	406.455	418.469	501.034
5%		336.472	276.192	367.282	447.806	478.333	528.899
10%		386.847	287.398	382.988	469.325	505.787	543.200
30%		451.688	304.861	407.230	489.289	530.746	569.488
50%		507.362	317.950	428.012	503.077	545.943	592.428
70%		566.763	331.888	452.570	517.204	563.271	620.986
90%		653.063	352.361	489.840	540.086	592.896	657.122
95%		665.474	360.000	500.025	546.236	603.751	605.798
EBP		675.362	367.326	509.388	556.737	616.348	670.130
ASTM D1160 at 760 MM HG (LV)	C						
IBP		254.435	257.362	346.303	420.028	431.361	517.058
5%		339.121	274.012	369.896	456.370	487.576	540.588
10%		384.460	283.026	382.600	475.317	513.342	552.699
30%		464.400	310.236	418.144	505.839	549.996	591.024
50%		528.559	327.494	444.050	523.987	569.772	619.523
70%		593.336	347.224	475.445	543.936	593.358	655.053
90%		669.583	372.957	515.896	571.924	627.372	694.571
95%		692.158	383.897	533.202	582.055	648 801	704.956
IBP		710.142	394.389	549.110	599.351	673.672	710.142

1.4.5.2 Considerations for New Design of a Crude–Vacuum Unit

Crude column

Efficiency above wash zone: 65%–70%
Efficiency of wash zone: 30%–40%
Efficiency of bottom stripping zone: 30%
Side stripper tray efficiency: 50%
System factor: 0.85–0.9 (some cheaper heavy crudes can be foaming)

ASTM gap and overlap specifications will influence the product draw location from the main column to the side strippers. Three stages are preferred in side strippers. The kerosene stripper may have a higher number of stages for meeting the flashpoint of kerosene.

Typical simulation inputs for crude column

1. Crude true boiling point (TBP) data (essential)
2. Crude density or API gravity $= 141.5 - (131.5/\text{sp. gr.})$ or density (essential)
3. Crude molecular weight (optional)
4. Viscosity (optional)
5. Thermodynamics, VLE (K-value): GS/SRK
6. Transport property: PETRO/SIMCI heavy oil correlation. The viscosity of crude and heavy fractions can be estimated from a mid-percent viscosity curve given along with crude TBP, and the data can be entered as input in the simulation.
7. Density evaluation: by API method

Pumparound (PA) differential temperatures in crude column

- Differential temperature (ΔT) across top PA: around 25°C–30°C, return temperature must be higher than the column top water dew point
- Differential temperature (ΔT) across kerosene PA: 55°C–60°C
- Differential temperature (ΔT) across LGO PA: 55°C–60°C
- Differential temperature (ΔT) across HGO: 55°C–60°C

Typical reflux ratio: 1.5–1.8

Vacuum section (fuel type)

Typical stages in vacuum column (fuel type)
Top PA or vacuum diesel PA $= 1–2$ stages
Between heavy diesel draw and LVGO draw $= 2–5$ stages

LVGO PA $= 1–2$ stages
LVGO–HVGO $= 1–2$ stages
HVGO PA $= 1$ stage
Wash section $= 2–3$ stages, bottom stripping section $= 2–3$ stages

Packing is used in most of the vacuum columns but for the bottom stripping section. In the stripping zone, trays are normally used. The efficiency of the stripping trays can be taken as 30%. Counter-flow type trays are preferred.

Typical simulation inputs for standalone vacuum distillation unit

1. For standalone vacuum unit, feed RCO, TBP, and density need to be supplied.
   - Differential temperature (ΔT) across heavy diesel PA (ΔT): 70°C–90°C or return temperature 55°C to 60°C
   - Differential temperature (ΔT) across light vacuum gas oil (LVGO) PA: 55°C
   - Differential temperature (ΔT) across HVGO PA: 55°C, where ΔT is the differential temperature between pumparound supply and return
   - Wash liquid at the bottom of wash bed is typically ~0.8–1 M3/h/M2 Column C.S. area
2. Thermodynamics (K-value): BK10/GS.
3. Transport property option: PETRO (method)/SIMCI heavy oil correlation. For standalone vacuum unit, the feed RCO viscosity may be entered from a mid-percent viscosity curve given along with crude TBP.
4. Density option: API (method).

Fuel-type vacuum columns are designed for a far lower number of stages (typically 9–10 stages above the flash zone) than lube-type vacuum columns. Typically, 2 stages for wash zone and 2 to 3 stages are provided in the stripping zone. Fewer stripping stages may reduce VGO recovery and would need higher stripping steam or higher vacuum. Counter-flow trays are preferred in stripping zones to avoid tray damage by occasional level buildup and quick pump out (due to *vapor gap*).

Common challenges/processing problems experienced in crude vacuum units while processing heavy crudes are as follows:

- High crude side pressure drops due to higher crude viscosity
- Poor desalter performance due to lower temperature achievable and often due to presence of mud and silica along with crude
- Higher acid corrosion in crude column overhead circuit and condensers
- Fouling in top to kerosene zone in crude column due to salting
- Low diesel yield
- Higher vacuum heater duty
- Vacuum heater coking
- Fouling in vacuum column
- Low HVGO product yield (due to dropping of VGO to contain contaminants)
- High-sulfur and metals in VGO
- High-sulfur and metals in VR
- Higher corrosion in general

1.4.5.3 Features Needed in Crude Units for Processing of Heavy Crudes

1. Use of efficient filtering media in crude booster pump suction to arrest mucks entering the preheat train and resultant fouling.
2. A configuration with pre-flash drum (Figure 1.27) may be necessary to avoid water ingress in the high-temperature heat exchanger train. A horizontal pre-flash drum is preferred as the service may be foaming.
3. Provision of higher desalter temperature (140°C–145°C) to reduce viscosity of crude to facilitate better desalting.
4. Parallel train of heat exchangers to reduce pressure drop on account of higher crude viscosity.
5. Problem of low pre-heat/desalter temp can be due to following:
   - Low heat transfer coefficient U_D because of high viscosity of the streams necessitating higher heat transfer area.
   - Low pumparound duty in the crude distillation unit (CDU) due to lower potential of straight run distillates.
   - Less heat recovery from VR as VR stream cannot be cooled to lower temperature due to its high viscosity. At lower temperature, the pressure drop issue in the VR rundown circuit is likely.
6. The cold heat exchanger train (before desalter) may have less heat available at a low temperature to heat crude to desalter temperature, as the heat of VR cannot be fully recovered and imparted to crude in cold train. The short fall of heat may need to be supplied with column overhead heat integration with crude or top pumparound heat integration to crude to deliver the shortfall of heat to crude in the cold train.
7. Additional surface area in preheat exchangers, particularly in the cold heat exchanger train, to achieve and desalter temperature due to higher viscosity at lower temperatures of the streams in the cold train.
8. Better desalting facilities to take care of increased salt content in crude. Crudes from Tar Sands have high sediment and clay content, and some blends also have high viscosity thereby creating more difficulties in desalting.
9. Efficient caustic injection facility at desalter downstream to reduce chloride corrosion at crude and vacuum column overhead section, as shown in Figure 1.38 and Figure 1.40.
10. High velocity in heat exchangers needs to be maintained to avoid asphaltene precipitation and may result in higher pressure drop and may need higher power for pumping.
11. Recycling of naphtha/heavy naphtha along with heavy crude to reduce its viscosity, if required.
12. Special metallurgy in section of column based on naphthenic acid distribution in the crude. Special metallurgy in furnace tubes/transfer lines (e.g., SS 316L and SS 317, depending upon TAN of the crude processed).

Other issues include:

1. Refiners often increase atmospheric tower bottoms (ATB) initial boiling point (IBP) to stay within vacuum column diameter limits. An additional heavy diesel cut withdrawal from crude column is useful to reduce vacuum furnace duty and column wash zone velocity in vacuum tower. The heavy diesel cut from crude column may join the HVGO PA return of the vacuum column. The author provided the arrangement in the design of a vacuum column to lower the wash zone load (C_{SB}) of the vacuum column processing RCO derived from heavy crude.
2. Higher viscosity of RCO and VR may need splitting of the streams and laying parallel heat exchanger trains, particularly in revamp applications. Additional pumps may be required in series to provide battery limit pressure for sending residues to storage. The author provided a booster pump for the RCO at battery limit, in series, in a crude unit (processing heavy crude) for sending the RCO to the tank.
3. Reflux drum arrangements.

There are two schemes/alternatives of a reflux drum arrangement:

- Dry reflux drum (temperature higher than water dew point)
- Wet reflux drum (forced condensation)

In dry reflux drums, there is the likelihood of reaching the water dew point anywhere and that can result in severe initial condensation point (ICP) corrosion. When corrosion in the overhead circuit or overhead/crude exchanger is a problem, then the wet reflux drum may be a preferred option.

As desalting of heavy crude is a problem, the salt content of crude oil at the outlet of the desalter is usually higher than 1–2 pounds per thousand barrels (>1 ptb). The overhead chlorides are likely to be higher and can result in higher overhead corrosion. Further, due to availability of lower heat in the cold heat exchanger train, an overhead heat integration or top pumparound heat integration are common. This needs higher attention to avoid acid corrosion in the overhead/crude heat exchangers.

The most common reflux drum used is a wet or forced condensation type. The water quantity must be a minimum 30% excess over the quantity required for saturation, and sufficient mixing length ahead of the condenser needs to be ensured. Single-reflux drums are, however, more common. But when overhead/crude exchangers are used, then two drums is the preferred option as higher differential temperature is available between crude and overhead. The two-reflux drum arrangement with forced condensation is shown in Figure 1.30. The overhead vapor heat is partly transferred to crude in the heat exchanger and the condensed vapor is routed into the hot reflux drum, typically at a temperature of 85°C–90°C. The vapor from the hot reflux drum is further cooled and routed to the product

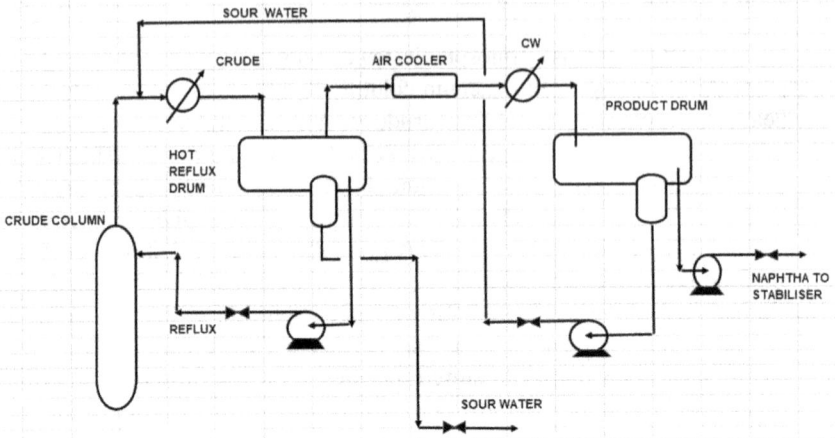

FIGURE 1.30 Typical two reflux drums in overhead with forced condensation.

drum. The sour water from the second drum is recirculated at the inlet of the over-head/crude heat exchanger for forced condensation. The reflux is returned to the column from the hot reflux drum. An air cooler in most of the cases is provided at the upstream of the hot reflux drum (at the downstream of the overhead/crude heat exchanger). This is to maintain the hot reflux drum temperature. An air cooler followed by a water cooler is often provided to cool the vapors from the hot reflux drum going to the cold drum. Provision of pumping of the hot reflux drum liquid is often provided to maintain the level of the hot reflux drum.

1.4.5.4 Features Needed in Vacuum Distillation Units for Heavy Crudes

The processing of heavy crudes in the CDU usually results in a higher feed rate and vapor velocity in the vacuum column. There may be some modification needed if it is not designed to process a higher feed rate and handle higher VGO content in the feed. Some special design and operation approaches in the vacuum unit may be required for heavy oil processing as described next:

- Efficient vapor horn to be installed at the column inlet to reduce entrainment of heavier ends for limiting load in the bottom of the wash bed and metal content of the VGO stream.
- Very strong wash zone required, i.e., higher irrigation rate (wash oil) to prevent entrainment and wash zone bed coking.
- Additional bed for wash oil circulation facility may be provided below HVGO wash, as in Figure 1.31.
- VGO cut point may need reduction to maintain feed quality for the downstream VGO treater with regard to N2, metal, asphaltene (C7 insoluble), and CCR to meet VGO HT feed specifications.

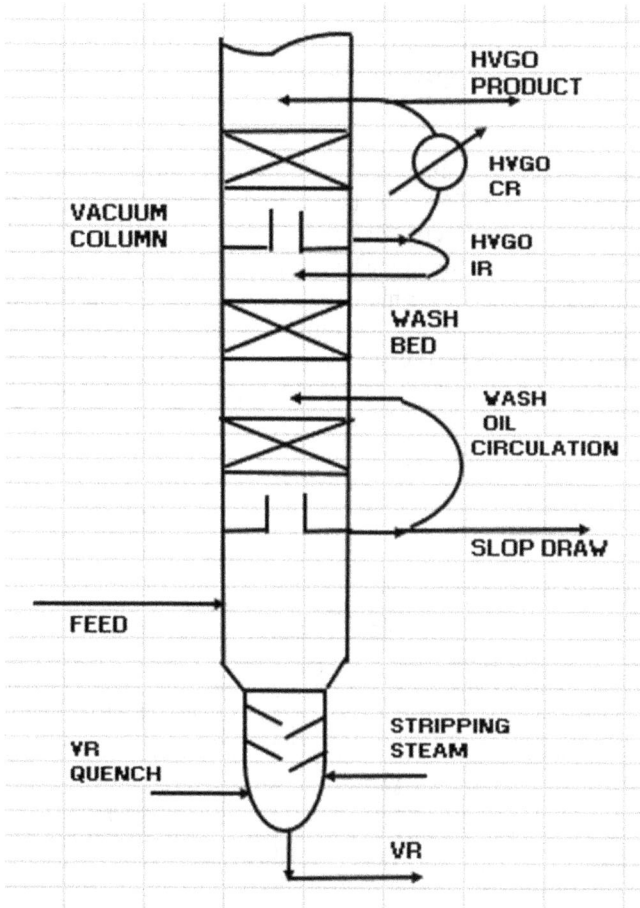

FIGURE 1.31 Wash oil circulation below HVGO wash bed.

- XVGO/HHVGO withdrawal facility to draw extra-heavy VGO fraction and divert to the FCCU feed can be provided. Lighter VGO will be fed to the VGO treater/hydrocracker to avoid nitrogen, asphaltene, Ni, and V in the feed, and heavier VGO can be routed to the FCCU. The FCCU can tolerate a little higher metal and CCR than fixed-bed hydrotreaters, as metal-laden catalyst is rejected on a regular basis from the inventory and fresh catalyst is made up.
- Metallurgy of selected sections of the column need to be upgraded depending on the naphthenic acid distribution in the crude.
- The vacuum column bottom quench arrangement needs special attention.
- A coke strainer with tar pot arrangement may be preferred to further reduce residence time at the vacuum column bottom to lower cracking and production of the lower amount of decomposition gases, as shown in Figure 1.33, while processing heavy crudes.

- A higher quantity of coil steam may be required to lower the residence time in heater coils and reduce production of decomposition/cracked gases.
- The hot well may have special features like off gas amine treatment, shown in Figure 1.35.

The author has revised the configuration of a vacuum unit, as shown in Figure 1.31, processing low sulfur, high TAN, and heavy crude with high nitrogen, deployed for production of feed to the hydrocracker, and introduced an additional HHVGO drawn by eliminating the wash oil circulation bed and using the space as in Figure 1.32. The unit has operated satisfactorily for more than 20 years. LVGO and HVGO together is routed to hydrocracker maintaining the nitrogen content of 800 ppm max. (to meet HCU feed specification), and HHVGO is routed to the FCCU. This arrangement has increased overall VGO recovery, as the FCCU unit was in position to process feed

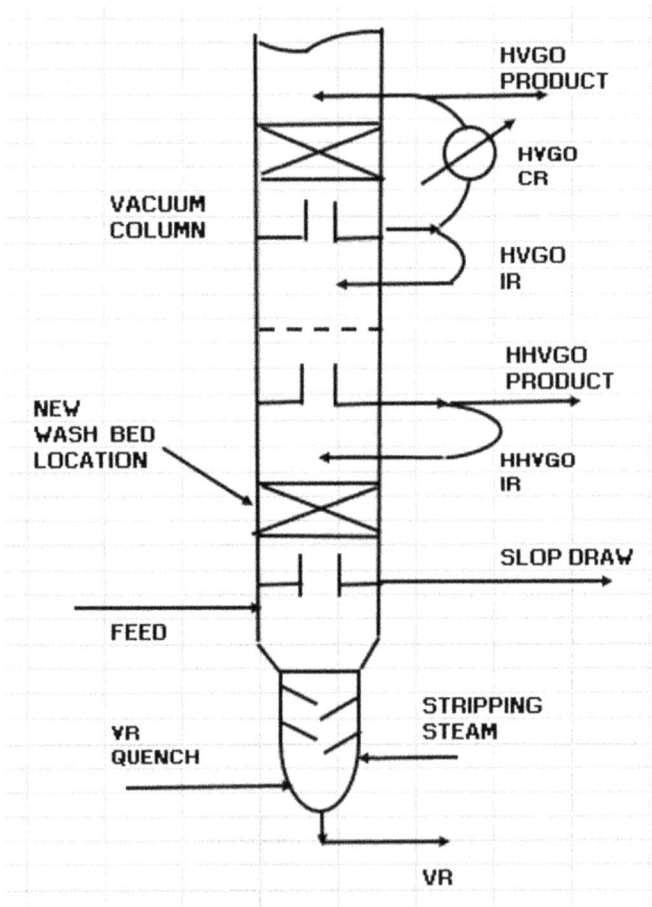

FIGURE 1.32   HHVGO draw by relocating wash bed below HHVGO draw.

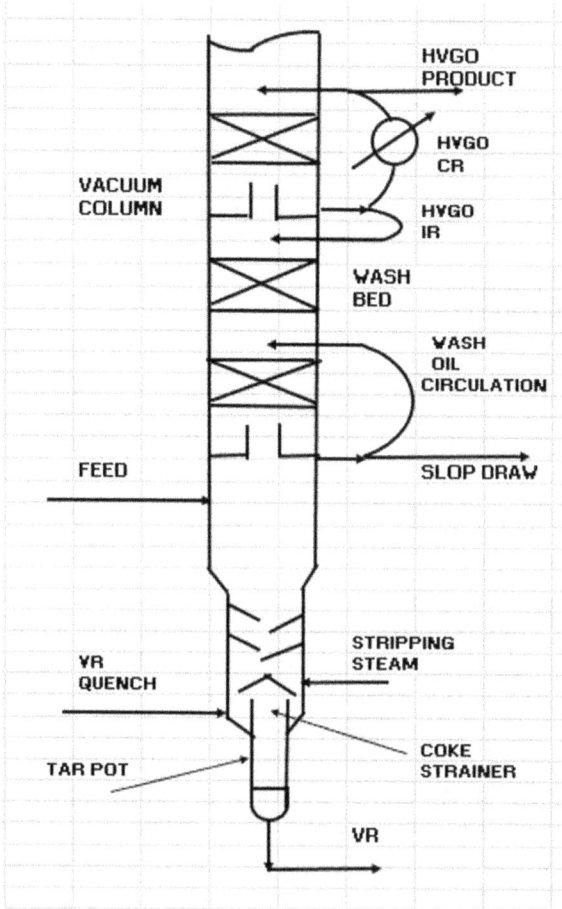

FIGURE 1.33 Vacuum column bottom with tar pot to lower residence time of VR.

with higher nitrogen (it was primarily processing VGO with low nitrogen and used a fresh catalyst make up of about 2 MT/day).

The vacuum column bottom with tar pot to reduce the bottom residence time is shown in Figure 1.33 to reduce cracking in the bottom and reduce production of decomposition gases.

The VGO quality and the performance from a vacuum unit is highly influenced by the following:

1. Wash zone efficiency of vacuum column

In refinery vacuum tower services, vapor volumes are very large for the feed rate. The tower cross-section area provided is based on the vapor handling requirements. The *C-factor* derived from the Souders-Brown equation is traditionally used for limiting the maximum velocity of vapor through the flash zone. The wash zone vapor velocity has a maximum limit expressed by C_{SB} or C-factor, defined as

$$C_{SB} = V/A_{CS} \sqrt{\left(\rho_V/\left(\rho_L - \rho_V\right)\right)}$$

where C_{SB} is the Souders–Brown C-factor in ft/sec (often shortened to C-factor and not using the subscript), V is the velocity of rising vapor in ft³/sec, A_{CS} is the cross-sectional area open to flow in ft², and ρ_V and ρ_L are the vapor density and liquid density, respectively, in consistent units.

The maximum C-factor limit for an open-structured packing or grid in a refinery vacuum tower is 0.45 ft/sec. The vapor leaving the wash bed becomes the vapor entering the HVGO bed.

A wash bed is located just below the HVGO draw, and a combination structured packing and grid is used in the wash beds. Combined beds have a sparse grid at the bottom and denser structured packing on top. In the wash section of the vacuum column, the liquid flow is very low compared to the very high vapor flow rate. Wash liquid from the HVGO tray is introduced on top of wash bed with the distributor to evenly distribute the wash liquid. The wash bed maintains the HVGO quality in respect of CCR, asphaltene, metal content, and the end point. Typically, two stages are provided in the wash zone. The dried wash liquid from the wash bed together with the entrained liquid from the flash zone bottom collect in the pan below the wash bed and are mostly drawn as vacuum slop or recirculated with feed to the furnace inlet (as furnace recycle to eventually work as wash liquid) or simply dropped to the column bottom stripping section through internal pipe downcomers.

The main reason for wash bed failure is coking due to insufficient wash oil or pressure-surge-induced damage caused during startup while pulling vacuum. A metered withdrawal of vacuum slop under level control of the pan tray or an outside-located draw-off vessel (not to allow overflow) provides a good indication of wash liquid at the bottom of the bed and help in evaluating the wash liquid at the middle of wash bed where liquid is likely to be minimum and may be the cause of coking at the location.

It may be remembered that at the bottom of the bed entrained liquid from flash zone may provide some additional irrigation, but at the middle of the bed the liquid irrigation is minimum and is critical in arresting coking of the bed. A proportion of about 25%–30% of the slop oil flow can be assumed to be contributed by entrainment. A sodium content analysis of slop oil can help in estimating the entrainment of heavies from the flash zone.

In addition, the slop wax collector tray should have a sloped construction, and liquid from the collector chimney tray may flow freely without accumulation into a slop oil drum provided outside and is quenched by cooled recirculated slop oil to lower the temperature to arrest coking in the drum. Level in the drum is maintained by manipulating the slop oil draw. The arrangements help reduce the probability of the slop wax collector coking.

2. Efficiency of de-entrainment devices

The vacuum tower feed enters in the tower flash zone as a mixture of vapor and liquid. The high-velocity feed creates small liquid droplets. The feed entry device allows the larger droplets to fall toward the tower bottom. Commercial units operate at much higher C-factors. Some entrainment typically carries into the bottom of the wash bed. An efficient de-entrainment device at the flash zone is essential to contain the entrainment and heavies loading the bottom of the wash bed.

Efficient de-entrainment devices are very much required for the processing of feed derived from heavy crudes. Different types of flash zone disengagers in flash zone (two-phase feeds) in use are described next:

- Radial: Feed may enter the vacuum tower radially (aimed across the center of the tower). Modern radial entries use vanes to direct vapor to both sides of the entry device. The vanes also provide an impingement surface for liquid collection. The liquid flows sideways along the vane and falls off the end. The leading edge of the vanes move closer as the feed crosses the tower. Each vane cuts away a portion of the vapor at that point and directs it to the tower sides.
- Circumferential: Feed may enter tangentially (aimed along a tangent to the tower circumference). Modern tangential entries use vanes to direct vapor into a core region of the tower. The vanes are located at stepped elevations around the torus. Each vane cuts away a portion of the vapor at that point and directs it to the tower center. The author has experience of repeated damage of a radial device and vanes, and thinks that a traditional circumferential device (tangential entry) may be sturdier one. He does not have any experience with damage to the circumferential device.

3. Dry-out ratios
Wash oil dry-out ratio is defined as the actual volume wash liquid introduced as internal reflux at the top of the bed divided by the actual volume at the bottom of the bed. Operating units have been known to exhibit a dry-out ratio as high as 5:1 or higher for deep cut units. Typical may be 3 to 4:1.

4. Minimum wetting values
A safe minimum wetting value for the wash bed may be 0.7 m3/Hr-m2 of distillate liquid at the middle of the bed or 0.5 M³/Hr at the bottom (without consideration of the entrainment). This value is based on liquid volume at actual conditions and the tower cross-section area for the wash bed. Wash liquid means that this is the liquid rate at the bottom of the bed without adding any amount for entrainment in the calculation. This minimum liquid rate specified is based on experts' experience with many units over the last two decades.

5. Barometric legs and hot well
Another important aspect in the vacuum overhead system is that the barometric legs should be inclined at a maximum angle of 45 degrees and preferentially should not

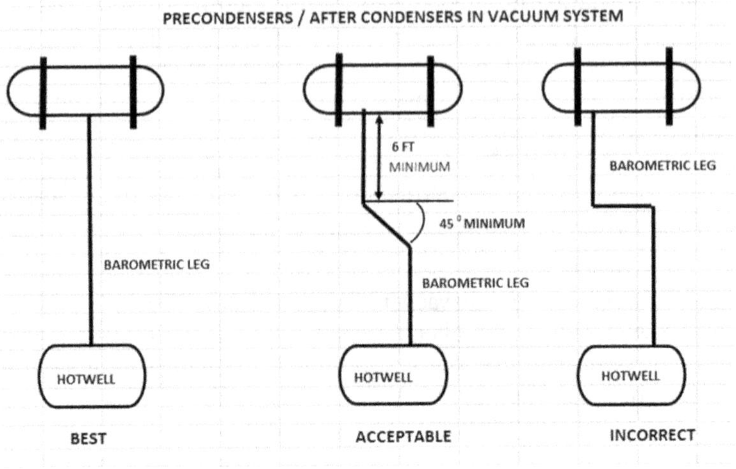

PRECONDENSERS / AFTER CONDENSERS IN VACUUM SYSTEM

6 FT
MINIMUM

BAROMETRIC LEG

45° MINIMUM

BAROMETRIC LEG

BAROMETRIC LEG

HOTWELL

HOTWELL

HOTWELL

BEST

ACCEPTABLE

INCORRECT

FIGURE 1.34   Inclination of barometric legs.

enter the hot well through the bottom with a U loop as the mucks can collect and eventually partially or fully block the legs, and may result in buildup of condensate in the condensers and resultant loss of vacuum. The preferred barometric leg inclinations are shown in Figure 1.34.

The cracked gases produced in the vacuum unit processing RCO of heavy crude may have a very high proportion of H2S and may need an amine wash for of the ejector off-gases (non-condensables) before routing it to furnace firebox or other safe location. A low pressure drop meter (typically a venture meter) measures the flow of non-condensables to estimate the generation of off-gas. There should be a sampling provision to collect a sample of the sweet off-gas for analysis to further estimate the air leak and cracked gas generation. A sketch of the hot well is given in Figure 1.35.

Barometric legs combined and entering with a U loop has been seen in number units particularly in bigger units (having multiple pre-condensers), as in Figure 1.36, is not advisable if can be avoided.

6. Vacuum furnace transfer line pressure drop

The vacuum furnace transfer line pressure drop is a very important parameter for processing of RCO feed derived from heavy crudes. A lower pressure drop is desired in the transfer line to the column. A lower pressure drop in the transfer line reduces the film temperature of the oil inside the furnace tubes and reduces coking, cracking, and production of cracked gas. The typical design pressure drop is 150 mm Hg. The preferred velocity in the column feed nozzle is 0.6 times sonic velocity. The sonic velocity can be estimated as

$$Vs = 6160 \left(K/1.1 \right)^{0.5} \times \left(T/M \right)^{0.5}$$

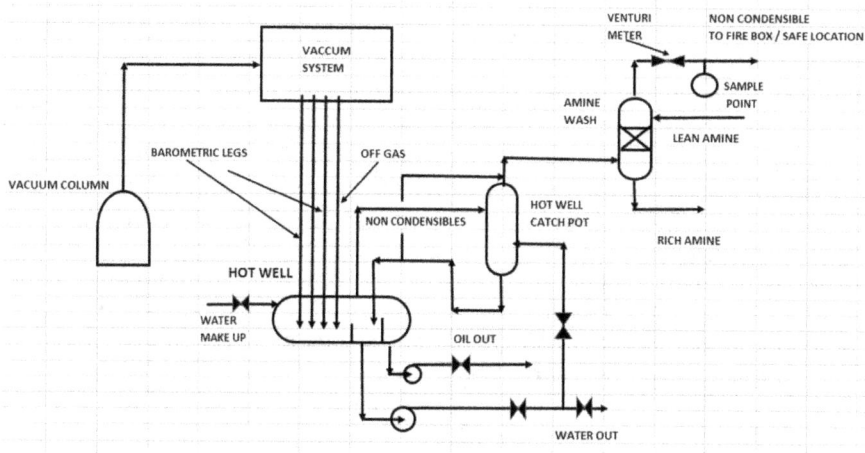

FIGURE 1.35   Modified hot well arrangement with ejector off-gas amine wash.

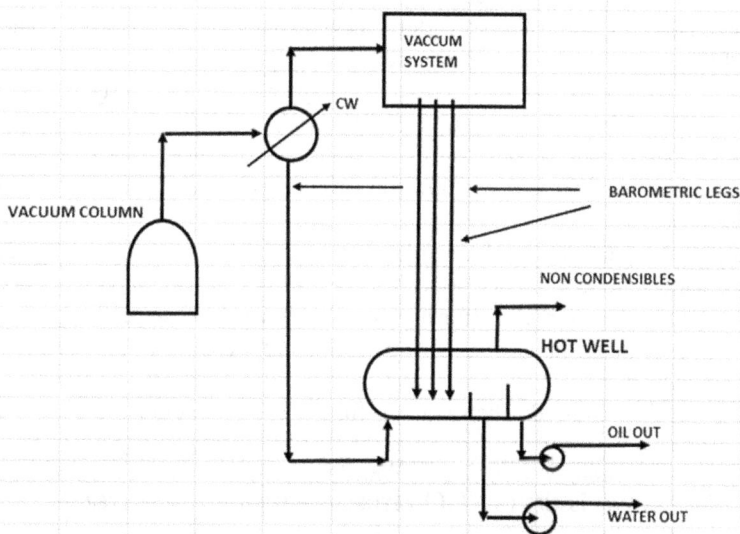

FIGURE 1.36   Barometric leg with a U loop.

where Vs is velocity of sound in meter per minute, K is C_p/C_v, T is absolute temperature in kelvins, and M is the molecular weight.

7. Correct location of introducing VR quench

Units operating with a high furnace coil outlet temperature should have the cooled VR quenched immediately at the outlet of the stripped liquid from the bottommost stripping tray, preferably in a pan at the outlet of the tray.

8. Upgrades in metallurgy

Metallurgical upgrades for high-TAN crudes may be required. Metallurgical upgrades are required for vacuum gas oil circuits that operate between 260°C and 345°C. Piping and column internal components such a beam, packing supports, and tower attachments commonly use 317L stainless steel. 904 stainless steel (high molybdenum content) for cladding and internals, or 316 L cladding is used in the vacuum columns for better resistance to naphthenic acid attack. It may be noted that the corrosion products can also find way to the VGO and affect the quality.

1.4.6　Processing Issues in Coker Units for Residues Derived from Heavy Crudes

- The unit capacity is likely to go higher due to higher yield of VR from heavy crudes. May need capacity augmentation/debottlenecking to handle higher feed rate, if required cushion in capacity is not available.
- Coke production is likely to be higher due to higher CCR of feed. Cycle time reduction may be necessary, if adequate clearance above maximum coke height is not available with the designed cycle time. Refer to Chapter 3 for further detail on cycle time reduction.
- May produce more of shot coke due to higher asphaltene in feedstock and face the problems associated with higher shot coke elaborated in Chapter 3 on coker.
- May result in more coke laydown in furnace tubes needing more frequent decoking.
- Higher heat transfer surface area would be necessary for attaining fresh feed preheat due to higher viscosity of VR.
- Metallurgy to suit high sulfur in VR of heavy crudes is necessary.
- SS 316 lining at the flash zone and some other locations in the coker main fractionator may be necessary and Monel lining in the top of the column may be also required. The subject is further elaborated in Chapter 3.

1.4.7　Processing Issues in VGO Hydrotreater/Hydrocracker Units for Feed Derived from Heavy Crudes

- Feed rate is likely to increase due to higher production of VGO components both from the vacuum unit and from the coker unit (VGO and HCGO).
- Extensive filtering of feed is required to arrest particulates. May need augmentation of feed filters (backwash filters) and higher temperature of operation of filters may be necessary.
- May need higher quantity of metal guard catalyst in reactors at the beginning to arrest all the metals like Ni, V, Si, As, and phosphorus. Arsenic is very detrimental for hydrotreating catalyst. It is present in West African and some Russian crudes and particularly in synthetic crudes from Canada. The

effect of metal on hydrotreating catalyst is described at length in Chapter 4, Section 4.1.

- Feed cut point (end point) may need reduction to limit feed contaminants like N2, metals, asphaltene, and CCR in line with feed specifications.
- Hydrogen consumption may increase and hydrogen makeup compressor capacity may be a constraint.
- Recycle gas purity may reduce due to more thermal cracking due to likely heavier (higher CCR) feedstock.
- The feed is likely to have higher chlorine compounds that will convert to HCl in the reactor. The HCl and NH_3 formed may react together to deposit NH4Cl salt in the reactor effluent circuit in a colder region (200°C–210°C) and increase the pressure drop in the circuit. The deposition temperature is a function of partial pressures of HCl and NH_3. It may be noted that NH_4Cl deposits at higher temperature than NH_4HS and is likely to deposit at the upstream of water injection point at the inlet of REAC. This has been experienced in several hydrotreaters (VGOHT and DHDT, NHT). This is elaborated in Chapter 4.
- Run length may reduce due to metal laydown and coking on catalyst due to higher asphaltene, CCR, and metals in the feed.
- Conversion may reduce at the same weighted average bed temperature (WABT) or catalyst average bed temperature (CAT). The subject is further elaborated in Chapter 4.

1.4.8 Processing Issues in FCCU for Feed Derived from Heavy Crudes

- Feed hydrotreatment may be necessary to enhance UOP "K" for better conversion. The VGO components (VGO + HCGO) derived from heavy crudes are likely to have high sulfur, nitrogen, and lower content of paraffins. Thus, hydrotreated VGO is usually fed to the FCCU to have low sulfur in products and achieve higher conversion.
- Higher catalyst rejection and fresh catalyst make up may be necessary to contain metal buildup and consequent catalyst deactivation if part raw VGO (not hydrotreated) is fed to the unit.
- May need higher CAT:OIL ratio if raw VGO (without hydrotreating) is processed in the unit. This is required for maintaining conversion and consequently results in higher catalyst circulation. Standpipe sizes, slide valves and cyclone adequacy need checking.
- Much increase in feed CCR is not likely, as the feed that is normally hydrotreated may not require additional heat removal in catalyst coolers.
- Air blower capacity may be limited with higher carbon laydown on catalyst on account of higher CAT:OIL.
- Regenerator superficial velocity needs to be checked as a result of higher air flow to the regenerator, if part raw VGO is processed.
- Regenerator cyclone inlet velocity may hit maximum and need to be checked. The aforementioned concerns are covered further in detail in Chapter 2.

1.4.9 Effect on Other Units

The additional hydrogen requirement for hydrotreaters can be supplied through higher recovery of hydrogen from hydrogen-rich streams (elaborated in Chapter 6, Section 6.6), or small-scale augmentation of hydrogen unit capacity like an additional fixed-bed reactor, downstream tubular reformer for reforming heated by hot flue gas (convection reformer), or heat transfer exchanger reactor (HTER) (discussed in Chapter 6, Section 6.7).

In the sulfur unit, however, the addition of new trains may be required to handle the higher quantities of acid gases produced from processing high-sulfur heavy crudes. Oxygen enrichment of air can be an alternative to increase capacity of sulfur units (described in Chapter 7, Section 7.5). Other product treatment units like amine absorbers, amine regenerators, sour water strippers, and sweetening units may need some small modifications to handle streams with higher sulfur content and sometimes in higher quantities.

1.5 CORROSION IN CRUDE VACUUM UNITS WITH HEAVY CRUDES

1.5.1 Acid Corrosion

In general, corrosion of the upper section of the crude column is caused by acid attack (shown in dark in Figure 1.37), and corrosion caused by naphthenic acid at locations shown in dark in the middle section of the column in Figure 1.37 while processing high-TAN crudes. The bottom section is prone to sulfur attack.

The most common problem faced by refineries is corrosion due to hydrochloric acid attack of the overhead system of crude and vacuum distillation units. The problem of acid corrosion becomes more pronounced while processing heavy crudes as a result of usual inefficient desalting. The problem originates with the presence of salt water in crude oils. Most of the salt content of the brine dispersed in crudes is present in the form of sodium chloride (NaCl). Typically, about 10% of the salt is present as magnesium chloride ($MgCl_2$) and calcium chloride ($CaCl_2$). In the presence of water and heat, these chloride salts hydrolyze to form HCl and lead to corrosion. NaCl is the most heat-stable. $CaCl_2$ is moderately heat-stable, and $MgCl_2$ is the least heat-stable. Most of the $MgCl_2$ will hydrolyze at typical crude-tower flash zone temperatures (about 365°C). Most of the NaCl will remain unaffected at these conditions. Typically, two main mechanisms contribute to the corrosion:

- Acid attack at the initial condensation point (ICP) resulting in low pH and metal corrosion
- Deposition of amine chloride salt resulting in under deposit corrosion as a result of injection of an amine neutralizer and its combination with HCl.

If water condensation takes place inside the column or liquid water inadvertently gets into the column, then acid corrosion is likely to occur inside the column at the

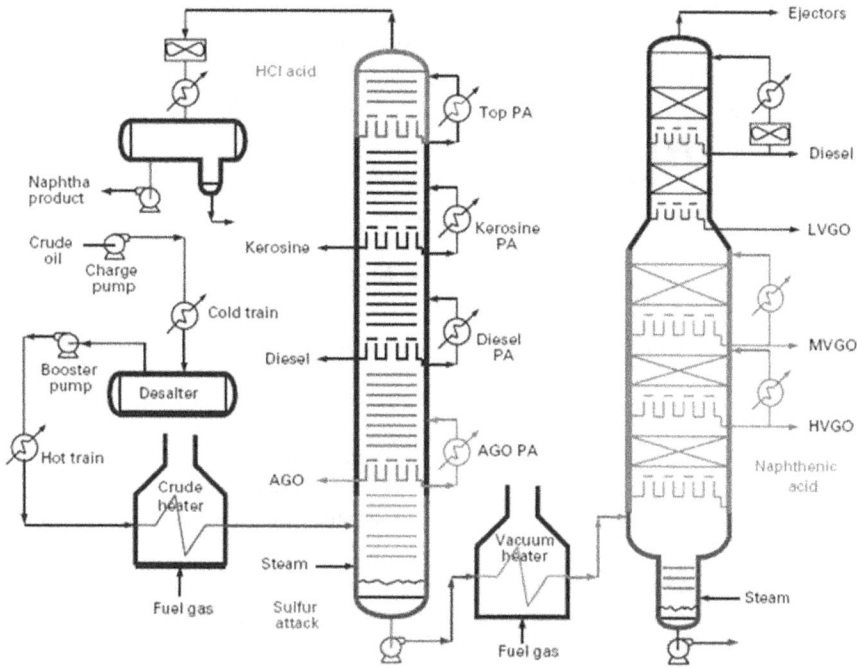

FIGURE 1.37   Corrosion prone locations in crude vacuum unit.

upper section. The top temperature of the column must be higher than the water dew point to avoid water condensation inside the column. Due to improper separation of water and oil or an incorrect water hydrocarbon interface level indication, water can enter in the column along with reflux and may result in acid corrosion inside.

Typically, the top five to six trays in the crude column are made of Monel and the column shell in the upper section is strip-lined with Monel to reduce acid corrosion. Acid corrosion is also common in overhead circuits, in overhead lines, and over-head coolers/heat exchangers. This results in thinning of overhead lines and frequent leaks in overhead crude heat exchangers, air coolers, water coolers.

The overhead chlorides are traditionally neutralized by injection of ammonia or amines to maintain the reflux drum water pH typically around 6.5. Amine chlorides formed are expected to exit the system along with sour water from the boot of the reflux drum. But, if amine chlorides do not adequately partition in water in the reflux drum, the same can get recirculated along with reflux or get in the column along with reflux. This will increase the partial pressure of amine and hydrogen chlorides and may result in deposition inside (in the upper section of the column) and in the over-head circuit, giving rise to a higher pressure drop in the column and can also result in *under deposit corrosion* inside the column and in the overhead circuit.

Induction of cheaper heavy crudes in the crude mix increases the problem. The cheaper heavy crudes are characterized by high density, high pour point, high viscos-ity, low yield of straight run distillates, high naphthenic acid content (higher TAN), and high metal content. Efficient desalting may not be possible for some crudes; some crudes (few crudes of Canadian origin) cannot be desalted particularly due to their high silica content. Thus, the desalted crude chloride level and basic sludge and water (BS&W) are likely to go higher. Pre-desalter heat exchanger trains are modified to deliver higher desalter temperature; desalters are being modified for efficient desalting of heavy crudes to lower desalted crude salt content. The hydrolysis reactions contributing to HCl liberation are

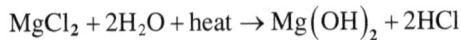

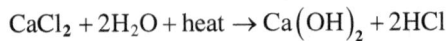

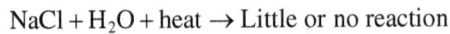

$$MgCl_2 + 2H_2O + heat \rightarrow Mg(OH)_2 + 2HCl$$

$$CaCl_2 + 2H_2O + heat \rightarrow Ca(OH)_2 + 2HCl$$

$$NaCl + H_2O + heat \rightarrow \text{Little or no reaction}$$

When wet crude oil containing salts are heated, most of the $MgCl_2$ and a small amount of $CaCl_2$ begin to hydrolyze at about 120°C liberating HCl, and at 370°C about 85% of $MgCl_2$ and 15% $CaCl_2$ hydrolyze and produce HCl. NaCl does not hydrolyze to appreciable extent at crude unit conditions.

The salts in most crudes have a proportion of salts similar to that of seawater (70%, 10%, and 20% of Na, Ca, and Mg chlorides, respectively). An injection of dilute caustic to crude exiting the desalter results in a series of reactions to arrest HCl liberation and produce more stable NaCl. The following chloride-salt removal efficiencies were observed by experts, typically in single-stage desalters:

- 90% NaCl removal
- 50% $CaCl_2$ removal
- 40% $MgCl_2$ removal

The hydrolysis of $MgCl_2$ proceeds far more readily than that of NaCl and its potentially lower removal efficiency in the desalter will create a disproportionate downstream corrosion problem. *A double stage desalter can bring down the salt content to less than 1 ptb (pounds per thousand barrel) and typical BS&W content 0.5% at the desalter outlet.*

It may be noted that H2S is known to be an HCl corrosion accelerator, explained by the following reactions:

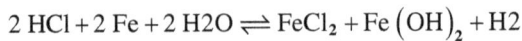

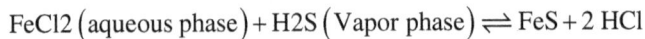

$$2\,HCl + 2\,Fe + 2\,H2O \rightleftharpoons FeCl_2 + Fe(OH)_2 + H2$$

$$FeCl2\,(\text{aqueous phase}) + H2S\,(\text{Vapor phase}) \rightleftharpoons FeS + 2\,HCl$$

Thus, HCl is regenerated by the second reaction and initiates a fresh attack on metal, and FeS is the product of corrosion. Reduction in column pressure (by use of compressors) to pick up the gas produced and recontact with naphtha at a higher pressure around 6 Kg/cm2 are often practiced. The recontacted naphtha is then fed to stabilizer for recovery of LPG. The reduction in total pressure can reduce H2S partial pressure and may reduce the regeneration of HCl by the aforementioned second reaction.

1.5.2 CAUSTIC INJECTION

Traditionally, dilute caustic solution in water is injected in desalted crude just at the outlet of the desalter to convert the balance salts in desalted crude (convert hydro-lyzable $MgCl_2$ and $CaCl_2$ to less hydrolyzable NaCl) to reduce acid corrosion in the column overhead. Injection of sodium hydroxide downstream of the desalter is an effective method of moderating hydrochloric acid attack in the crude-tower overhead system. A rough rule of thumb based on observation in one plant is to inject 1 lb NaOH/lb salt in the desalter effluent. The exact amount of caustic injection is based on the chloride content of the boot water in the crude-column overhead reflux drum. A reasonable target is 10–20 ppm chlorides, measured in the reflux drum boot water.

Often different views are expressed regarding the strength of the caustic solution, the point of injection, and the methodology of mixing with crude to effectively to contain HCl release. Thus, the subject may need a thorough review and analysis. The contribution of different factors, based on views of the experts, are summed up as next.

Caustic is not directly soluble in crude. It requires a sodium naphthenate interme-diary. Typically, all the crudes have adequate naphthenic acid content to react with caustic unless the TAN value is significantly lower than 0.1 mg KOH/g crude. The caustic should be added as 3–5 wt.% aqueous solution immediately after the desalter to allow the maximum time for adequate mixing with the crude oil before hot zones are reached. Use of a higher strength of caustic is known to lower efficiency for chlo-ride reduction and may need higher caustic injection (to lower chloride in the reflux drum boot water), and may in turn result in a high sodium level in the vacuum resi-due (coker unit feed) and can result in a higher coke deposition in coker heater tubes.

If injection further downstream in the hotter zone is done, then the small amount of water present in the caustic solution may go into the solution in the crude oil. Solid caustic may drop out in the exchanger train or crude furnace tubes and foul the heat exchangers and furnaces without effectively lowering HCl liberation. The caustic dropout can also result in caustic embrittlement. A proper mixing of caustic solution with crude is very important. The injection rate varies typically from 0.25 to 2 PTB (pounds of NaOH/1000 barrels of crude), finally adjusted based on the chloride level in the reflux drum boot water. The maximum caustic injection rate may be less than 50% of the total moles of naphthenic acid present in crude.

It is believed that caustic can react with naphthenic acid corrosion inhibitors and reduce the effectiveness of both the corrosion inhibitor and caustic. Caustic is also known to react with many dispersant type antifoulants. Thus, the caustic injection point could be well away from any antifoulant and corrosion inhibitor injection point.

Injection of caustic at the upstream of the desalter is not advised as high desalter water pH can result in formation of emulsion and can result in improper water sep-aration. Further, due to higher water pH the ammonium chloride/amine chlorides will partition more into crude and travel to the column top section and increase

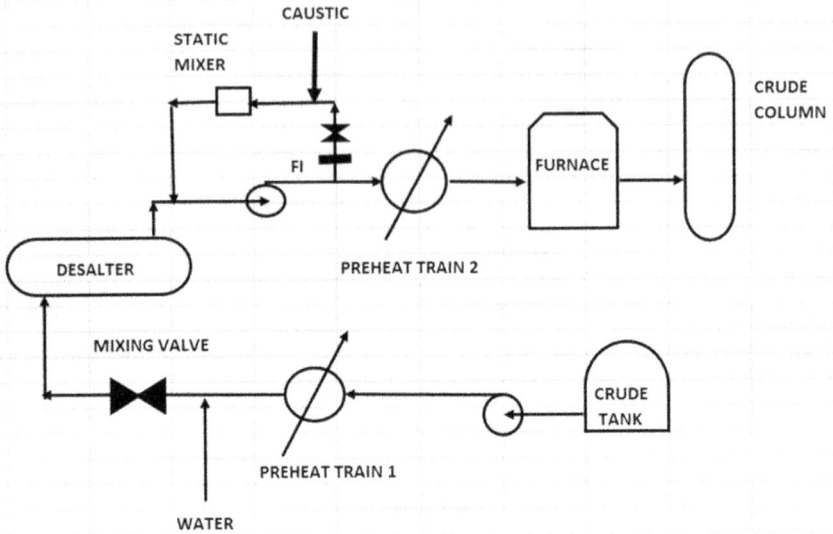

FIGURE 1.38 Caustic injection facility with slip stream.

the likelihood of ammonium chloride deposition in the overhead and under deposit corrosion. Further, injection at the desalter upstream may result in removal of all the caustic with brine and may not be available to neutralize where the substantial hydrolysis of $MgCl_2$ and $CaCl_2$ will take place and HCl will be liberated.

For the units without a desalter, caustic is advised to be added to the preheat train almost at the same temperature, at a point where the temperature is less than 140°C. Injection at a location having a higher temperature may result in migration of water to crude oil and deposition of caustic in equipment.

The more efficient method may be to pump the solution into the desalted crude line at the suction line of the desalted crude pump. The method involves taking a small slip stream of pump discharge and adding caustic solution, flow the slip stream through a static mixer, and return to suction of the pump with a quill, as in Figure 1.38 and Figure 1.39. The ratio of caustic solution to crude slip stream may be in the range of 1:2 to 1:3.

If adequate space at suction is not available, the caustic solution can be injected at the suction of the pump through an injection quill, as shown in Figure 1.40 and Figure 1.41.

1.5.3 Chemistry of Caustic Reaction

The basic chemistry of caustic reacting with the hydrolysable chlorides are as follows.

The naphthenic acids present in crude react with basic sodium hydroxide to form sodium naphthenate, which will be soluble in oil.

$$RCOO - H + NaOH \rightarrow RCOO - Na + H2O$$

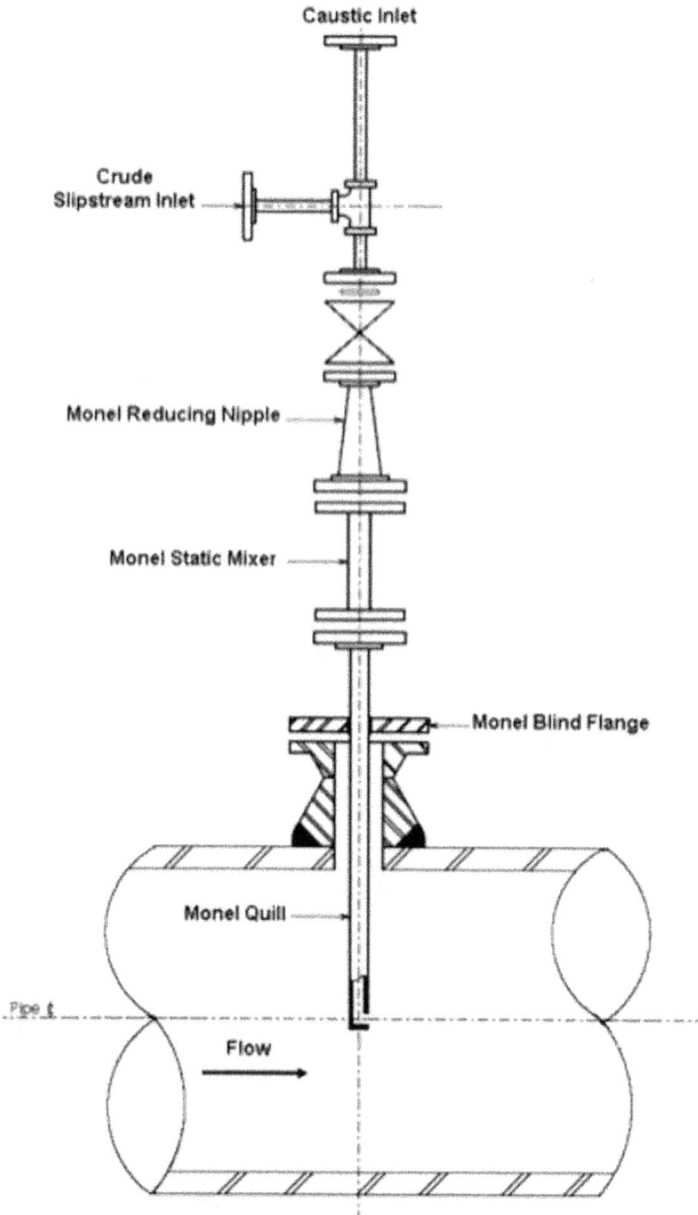

FIGURE 1.39 Caustic injection device with crude slip stream.

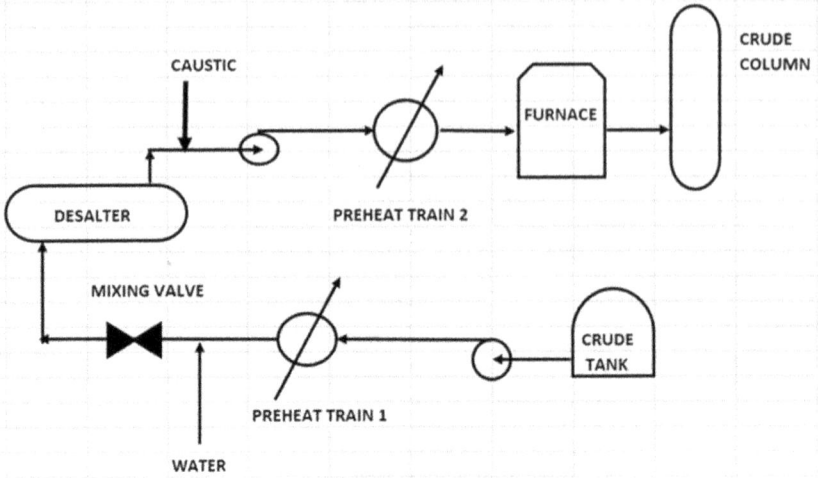

FIGURE 1.40   Caustic injection facility with injection quill.

INJECTION QUILL

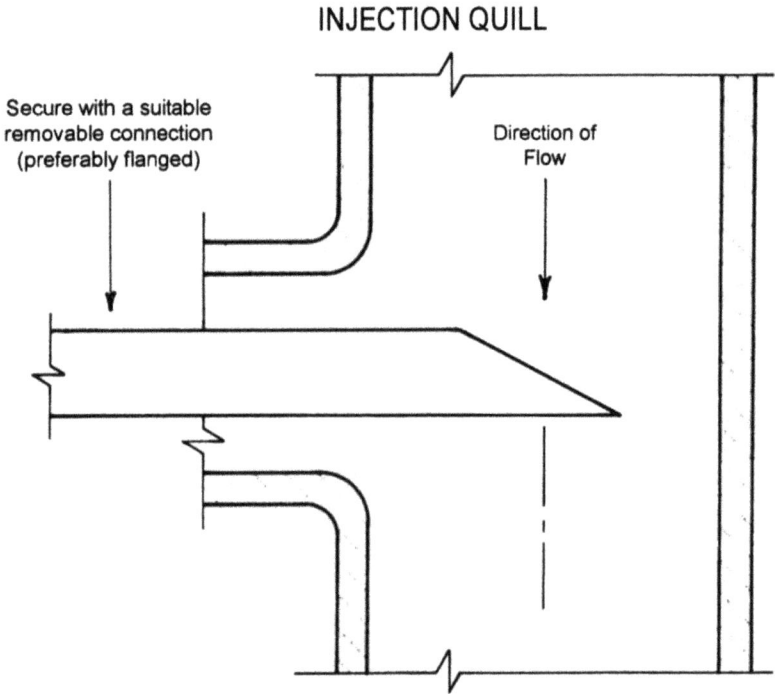

FIGURE 1.41   Caustic injection point arrangement with quill.

Magnesium and calcium chloride being hygroscopic in nature will retain adequate water for hydrolysis. After the crude is heated to a sufficient temperature of 121°C–204°C, the hydrolysis of magnesium chloride and calcium chloride begins following these reactions:

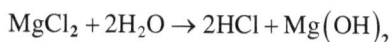

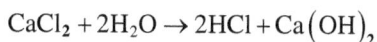

$$MgCl_2 + 2H_2O \rightarrow 2HCl + Mg(OH)_2$$

$$CaCl_2 + 2H_2O \rightarrow 2HCl + Ca(OH)_2$$

Then reaction between strong acid (HCl) and weak base (RCOO-Na) will follow to produce sodium chloride by the following equation:

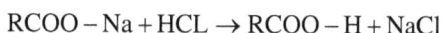

$$RCOO - Na + HCL \rightarrow RCOO - H + NaCl$$

The sequence of reaction is very important. If the first reaction does not take place significantly, the overall net efficiency will not be achieved. Thus, thorough mixing of caustic with crude is most essential to initiate the set of reactions.

Ammonia, morpholine, or higher-boiling-point neutralizing amines can be injected into the atmospheric column overhead, and judicious amounts of caustic can be added to the crude at the desalter outlet to control acid corrosion. Acid corrosion at the ICP is controlled by injection of amines, ammonia, etc. in the overhead. The receiver water pH acts a guide for controlling the rate of injection of neutralizer.

Ammonia is often not considered as an effective neutralizing agent due to its poor solubility in water at dew point temperature. Corrosion due to deposition of amine salts are often mitigated by water wash provided as a means to dilute/wash the corrosive salts. Lower pressure operation and a higher reflux ratio is likely to reduce H2S-accelerated HCl corrosion. Corrosion inhibitors/filmers are often used for protection from both types of reactions and corrosion.

Most refiners use a water wash upstream of heat exchangers/ condensers in column overhead line to achieve forced condensation of water to reduce acid corrosion at the ICP and salt deposits. This does not protect against amine chloride salt fouling, if the same forms at a lower temperature and deposits ahead of the water injection point. The corrosion risk increases if the amine gets into the column along with reflux and can induce an amine recycle loop. The ideal neutralizer will form its amine chloride at 8°C lower than the water dew point. To protect deposition of amine chloride salts in the column, the neutralizer salt point should be 14°C lower than the tower top temperature.

1.5.4 Tramp Amines

Tramp amines are amines other than appropriate amines found recycling in the system. Tramp amines affect overhead water pH and have a very high salt point.

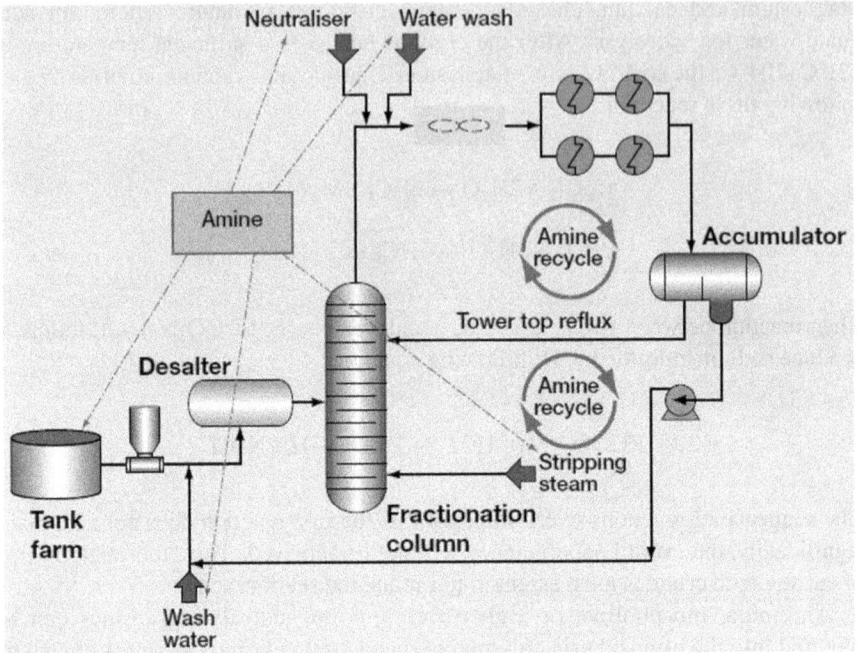

FIGURE 1.42   Tramp amine cycle. (Reprint of *PTQ*, 3rd quarter issue, 2012, pp. 75–81.)

They come from crude, neutralizers used, steam, stripped sour water, wet reflux, amine from absorbers, H2S scavengers, etc. It becomes impossible to keep the salt point below the column top temperature and water dew point. The situation can cause the salt point to exceed the column top temperature and cause deposition inside the column. A figure demonstrating the amine cycle is presented in Figure 1.42. It is difficult to maintain tramp amines at a lower level compared to controlling chlorides.

Some special care can be introduced to reduce corrosion:

- Sufficient water needs to be added in the overhead to ensure at least 10%–15% in the liquid phase after saturating the gas.
- The reflux drum should be adequately sized to handle the extra water without carry over along with reflux.
- Overhead line vapor velocity should be maintained within 30–60 ft/sec to avoid dislodgement of protective FeS film.

    The line from the reflux drum to the reflux pump should extend up by 6 inches inside the reflux drum to eliminate water in the reflux (as the boot water is likely to contain an appreciable amount of amine chlorides).

The tendency of these amines to cycle up is largely driven by the overhead receiver water pH in the overhead drum and the desalter effluent brine pH at the outlet of the desalter. As the pH rises above 5.5, the tendency to cycle up increases due to a lower partitioning rate from hydrocarbon to water. Amine partitioning depends on the type of amine used, polarity of the hydrocarbon, and water pH. A lower pH will protonate amine (NH3 to NH4+) and drive the ionic compound to the water phase. A higher pH will deprotonate the amine and drive the non-ionic compound into the hydrocarbon phase. Amine partitioning also depends on the type of amine. As the amine becomes heavier (more carbons are added to amine), its partitioning becomes less pronounced with pH. Ammonia is easily partitioned into water. MEA partitions to a lesser extent and so on.

In crude units, keeping the overhead receiver water slightly acidic helps in break-ing the amine recycle loop. The pH of the receiver water is maintained at 6–6.5 to control the pH at the ICP at 5–5.5 and to control HCl corrosion at ICP, neutralizer is added. A lower pH of receiver water will increase the HCl corrosion at the ICP and will be reflected by a higher ppm Fe (>1 ppm) in the water. As the pH at the ICP rises above 6.5, the tendency for deposition of amine chloride increases resulting in under deposit corrosion. Further, destruction of the protecting layer of iron sulfide becomes a concern at higher pH. The protective film is weakened as the pH increases. A higher pH may eventually result in higher iron (>1 ppm) in the water phase with the likely damage of the FeS protective layer.

Thus, both the upper and lower level of pH are hard limits not to be exceeded. At the desalter, a slightly acidic pH will help in partitioning of amines into desalter brine. Using nonvolatile weak acid (e.g., acetic acid) is a good way to maintain desalter brine pH. The acid decomposes into an inert substance at the crude unit heater.

Most amine chloride salts are too light and does not get out through the atmo-spheric tower bottom. Hence, they accumulate as a solid at an intermediate point in the crude column unless removed through the top. The jet fuel draw-off nozzles on several units have been partially plugged with these solids. A sample analysis drawn from a low-point bleeder on the jet draw-off line reportedly showed 85% dry amine chloride salt and around 15% ferric chloride.

If a neutralizing agent (amine or morpholine) is refluxed back to the atmo-spheric tower, the amine chloride salts will be heavy enough to sublime on the upper trays of the column. Ammonia, morpholine, or higher-boiling-point neutralizing amines can be injected into the atmospheric-column overhead. Judicious amounts of caustic can be added to the crude at the desalter outlet. The HCl corrosion can manifest into

- Flooding of the atmospheric-tower top trays
- Condenser-tube leaks
- High pressure drop and low heat-transfer coefficients in the atmospheric-tower overhead condensers
- Plugged kerosene draw-off nozzle

- High iron content (>1 ppm) and erratically low pH in the reflux drum water
- Failure or leak in the overhead vapor line
- Flooding of the atmospheric-tower top trays

Vacuum tower overhead corrosion can be attributable to liberation of HCl in the vacuum tower. The source of the HCl in the vacuum system overhead is primarily due to the decomposition of $CaCl_2$ and the residual $MgCl_2$ and the vacuum-tower overhead corrosion is normally less severe than atmospheric-tower overhead corrosion because of the large amount of steam that condenses along with the HCl in vacuum towers and dilutes the acid. This steam introduced is either as bottom-stripping steam or as motive steam to the first-stage ejectors or stripping steam to side stripper for lube type vacuum columns.

1.5.5   USE OF FILMER

Filmer works by coating the metal surface with hydrophobic barriers that prevent corrosive species from coming into contact with metal surfaces and preventing corrosion.

It is possible to have protection of 90%–95% of the coated surface with adequate dosage of the filmer. Traditional filmers need a pH above 4 to maintain optimum film stability. Special ones are known that can handle up to a pH of 2 and less. Certain filmers are known to have the capacity to act as salt dispersants and reduce risk associated with salt fouling due to amine-based neutralizers and tramp amines. Filmers have been used with success to disperse the deposited salts in the areas where no water wash is present, like pumparound circuits and tower trays.

1.6   PROCESSING OF HIGH-TAN CRUDES

1.6.1   INTRODUCTION

Worldwide production of crudes with high naphthenic acids have increased several-fold in recent years. The high-total acid number (TAN) crudes are priced at a significant discount compared to benchmark crudes due to their inherent corrosivity. The high-TAN crudes contain high levels of naphthenic acids and mostly are heavy crudes. Incorporation of high-TAN crudes in feed provides significant profit improvement opportunities. Refiners are changing their perspective on the best way to take advantage of low-cost crudes with high acid numbers. Some have relied solely on metallurgy to protect against naphthenic acid corrosion but are now started to depend on chemical programs. Others are reevaluating existing chemical programs to ensure safety and reduce risks, and waiting to introduce high-acid crudes until installation of improved metallurgy. It is thought that most of that potential profit can be realized using chemical programs. Refiners are gaining confidence in new chemical treatment technology. Capital required for metallurgy upgradation versus operating costs are being critically evaluated. The typical metallurgy used in a crude and vacuum unit are capable of handling a maximum TAN of 0.5 (mg KOH/gm crude). There are ways to manage the risk to capture the economic benefit of high-TAN crudes (suggested by experts and being practiced) and as explained in the following sections.

1.6.2 METHODOLOGIES FOLLOWED

1. *Careful evaluation of the TAN limits* of existing metallurgy and operation carried out at the safe limits without equipment modifications or use of corrosion inhibitors.
2. *Blend* to the TAN limits of the existing metallurgy, i.e. blend high-TAN and low-TAN crudes to remain within the limits in the blend and operate at a safe limit without metallurgical modifications or corrosion inhibitors.
3. *Upgrade metallurgy*, Investing in improved metallurgy in a limited way to further increase TAN and operate without corrosion inhibitors.
4. *Use corrosion inhibitors and partially upgrade metallurgy.* Some refiners have started introducing high-TAN crudes to a limited extent and implementing corrosion inhibitor programs for the same. The refiners are carefully evaluating the corrosion in critical locations and upgrading metallurgy in most corrosion-prone locations and circuits in a limited way and increasing the TAN value of the crude further to maximize profit.
5. The *incremental investment in metallurgy* involves significant capital. The opportunity cost of waiting for the shutdown to install the metallurgy can be huge for a refinery. Moreover, some refiners are apprehensive about future discounts on high-acid crudes and, accordingly, have reservations about metallurgy upgradation and resumed high-TAN crudes in blends with use of corrosion inhibitors only. Chemical costs are known to vary based on many factors, but they are known to be far less than the typical discount on crudes.

1.6.3 CONCERNS FOR USE OF CORROSION INHIBITORS FOR NA

Chemical treatments can be implemented quickly but require the formation of a protective layer. The protective layer is known to comprise of phosphorus, sulfur, iron, and oxygen with a thin, highly cross-linked organic layer at the surface. The chemistry of corrosion inhibitors currently in use are predominantly of a class of compounds that are called *phosphate esters*.

Traditional phosphate esters, particularly the older versions, were known to break into insoluble particulates that are not very effective in building a protective layer. Only a fraction of the traditional phosphate ester products remain soluble in oil to chemically react in the formation of a passivation layer. Precipitated phosphorous degradation products introduce potential for harming downstream equipment/units, including fouling and catalyst poisoning. The byproducts of both thermal degradation products and unreacted phosphate ester introduce high acidity in the system for potential damage.

High phosphorus content in the older version of phosphate ester may impair the downstream unit catalyst increasing fresh catalyst addition, and may potentially foul the hydrotreater, decreasing the run length of the unit. The risk of fouling is difficult to quantify but has been significant enough to cause potential problems over time and discourage some refiners from using corrosion inhibitors.

New versions of additives are known to control corrosion with up to 80% less phosphorous added, reduce the need for refiners to expose equipment to fouling risks. The amount of P (phosphorus) required by the newer generation is far less than the older generation to obtain equal corrosion protection.

Corrosion treatment programs may be carried out strictly under the guidance of an expert group. Due to the poor stability of iron phosphate scale formed by traditional phosphate-based chemistry, scale tends to get dislodged from metal surfaces when subjected to high velocities and wall shear stress. This dislodged scale can foul the heat exchangers. Accordingly, velocities of fluids are also taken as factors for dosing. Experts may install corrosion coupons at sensitive locations to monitor. In general, refiners are using corrosion inhibitors and incorporating high-TAN crudes without many problems.

However, the author has observed high pressure drops in sections of a crude column, suspected due to deposition of phosphorous compounds inside the crude column, using chemical treatments for TAN. The fixed opening dispersers in the section around kerosene tray were found completely blocked due to the deposition. The deposits were analyzed and phosphorus in the deposits were found. The affected tray segments were later changed with non-fouling types of trays. The problem is resolved to a great extent.

1.6.4 MECHANISM OF NAPHTHENIC ACID CORROSION

Naphthenic acid (NA) corrosion is heavily influenced by temperature. Corrosion at low temperatures is not strong. Once in a boiling state, especially at a high temperature and anhydrous environment, the corrosion will be the strongest. Most of the high-temperature naphthenic acid corrosions happens in liquid phase, but if naphthenic acid is condensed in the gas phase, the gas phase corrosion may happen and the corrosion extent will be influenced by acid value. In addition, the product of corrosion—$Fe(RCOO)_2$—is oil soluble and is formed from the direct reaction between naphthenic acid and the metal surface or iron sulfide scales at high temperatures. It can get easily dissociated from the metal surface and carried away by oil. The new bare metal surface is exposed and results in fresh corrosion deeper into the metal. For this reason, the metal surface corroded by naphthenic acid is found very smooth and clean and without dirt.

After prolonged contact with naphthenic acid and flushing by the acid stream, the metal surface may show corroded grooves (high-speed acid stream will form parallel sharp groove lines, while low-speed acid stream will form corrosion pits with sharp edges). These are indicators to identify this kind of corrosion.

1.6.5 Influence of Temperature on Naphthenic Acid Corrosion

- At <200°C, in the anhydrous case, there is no corrosion. In water environments the corrosion will increase with temperature.
- At 270°C~280°C, it reaches the acid boiling point, and results in appreciable corrosion.
- At 350°C~400°C, due to the high-temperature melting of FeS film, together with the elemental sulfur discomposed from crude oil sulfides, the corrosion on metal equipment becomes violently enhanced by the interaction between naphthenic acid, elemental sulfur, and H2S. So, vacuum furnace tubes and outlets to the column line may need stainless steel cladding inside for protection.
- At temperatures >400°C, the acid is decomposed and corrosion is weakened. Naphthenic acid when thermally decomposed will result in formation of CO2 and water, which have corrosivity similar to water condensate when cooled.

1.6.6 Factors Influencing High-Temperature Naphthenic Acid Corrosion (Flow Velocity and Flow Regime)

- The heavy corrosion is likely in elbows, tee joints, and in locations where the flow is turbulent for pumps and other equipment.
- If the proportion of gas/vapor in the stream is higher than 60% and the gas stream speed is higher than 60 m/s, the corrosion speed of some equipment (such as furnace tubes, elbows, vacuum distillation transfer lines) may nearly double.
- The generally controlled flow velocity for carbon steel is <25–30 m/s. In SS 316, the velocity may be <120 m/s. This is applicable for two-phase flows in furnace transfer lines.
- Acid value doesn't directly show the effect on naphthenic acid corrosivity.
- But as for high-acid crude oil, its acid value is used for defining its corrosivity.
- Naphthenic acid of light molecular weight is known to need lower activation energy during reaction with metals, and it has higher activity at low temperature, hence may result in higher corrosion compared to heavier ones.

1.7 PROCESSING OF CRUDE OILS WITH HIGH CALCIUM

Calcium is present in many crude oils in the form of brine-soluble calcium salts that are removed within the desalter. However, in recent years the content of oil-soluble calcium naphthenate in crude blends has increased in oils from the North Sea, West Africa, and China. Calcium naphthenate is not water soluble and releases oil-soluble naphthenic acids if treated with aqueous acid. The latter causes downstream corrosion problems in areas of high temperature. Crudes that contain calcium naphthenate include Doba, Kuito, Heidrun, Ciao Fao Dien, and Shengli. Calcium naphthenate is known to accumulate at the interface in the desalter. This often results in water carryover with desalted crude and increased oil in brine, leading to an adverse impact on the wastewater treatment plant. East African Doba crude sells at an attractive price, but it presents potential processing problems due to a calcium content averaging over 200 ppm.

Chemical dosing experts have developed a few new calcium removal additives that is known to demonstrate significantly higher calcium removal efficiency. The special formulations convert the calcium naphthenate to form water-soluble calcium salts, which are removed through the desalter brine. The additive is known to form a calcium compound that is water soluble, biodegradable, and noncorrosive. The author, however, has experienced some problems of liberation of H2S (from desalter brine) in effluent treatment plants with additive injections (possibly organic acid) at the inlet of the desalter along with wash water while processing Doba crude in admixture with regular crudes.

1.8   CASE STUDIES IN DISTILLATION UNITS

1.8.1   CASE STUDY 1

A crude column was having three side cuts, namely, kerosene, LGO, and HGO, with their respective pumparound refluxes. Additionally, a top pumparound (PA) reflux was provided from a fourth tray to the topmost tray. The entire overhead condensation duty was envisaged to be removed by the top pumparound. Only a provision of top reflux was there. The overhead system of a crude unit with top pumparound and compressor and recontact drum are shown in Figure 1.43. The system was experiencing higher pressure drop across the upper 15 trays. This was further experienced when lighter crude was processed resulting in capacity limitation of the unit. It was found that, due to the absence of a considerable amount of top reflux, the water dew point would increase and partial pressures of amine and HCl would be higher than a conventional tower with top reflux. Thus,

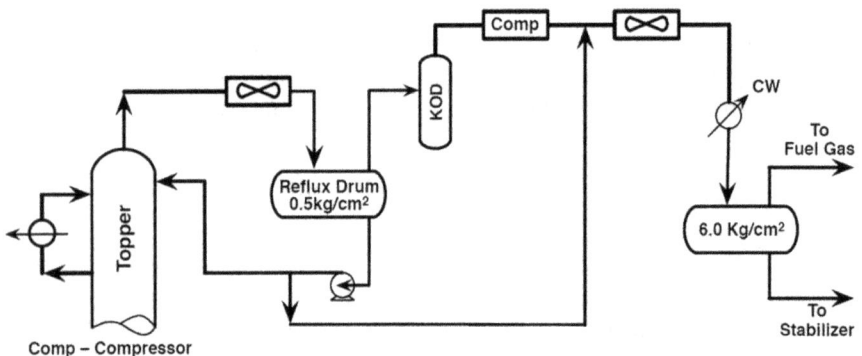

FIGURE 1.43   Overhead system of a crude unit with top pumparound and compressor and recontact drum.

the likelihood of water dew point reaching inside the column would increase and the possibility of deposition of amine chloride would also increase inside the column. The deposition of amine chloride inside the column can increase the pressure drop in the upper section of the column. Further, the top pumparound trays were rated. The four trays were provided with dynamic sealing in the downcomer (the downcomer inlet width was higher than the outlet to establish a dynamic seal and, accordingly, clearance under the downcomer was kept much higher than the weir height of 2 inches to handle high liquid flow). It was found that beyond a flow of the top pumparound of 1800 MT/Hr, the top pumparound trays are likely to flood. The refinery was advised to carry out the following:

- Lower the top pumparound return temperature by 10°C–15°C (maintaining the PA return temperature 5°C–10°C higher than the water dew point) so that they can reduce the flow of the top pumparound from the designed 2200 MT/Hr to around 1800 MT/Hr.
- They were also advised to increase the top temperature by 8°C–10°C, limited by the naphtha end point to lower likely deposition of amine salts inside the column.
- There were advised to resume the caustic injection downstream desalter (they discontinued its as boot water chloride was about 20 ppm) to further lower the requirement of amine injection to the overhead to maintain boot water pH and reduce likelihood of deposition (due to lower partial pressure of amine in the overhead).
- Maintain desalter brine pH between 6.5 and 6.8 by injection of weak acids, if required to partition the tramp amines into desalter brine and to reduce tramp amines entering the column along with crude.
- Maintain reflux drum boot water pH 6.5 to 6.8 again to help partition the amine chlorides into boot water.
- Ensure better performance of boot water interface level indication to avoid water in the reflux and reentry of amine chlorides to the column.
- Check the design of column trays in the top PA section by the trays supplier and if possible by changing to design to handle a higher flow rate.

The column is working all right and the refinery may also change the top pumparound trays to increase the liquid handling capacity of the top pumparound section and and will also install proprietary antifouling trays.

1.8.2   Case Study 2

In a very big crude vacuum unit, the feed to the stabilizer was unstabilized naphtha from the crude column after recontact of the naphtha with off-gas. The crude column had a compressor to compress the bleed gases from the reflux drum and recontact with unstabilized naphtha from the drum in a small absorber, and from the absorber top finally gas was bled (with pressure control of the drum) to the amine absorber and then to fuel gas.

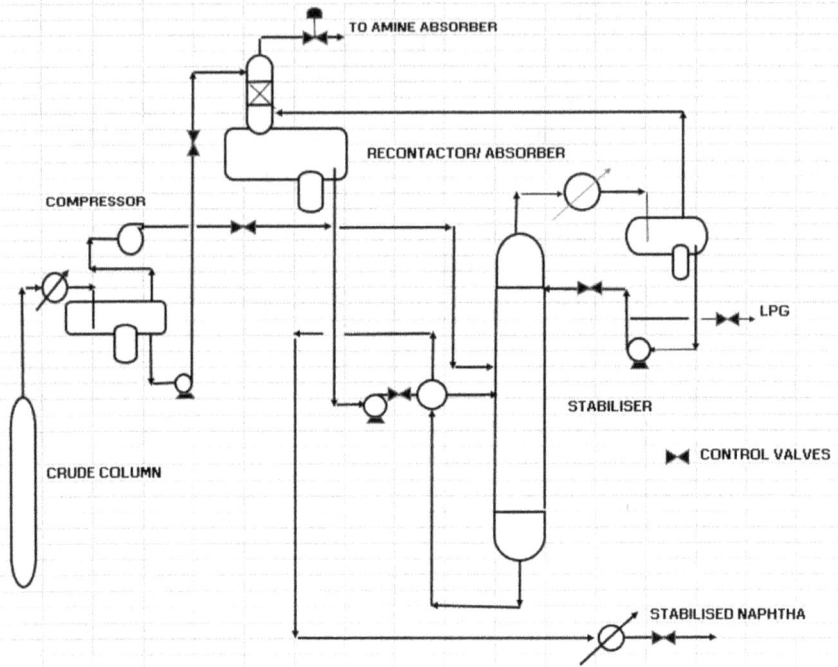

FIGURE 1.44   Original scheme with compressor discharge routed to stabilizer.

The crude unit stabilizer was not producing LPG up to the potential. The original scheme is shown in Figure 1.44.

There was lots of problem in the stabilizer and LPG recovery was low. The gas from the reflux drum was mixed with some hydrogen from the naphtha hydrotreater (NHT) unit in a drum, and further compressed and fed to just above the feed tray in the stabilizer. The refinery wanted to increase the capacity of the stabilizer commensurate with the capacity expansion of the crude distillation unit. Accordingly, the column internals were modified and fixed opening valve trays were provided for achieving higher capacity. The stabilizer had two reboilers: one heated by MP steam and the other by a pumparound (heavy kerosene pumparound) of the crude unit. The stabilized naphtha showed lot of butane slippage. Later it was also found that with the maximum available reboiler duty of around 13–14 G-Cal/Hr the trays are near dumping. Further, the gases (supercritical vapors) introduced in the column can cause problems of low LPG recovery. So, the gas stream from the compressor discharge was rerouted to the recontactor at the bottom of the packing instead of the column. It was also not possible to increase the reboiler duty further. So, a two-enthalpy scheme was proposed. The scheme envisages splitting the recontacted naphtha in almost two equal parts. The cold part is fed through the original feed point. The other part was heated by hot stabilized naphtha (of the column bottom) and further heated by MP steam (at 10 Kg/cm2 g) and fed to the tenth trays from

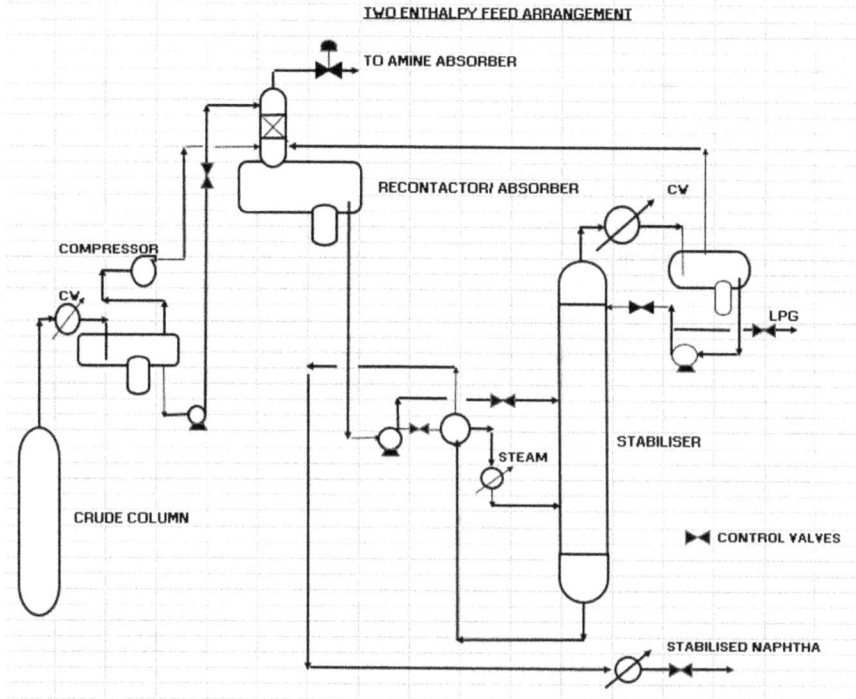

FIGURE 1.45 Modified scheme showing gas diversion to recontactor and use of two enthalpy feeds.

the bottom. The higher feed temperature increased the vapor load in the column and was able to arrest dumping. The cold feed was introduced to control the condenser duty. The butane slippage reduced to bare minimum. The LPG production increased by around 20%–25%. The unit has been operating for the last three years. The modified two-enthalpy scheme is shown in Figure 1.45.

1.8.3 CASE STUDY 3

In a refinery, a conventional naphtha stabilizer was installed to recover LPG and stabilize wild naphtha from the VGO hydrotreater, and a DHDT wild naphtha was installed. The scheme envisaged mixing the two streams and preheating by stabilized naphtha and feeding to a stabilizer with 45 trays. But it was not possible to produce on-grade LPG. It may be noted that wild naphtha streams have a considerable amount of hydrogen dissolved in it. Thus, it may not be possible to recover on-grade LPG by conventional schemes. The same was also corroborated by simulation. Later the scheme was changed as shown in Figure 4.22 of Chapter 4. A recontactor was installed for recontacting of the stabilizer off-gas with naphtha having three stages in the absorber. The configuration was validated with simulation. The modification was implemented.

It may be noted that in most cases a single-stage recontact usually suffices (a simple recontact may be adequate). But in some cases, an absorber with two to three stages may be necessary to increase the LPG yield.

1.8.4 CASE STUDY OF PUMPAROUND PROBLEM OF A CRUDE UNIT

A crude distillation unit processing heavy high-TAN crude experienced a unique problem. The kerosene pumparound reflux was often lost. The kerosene PA exchanges heat with crude in a heat exchanger located upstream of the desalter and used to help in maintaining the desalter temperature. With loss of kerosene pumparound flow, the desalter temperature dropped by almost 10°C, affecting desalter performance.

The unit was designed to process a blend of heavy crude and regular crude. But it was operating with heavy crude of Indian origin only. To identify the source of the problem, the column internals were examined and it was found that the kerosene PA and product was drawn from the 28th tray of the column and the static seal was absent from the draw off tray. However, there was a provision of dynamic seal in the central downcomer of the upper tray (tray 29). This not likely to work when the liquid flow rate through the downcomer is low. The vapor velocity through the valves was found almost same as the computed minimum velocity. Further, the unit was running at low coil outlet temperature due to refractory dislodgement in one charge furnace resulting in lower vapor generation. Thus, there is the likelihood of dumping of liquid and less liquid reaching the draw off box of tray 28 and loss of suction. In the absence of a static seal, vapor is likely to blow through the downcomer further aggravating the problem. Figure 1.46 represents the plan of the draw off tray of kerosene PA + kerosene product.

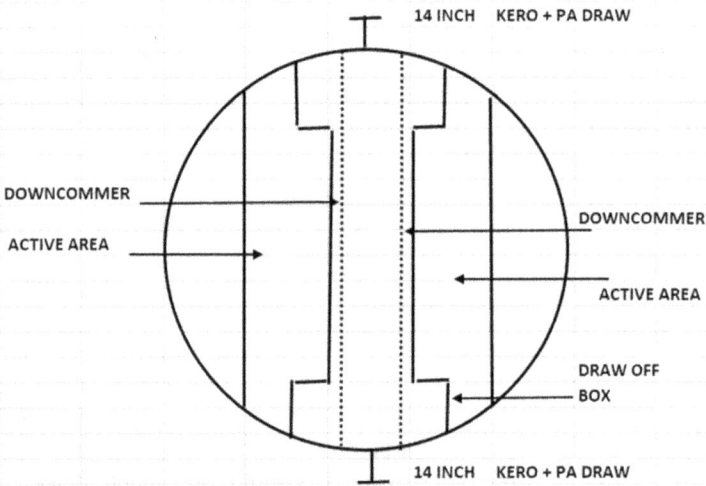

FIGURE 1.46 Plan of the draw off tray of kerosene PA + kerosene product (tray 28).

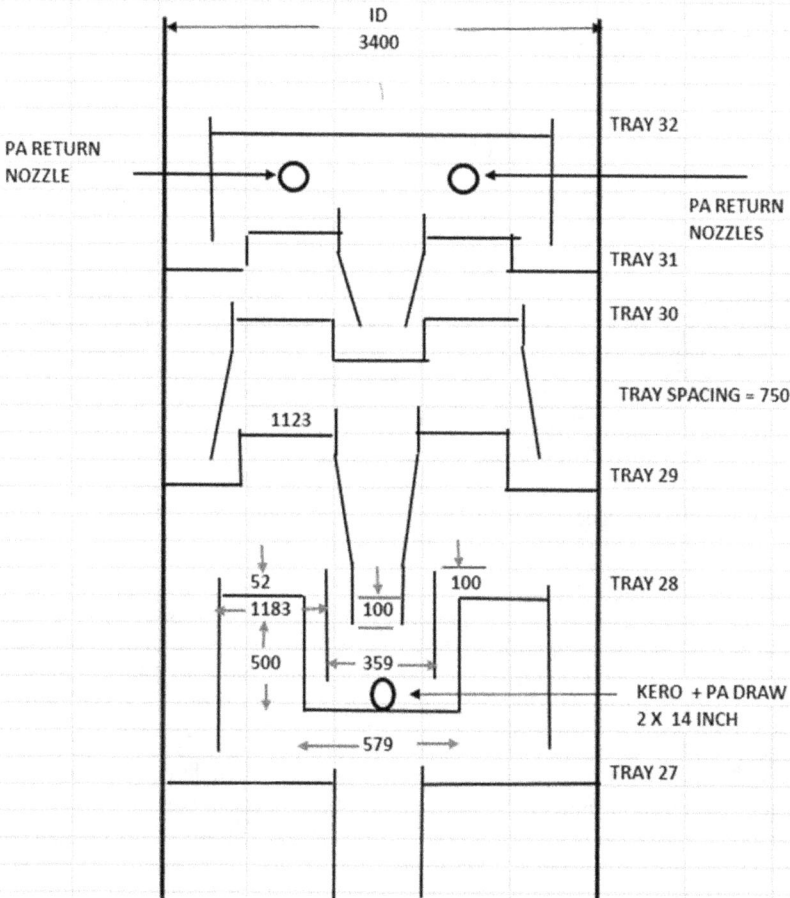

FIGURE 1.47 Elevation drawing of draw off tray.

Figure 1.47 shows the elevation of the draw off tray of kerosene PA + kerosene product.

1.8.5 CASE STUDY OF A VACUUM UNIT

In a vacuum unit, the barometric legs from condensers following booster ejectors (in parallel) were combined and introduced with a loop to the hot well, as shown earlier in Figure 1.36. Products of corrosion used to accumulate at the horizontal run of the barometric leg and in the loop. These accumulated mucks used to offer resistance to flow of condensate and eventually used to flood the condensers and increase the condensate temperature, and that in turn used to affect the vacuum adversely.

This kind of loop saves investment on the hot well, but in the long run may result in problems for getting the required vacuum. The author has seen this arrangement in two very big vacuum units. It may be better to dip the leg in another vessel (with a seal) and interconnect the vessels. A flow meter (low pressure drop) is always advisable on the non-condensable vent to estimate the quantity of leaks and cracked gas produced.

1.8.6 CASE STUDY OF VACUUM SYSTEM BOTTOM COKING

The vacuum column of a very big crude vacuum unit (AVU) had a bottom section as in Figure 1.33. It had six stripping trays at the bottom of the column. The bottom section had a coke strainer and cooled vacuum residue (VR) quench is introduced in the section. Below the strainer there was a tar pot or boot as in Figure 1.31 (having a lower diameter than the bottom section of the column) for lowering the residence time of the VR and consequent thermal cracking/coking. Level transmitters were provided in the upper wider section as well as in the smaller diameter section (tar pot). A level below the stripping tray appeared in the wider section. Increased pump out to bring down the level resulted in a lower level in the tar pot and the bottom vacuum residue pump suction starved, but the level in the upper section (upstream coke strainer) did not get low, indicating flow restriction in the coke strainer.

A pressure gauge was fitted to note the pressure at the suction of the pump. Delta pressure across the internal coke strainer was also measured as ΔP_s. Calculation of the level was performed as $H \times \rho = P2 - P1 + \Delta P_s$, where ρ is the density of the liquid (VR) at the operating temperature. The calculation revealed that the column bottom section is full with liquid. All stripping trays are submerged and the level is near the flash zone of the column. Finally, it was concluded that that the liquid from upstream of the coke strainer is unable to drop freely to the tar pot as the strainer may be choked with coke, resulting in liquid accumulation. Throughput of the unit was reduced. The coil outlet temperature reduced to have some respite.

The quench arrangement was also examined. It was found that the quench enters from one side through a nozzle, and there was no ring sparger inside to uniformly mix the quench liquid with the stripper outlet hot liquid. (Ideally, the quench liquid should be introduced right at the outlet of the liquid from bottommost stripping tray at the outlet of the seal pan.) So, when the level appeared in the upper wider section, the residence time increased, and due to absence of a sparger, hot locations remained in the pool of liquid and resulted in heavy coking around the coke strainer. Thus, the liquid from the wider section was unable to drop in the tar pot. The dimensions of the coke strainer were checked and compared with the standard and was found to be alright.

BIBLIOGRAPHY

Andrew W. Sloley. *Refinery Vacuum Towers, State of the Art* (Extended Abstract). Bellingham, WA: CH2M Hill, 2013.

Brandon Payne. Minimize corrosion while maximizing distillate. *PTQ*, Q3, 2012, pp. 75–81.

G. R. Martin, J. R. Lines, and S. W. Golden. Understand vacuum-system fundamentals. Properly operating ejectors and condensers is important in maximizing vacuum tower gas-oil yield. *Hydrocarbon Processing*, October 1994, pp. 1–7.

Henry Z. Kister. *Distillation Operation*. New York: McGraw Hill, Inc. 1990.

Henry Z. Kister. Flooded condenser controls: Principles and troubleshooting. *Chemical Engineering*, January 2016, pp. 37–49, www.chemengonline.com.

M. Dion, B. Pyne, and D. Gotewold. Operating philosophy can reduce overhead corrosion. *Hydrocarbon Processing*, March 2012, pp. 45–47.

Mahesh Subramaniam, James Ondyak, P. N. Ramaswamy, James Noland, and Parag Shah. Changing perspective. *Hydrocarbon Engineering*, March 2015, Dorf Ketal, USA, discuss the best way to take advantage of high acid crudes, pp. 1–4.

N. P. Hilton and G. L. Scattergoods. *Mitigate Corrosion in Your Crude Unit*. Nalco Engineering Services, September, 2010. Published as a corrosion control special report, reproduced from *Hydrocarbon Processing*, pp. 75–79.

Parag Shah, Mahesh Subramaniyam, James Ondyak, James Noland, and Nagi-Hanspal. Advances in processing high naphthenic acid crudes. *PTQ Journal*, Q4, 2015.

ABBREVIATIONS

API	American Petroleum Institute gravity, 141.5 − (131.5/sp. gr.), where sp. gr. is the specific gravity
ASTM gap	a positive difference of ASTM 95%-point temperature of upper product and ASTM 5%-point temperature of adjacent lower product
ASTM overlap	a negative difference of ASTM 95%-point temperature of upper product and ASTM 5%-point temperature of adjacent lower product
ASTM	American Society for Testing and Materials
ATF	aircraft turbine fuel
AVU	atmospheric-vacuum unit
COT	coil outlet temperature
CR	circulating reflux
CRU	catalytic reforming unit
CCRU	continuous catalytic reforming unit
DHDS	diesel hydrodesulfurization
DHDT	diesel hydrotreating/hydrotreater
EHCO	extra-heavy crude oil
EHGO	extra-heavy gas oil
FBP	final boiling point
FCCU	fluid catalytic cracking unit
HAGO	heavy atmospheric gas oil
HCU	hydrocracker unit
HGO	heavy gas oil
HO	heavy oil
HHVGO	extra-heavy vacuum gas oil
HVGO	heavy vacuum gas oil
IO	inter oil
IR	internal reflux
K	vapor–liquid equilibrium constant

LAB   linear alkylbenzene
LDO   light diesel oil
LGO   light gas oil
LO   light oil
LPG   liquefied petroleum gas
LVGO   light vacuum gas oil
MMTPA   million metric tons per annum
NG crude   North Gujarat crude (heavy crude of Indian origin)
PA   pumparound
PHT   preheat train
RCO   reduced crude oil
RON   research octane number
SCO   synthetic crude oil
SK   superior kerosene
SO   spindle oil
SRGO   straight-run gas oil
TAN   total acid number
TBP   true boiling point
UCO   upgraded crude oil
VGO   vacuum gas oil
VLE   vapor liquid equilibrium
VR   vacuum residue
XVGO   extra vacuum gas oil

2 Fluid Catalytic Cracking Unit (FCCU)

2.1 INTRODUCTION

The fluid catalytic cracking unit (FCCU) is deployed for upgradation of heavier petroleum stocks into lighter distillates or upgradation of the bottom of the barrel by converting heavier stock to lighter distillates. The feed typically can be vacuum gas oil (VGO), hydrotreated VGO, and heavy coker gas oil (HCGO) or an admixture of the components commonly referred to as distillate FCCU. The Conradson carbon residue (CCR) of the feed for distillate hydrocracker can range from 0.2% to 0.5% wt. A blend of VGO and reduced crude oil (RCO) with higher CCR can be also fed to the FCCU. Hydrotreated heavy residues like RCO, vacuum residue (VR), and dewaxed oil (DAO) can also be a feed for FCCU, separately or in admixture. The CCR of the feed usually ranges from 0.7% to 7% by weight. The units processing heavy feed stocks are known as residue FCCU (or RFCCU). The typical products are gas, liquified petroleum gas (LPG), FCCU gasoline, light cycle oil (LCO), clarified oil (CLO), and coke.

2.1.1 PROCESS DESCRIPTION

The FCCU deploys crystalline or amorphous zeolite catalysts for conversion of heavy oils to lighter distillates at elevated temperatures. Preheated oil feed is contacted with hot moving catalyst in the reactor riser (providing a stipulated residence time), and products of reaction are separated from catalyst by use of cyclones and fed to the fractionator for further separation into different products. The separated catalyst, soaked with oil, is routed to the catalyst stripper housed at the bottom of the reactor vessel for stripping with stripping steam to remove residual oil from the catalyst, and the strip out joins the products of reaction and finds its way to the fractionator. During reaction with oil, coke deposits on the catalyst and catalyst activity declines.

The spent catalyst from the reactor stripper flows to the regenerator through the spent catalyst standpipe (SCSP) controlled by the spent catalyst slide valve (SCSV) and the under level control of the stripper. In the regenerator, the coke on the catalyst is burnt with air and the catalyst is regenerated. The catalyst is separated from hot flue gas by the regenerator cyclones and the hot regenerated catalyst is again recirculated to the riser through the regenerated catalyst standpipe (RCSP) with a slide valve (RCSV) to control the flow of regenerated catalyst. The air to the regenerator is supplied by the main air blower (MAB), and flow of air to the regenerator is controlled based on the differential temperature between the dilute (catalyst lean phase) and catalyst bed dense phase (i.e., with control of afterburning).

DOI: 10.1201/9781003268246-2

The hot flue gas from the regenerator flows out through a double-disc slide valve (DDSV) and is sent to the carbon monoxide (CO) incinerator and flue gas cooler generating high pressure steam, and subsequently to the stack with or without flue gas treatment (FGD). The typical products of FCCU are gas, LPG, naphtha, LCO, HCO, CLO, and coke.

Recently, the FCCU has been deployed for production of olefins like ethylene and propylene and is also known to produce benzene and xylenes. The FCCU deployed for production of olefins, primarily propylene and ethylene, is termed the petrochemical FCCU. The petrochemical FCCU typically operates with a high reactor temperature of about 560°C, with hydrotreated VGO as feed, and is capable of producing around 40% wt. of LPG and around 16% wt. of propylene on feed. The sponge gas is known to contain 25% to 33% ethylene by volume. Ethylene is recovered from treated sponge gas of the FCCU. The processing scheme of a FCCU producing petrochemicals is shown in Figure 2.1, Figure 2.1A, and Figure 2.1B. Most petrochemical FCCU are deployed for production of propylene; some are for production of both ethylene and propylene.

Typical sponge gas ethylene concentration is around 18% by mole. But it is seen to be around 30% in some petrochemical FCCU operating with a very high catalyst-to-oil ratio. A typical scheme to recover the ethylene is presented Figure 2.1B. In a typical scheme, in the beginning the gas is washed with diglycolamine (DGA), then caustic washed. Oxygen is removed by use of dimethyl disulfide (DMDS) and H2 injection, and H2S is produced and washed again with caustic to remove traces. The metal guard removes mercury, arsine, etc. Then the gas is refrigerated in the cold box and fed to the demethanizer (DM) and dethanizer (DE) to get pure ethylene.

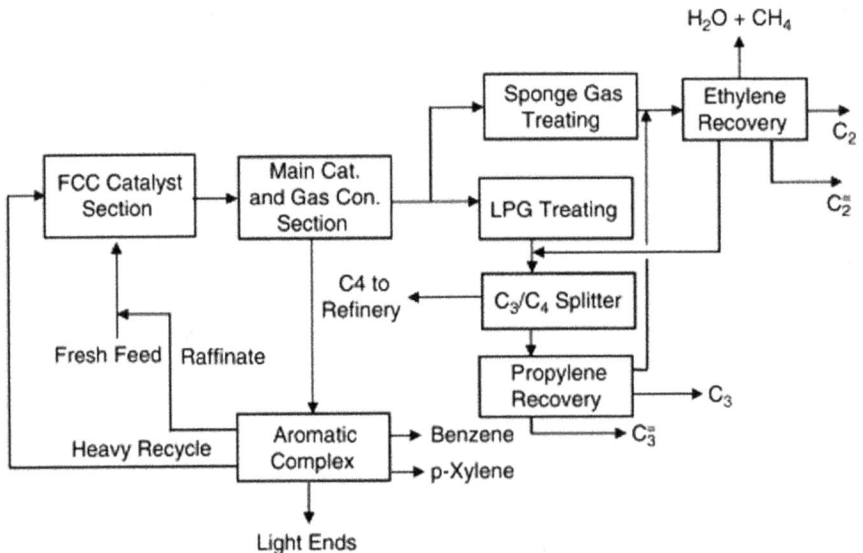

FIGURE 2.1   An overall FCCU processing scheme producing petrochemicals.

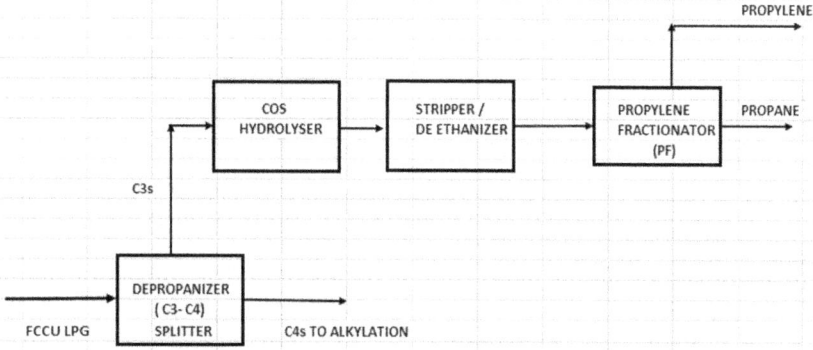

FIGURE 2.1A Simplified scheme for recovery of propylene and generation of feed stock for alkylation.

The FCCU can be divided in several sections:

1. Reactor–regenerator section (R-R section)
2. Regenerator flue gas handling section
3. Feed preheating section
4. Fractionation and gas concentration section

The main sections of an FCCU and their integration are shown in Figure 2.2. The overall flow scheme of an FCCU is shown in Figure. 2.3 and Figure. 2.4.

The R-R section is the heart of the FCCU because in this section fluidized catalyst reacts with hot oil and produces products and spent catalyst with carbonaceous matter is transferred to the regenerator to burn the carbonaceous matter and regenerate the catalyst, and then recirculates the catalyst again for reaction. The section provides almost the entire heat for the process through burning of coke of spent catalyst. A typical FCC reactor regenerator section is shown in Figure 2.5.

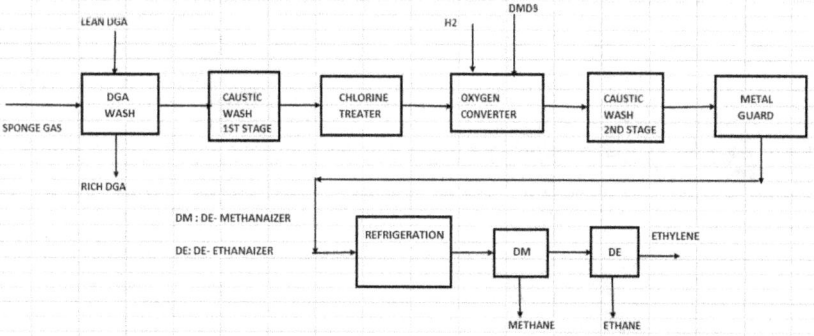

FIGURE 2.1B Tentative scheme for recovery of ethylene from FCCU sponge gas.

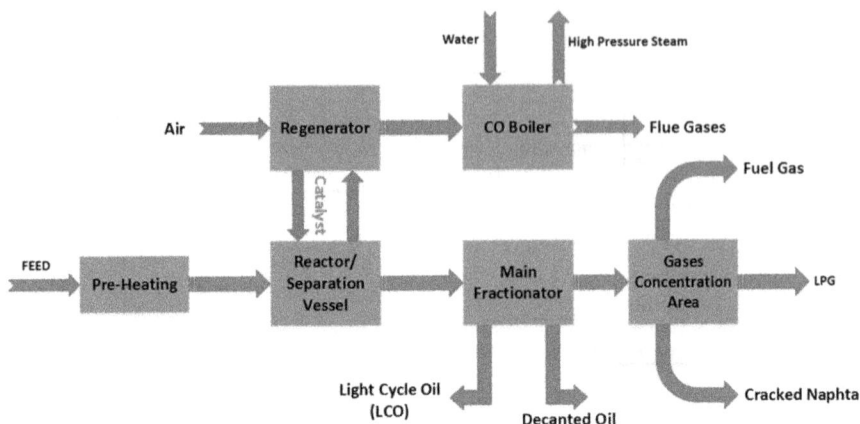

FIGURE 2.2   Integration of main sections of an FCCU.

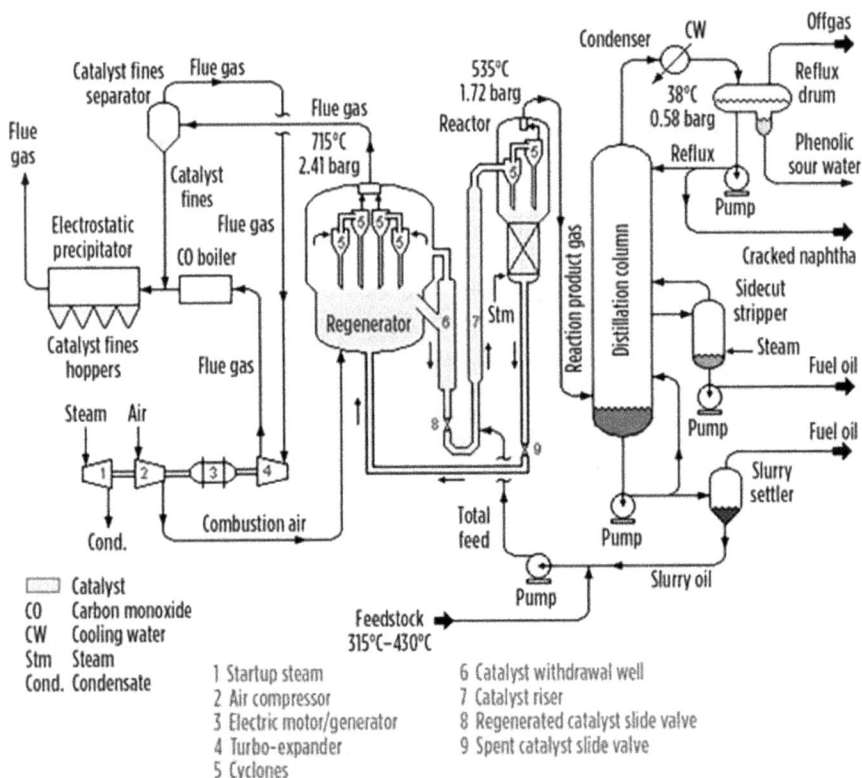

FIGURE 2.3   Typical FCCU flow scheme.

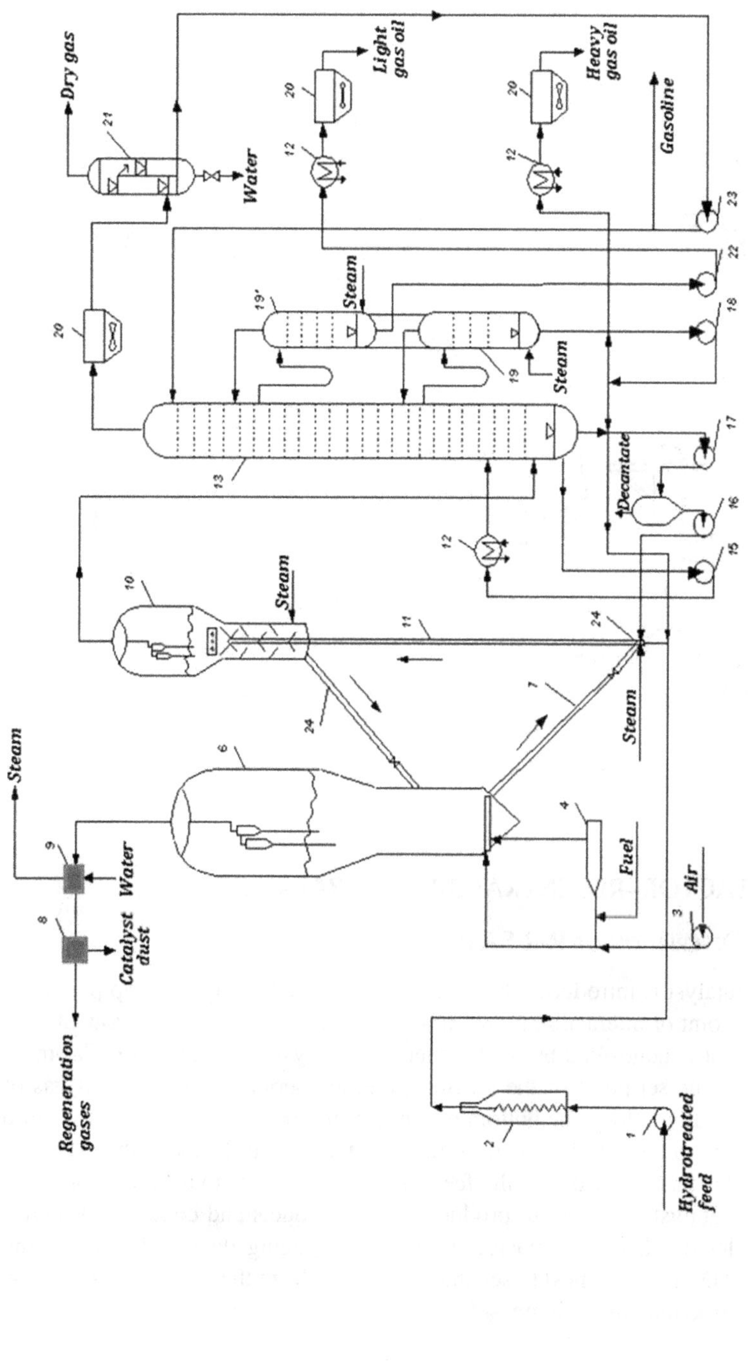

FIGURE 2.4 Typical FCCU major equipment/components.

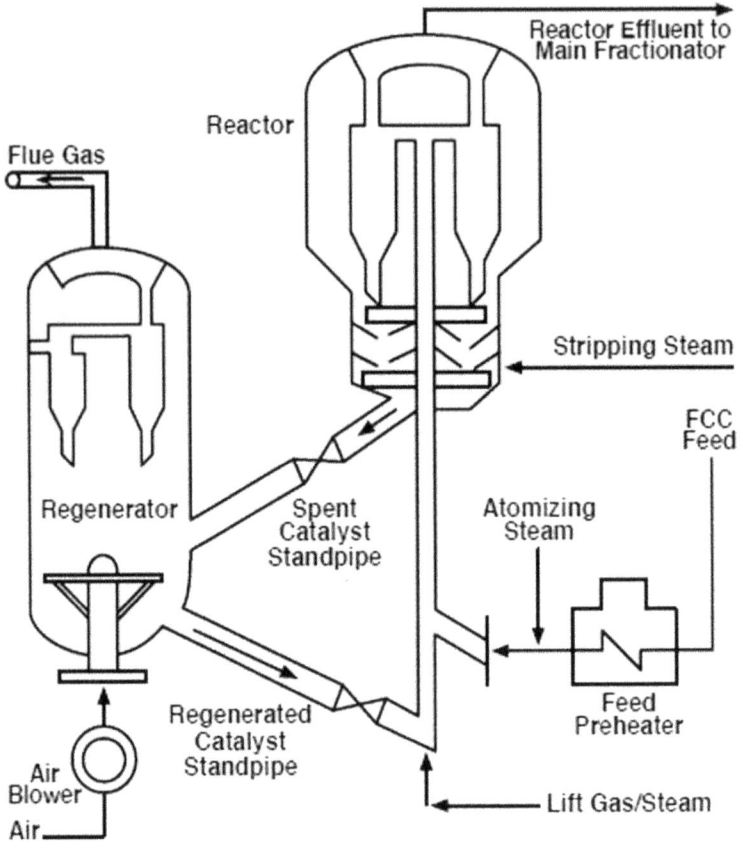

FIGURE 2.5 Typical FCC reactor regenerator section.

2.2   REACTOR–REGENERATOR HEAT BALANCE

2.2.1   DESCRIPTION OF R-R SECTION

The hot catalyst is introduced through the regenerated catalyst standpipe (RCSP) to the Wye (point of interaction between the RCSP and the riser). The flow of regenerated catalyst is controlled by the regenerated catalyst slide valve (RCSV) in accordance with the set point of the reactor outlet temperature (ROT). A lift gas or lift steam is usually used at the bottom of the riser to preaccelerate the catalyst through the riser. The preheated feed is introduced in the riser little above the Wye through feed nozzles that finely divide the feed into small droplets to achieve thorough contact with catalyst and react to produce lighter products and coke. The superheated vapor is let out from the reactor through disengaging devices [riser terminating device (RTD) and cyclones] to separate the vapors from the catalysts. The vapors are fed to the fractionating column.

The disengaged catalysts from cyclones are routed to the reactor stripper, which has trays or grids to efficiently strip off the hydrocarbons to reduce the hydrocarbon content of the spent catalyst going to the regenerator. The spent catalyst flows into the regenerator with coke and residual hydrocarbon through the spent catalyst stand pipe (SCSP), and its flow is regulated by the spent catalyst slide valve (SCSV) acting in accordance with the level of the reactor stripper. Air is injected in the regenerator through spargers for uniform distribution through the bed of catalyst. The coke burns and regenerates the catalyst and produces hot flue gas. The hot flue gas after separation of catalysts in regenerator cyclones is let out from the regenerator, typically through pressure control of the regenerator through the double-disc slide valve (DDSV). The regenerated catalyst again flows through the RCSP and RCSV to heat the feed in the riser and the cycle continues.

The heat produced during regeneration (burning of coke) is consumed in the endothermic cracking reactions. FCC therefore became an integrated reaction–regeneration process where heat is produced in the regeneration zone and transported by the catalyst and is used to heat the feed and vaporize the liquid feed to fracationator and to promote endothermic reactions. A part of the heat goes out of the regenerator along with the hot flue gas (and also provides the heat loss to surrounding). The flow of heat of combustion of coke is schematically presented in Figure 2.6.

2.2.2 Overall Heat Balance

Figure 2.6 illustrates the heat balance in the reactor–regenerator system.

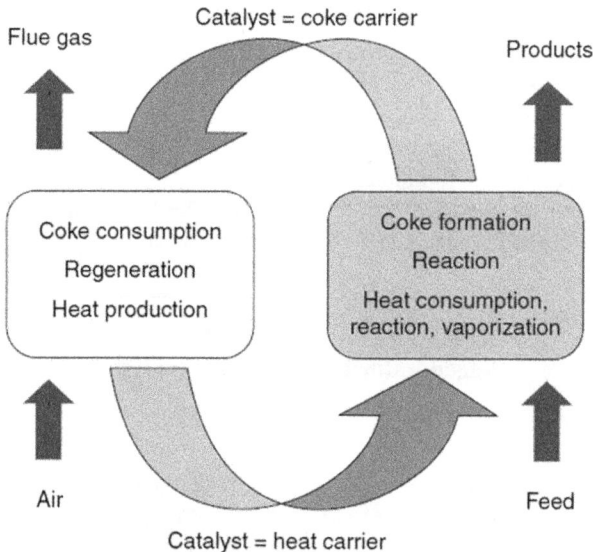

FIGURE 2.6   R-R heat balance in FCCU.

2.2.3   Regenerator Heat Removal

The coke produced in the reaction of hydrocarbon with hot catalyst needs to be burned to restore the activity of the catalyst for further reaction. The heat of combustion of coke is used in the following:

- Heat the air to flue gas temperature
- Heat the spent catalyst to regenerator dense phase temperature
- Heat the feed to reactor outlet temperature
- Provide the endothermic heat of reaction and heat of vaporization of products
- Provide the heat loss to the surrounding

Thus, the entire heat is supplied by the heat of coke combustion. This heat balance across the regenerator is also described in many books. But a heat balance once again may be necessary to go further and to examine the requirement of other facilities, like the catalyst cooler, and establish the effect of change of operating variables from the heat balance. Accordingly, the heat balance for the R-R section of an FCCU with a single regenerator is presented for an operating unit as follows.

The following data is from an actual plant using hydrotreated VGO as feed and a single regenerator. The dry flue gas analysis is presented in c.

Oil feed to the unit = 225 MT/Hr
Regenerator dense phase temperature = 670°C
Hot flue gas temperature = 709°C
Dry air flow to regenerator = 85000 NM3/Hr
Air temperature at blower outlet = 220°C
Basis: 100 ton moles (100,000 kg-mole) of flue gas
- Oxygen consumed for production of CO_2 = 15.02 ton moles
- Oxygen consumed for production of CO = 2.66/2 = 1.33 ton moles
- Oxygen excess = 1.25 ton moles
- Total oxygen consumed for CO_2, CO, and the balance = 15.02 + 1.33 + 1.25 = 17.6 ton moles
- Oxygen supplied with air = (100/79) 81.07 = 21.55 ton moles (air contains 21% of oxygen and 79% of nitrogen by volume)

TABLE 2.1
Regenerator Dry Flue Gas Analysis

Component	Volume %
CO_2	15.02
CO	2.66
O_2	1.25
N_2	81.07 (by balance)
Total	100

TABLE 2.2

Regenerator Wet Flue Gas Analysis

Component	Volume %
CO2	(15.02/108.04)=13.9
CO	(2.66/108.04)=2.46
O2	(1.25/108.04)=1.16
N2	(81.07/108.04)=75.04
H2O	(8.04/108.04)=7.44
Total	100

- Oxygen consumed for production of H2O = 21.55 − 17.6 = 4.02 ton moles
- Water formed = 2 × 4.02 = 8.04 ton moles
- Total weight flue gas quantity = (100 + 8.04) = 108.04 ton moles

Wet flue gas composition is worked out next and presented in Table 2.2:

Average molecular weight of wet flue gas = (13.9 × 44 + 2.46 × 28 + 1.16 × 32 + 75.04 × 28 + 7.44 × 18)/100 = 29.53

Air supplied 85000 NM3/Hr = 85000 × 29/(22.4 × 1000) = 110.04 MT/Hr

N2 supplied along with air = 110.04 × 0.77 = 84.7 MT/Hr(Air contains 77 % N2 by weight and humidity of the air is not considered in calculation for simplicity)

An N2 balance yields the following:

84.7 = WM × 0.7504 × 28, where WM is the wet flue gas quantity in T-moles/H

WM = 4.03 T-moles/Hr

Weight of flue gas = 4.03 × 29.53 = 119 MT/Hr

Carbon on coke = 12(13.9 + 2.46) × 4.03/100 = 7.91 MT/Hr

Hydrogen on coke = 2 × 7.44 × 4.03/100 = 0.6 MT/Hr

Total coke = 6.99 + 0.53 = 8.51 MT/Hr

% Carbon on coke = (7.91/8.51) × 100 = 93%

% Hydrogen on coke = (0.6/8.51) × 100 = 7%

Coke production on feed = (8.51/225) × 100 = 3.78%

Heat of formation of the flue gas components at 25°C is listed in Table 2.3.

Heat liberation for generating 100 ton moles of wet flue gas is computed as follows:

13.9 × 94.05 + 2.46 × 26.4 + 8.04 × 57.8 = 1836.95 G-Cal/Hr

Total heat liberation for 119 MT/Hr of flue gas production = 1836.95 × 119/(29.53 × 100) = 74.03 G-Cal/Hr

TABLE 2.3

Heat of Formation of Flue Gas Components

Component	Heat of Formation (Kcal/g mole)
CO_2	94.05
CO	26.4
H_2O	57.8
SO_2	71

Heat inputs

- Heat in along with air $= 110.04 \, (220 - 25) \times 0.3 = 6440.85$ T-Cal/Hr
- Heat in along with spent catalyst $= W \, (500 - 25) \times 0.4 = 190W$ T-Cal/Hr
- Heat in along with coke on spent catalyst $= 8.51 \, (500 - 25) \times 0.3 = 1212.7$ T- Cal/Hr
- Heat liberation inside regenerator $= 74030$ T-Cal/Hr

Total heat input $= 190W + 81683.6$ T-Cal/Hr, where W is the mass flow rate of the catalyst.

Heat outputs

- Heat out with flue gas $= 119 \, (709 - 25) \times 0.3 = 24418.8$ T-Cal/Hr
- Heat out with regenerated catalyst $= W \, (670 - 25) \times 0.4 = 258W$ Ton Cal/Hr
- Heat loss $= 0.02 \times 66300 = 14806$ T-Cal/Hr (assuming 2% heat loss on heat liberation)

$$\text{Total heat output} = 39224.8 + 258W \, \text{T} - \text{Cals/Hr}$$

Equating heat input and heat output around regenerator it gives the following:

$190W + 81683.6 = 39224.8 + 258W$
$68W = 42458.75$ MT/Hr
$W = 624.4$ MT/Hr: Catalyst/oil ratio $= 624.4/225 = 2.8$

A summary of the results of heat balance is presented in Table 2.4.

The hydrogen percent on coke is found on the higher side indicating oil ingress to the regenerator from the reactor stripper.

2.2.4 REVIEW OF REQUIREMENT OF CATALYST COOLER

Another example for a single-stage regenerator with higher coke make is considered to examine the requirement of a catalyst cooler. This is presented in the problem

TABLE 2.4
Summary of Results of Heat Balance

Attribute	Unit	Result
Coke production	MT/Hr	8.51
Coke production on feed	%	3.78
Carbon on coke	%	93
Hydrogen on coke	%	7
Catalyst circulation	MT/Hr	624.4
Cat/oil ratio	Ratio	2.8
Delta coke	%	

TABLE 2.5
Regenerator Flue Gas Analysis Dry and Wet

Components	Dry Flue Gas Composition (Volume %)	Wet Flue Gas Composition (Volume %)
CO_2	15.02	(15.02/108.04) = 13.91
CO	2.66	(2.66/108.04) = 2.46
O_2	1.25	(1.25/108.04) = 1.16
N_2	81.07	(81.07/108.04) = 75.03
H_2O	—	(8.04/108.04) = 7.44
Total	100	100
Flue gas mol. wt.		29.52

table in Table 2.5. The dry flue gas composition is considered as earlier, and the wet flue gas composition worked out and presented again for convenience.

The operating data and computed important indicators are presented in Table 2.6.

Basis: 100 T-moles (100,000 kg moles) of wet flue gas.

Computation of oxygen required to burn carbon

- O_2 required to produce CO_2 = 13.91 T-moles
- O_2 required to produce CO = 2.46/2 = 1.23 T-moles

O_2 require to burn carbon to the extent as flue gas = 13.91 + 1,23 = 15.14 T-moles

Carbon balance

Carbon burnt to CO_2 = 13.91 T-moles
Carbon burnt to CO = 2.46 T-moles
Carbon burnt to CO_2 + CO = (13.91 + 2.460) = 16.37 T-moles

TABLE 2.6
Operating data of FCCU

Attribute	Unit	Value
Unit capacity	MT/Hr	212.5
Catalyst circulation rate	MT/Hr	1200
Feed CCR	Wt.%	Nearly 5.0
Coke burnt	MT/Hr	15.52
Wt.% coke on feed	%Wt.	7.3
Wt.% carbon on coke	%Wt.	93
Wt.% hydrogen on coke	%Wt.	7
Cat/oil ratio	Ratio	5.6
Reactor temperature	°C	510
Maximum regenerator flue gas temperature	°C	730
Maximum regenerator dense phase temperature	°C	700
Air temperature at blower discharge	°C	181

Actual O2 requirement calculation

Carbon in coke $= 15.52 \times 0.93/12 = 1.2$ T-moles
Hydrogen in coke $= 15.52 \times 0.07/2 = 0.54$ T-moles

- O2 required for combustion of 1.2 T-moles of carbon $= 15.14 \times 1.2/16.37 = 1.11$ T-moles
- O2 required for combustion of 0.54 T-moles of hydrogen $= 3.72 \times 0.54/7.44 = 0.27$ T-moles
- Total O2 required for reaction with carbon and hydrogen $= (1.11 + 0.27) = 1.38$ T-moles
- O2 balance X evaluated through $X/(1.38 + X) = 1.15/100 = 0.016$ T-moles
- Total O2 supplied to burn the carbon on catalyst + free O2 $= (1.38 + 0.016) = 1.396 = 1.4$ T-moles (nearly)

Equivalent dry air to supply 1.4 T-moles of O2 $= 1.4 \times 100/21 = 6.7$ T-moles $= 6.7 \times 29 = 194.3$ MT/Hr

N2 balance calculation to evaluate actual wet flue quantity WM, in T-mole/Hr

$WM \times 0.7503 = 6.7 \times 0.79$, $WM = 6.7 \times 0.79/0.7503 = 7.06$ T-moles

Therefore, flue gas quantity $= 7.06 \times 29.52 = 208.3$ MT/Hr

Heat liberation calculation

Heat liberation equivalent to 100 T-moles of wet flue gas (computed earlier) $= 1802.3$ G-Cal/Hr

So, heat liberation from 7.06 T-moles/Hr of flue gas = $1802.3 \times 7.06/100 = 127.2$ G-Cal/Hr

Heat balance around regenerator calculation

Heat inputs
Heat in through air = $(181 - 25) \times 0.3 \times 194.3 = 9.1$ G-Cal/Hr
Heat in through spent catalyst = $1200 (510 - 25) \times 0.4 = 232.8$ G-Cal/Hr
Heat of reaction = 127.2 G-Cal/Hr
Total heat in = $9.1 + 232.8 + 127.2 = 369.1$ G-Cal/Hr

Heat outputs
Heat out with regenerated catalyst = $1200 (700 - 25) \times 0.3 \times 208.3 = 324$ G-Cal/Hr
Heat out with hot flue gas = $208.3(730 - 25) \times 0.3 = 44$ G-Cal/Hr
Heat loss assumed 1% on liberation = $0.0 1 \times 127.2 = 1.25$ G-Cal/Hr
Total heat out = $324 + 44.1 + 1.25 = 369.35$ G-Cal/Hr

The heat input and output are nearly equal. Thus, in this case there is *no requirement for a catalyst cooler.* Any imbalance can be controlled by variation of the preheat temperature or higher stripping of the spent catalyst in the reactor stripper. However, if heavier stock is processed the carbon lay down will increase, increase the heat liberation, and consequently may call for a catalyst cooler to balance the heat.

Calculation to evaluate the endothermic heat of reaction in the reactor for the afore-mentioned example (Tables 2.5 and 2.6)

A heat balance around the entire reactor–regenerator envelope is performed as follows:

Heat inputs
- Heat in with air = 9.1 G-Cal/Hr
- Heat in with feed = $212.5 (200 - 25) \times 0.64 = 23.8$ G-Cal/Hr
- Heat in with steam (10 MT/Hr MP steam (12 kg/cm2 (a) temperature 250°C) = 7.01 G-Cal/Hr
- Heat liberation = 127.2 G-Cal/Hr
- Total = 167.1 G-Cal/Hr

Heat outputs
- Heat out with flue gas = 44.1 G-Cal/H
- Heat out with Rx vapor = $212.5 \times 72 + 212.5 (510 - 25) \times 0.77 = 94.6$ G-Cal/Hr
- Heat out through hot steam (10 MT/Hr and 510°C) from reactor around 8.35 G-Cal/Hr
- Total heat loss on liberation assumed 4% = $127.2 \times 0.04 = 5.0$ G-Cal/Hr
- Heat of reaction in reactor = HXR G-Cal/Hr
- Total heat out = $44.1 + 94.6 + 8.35 + 5 = 152.1 + $ HXR G-Cal/Hr

$152.1 + $ HXR $= 167.1$ or HXR $= 15.01$ G-Cal/Hr
Heat of reaction per unit of feed = $150100/212.5 = 70.63$ Kcal/kg feed

So, endothermic heat of reaction in the reactor is around 71 Kcal/kg. This is typical of a VGO FCCU.

2.3   COMBUSTION MODES OF REGENERATORS

Oxygen-lean regeneration (partial CO combustion) is most preferred for use with heavy residuals where heat release in the regenerator and air required for coke burn is high due to high coke yield. In addition, the oxygen-lean regeneration environment reduces the likelihood of catalyst activity loss at high catalyst vanadium levels, due to the lower possibility of formation of V_2O_5 and vanadic acid at lower oxygen levels in flue gas. In grassroots applications, therefore, oxygen-lean regeneration is preferred for heavy residual operations with likely high vanadium loadings on catalyst. A two-regenerator arrangement (R2R) can also be an option. On the other hand, for better-quality residues as feed and gas oil feedstocks, complete CO combustion is preferred for its simplicity of operation. Other factors in the selection of regeneration mode are described next.

A unit designed to operate in an oxygen-lean mode for regeneration must have a CO incinerator for conversion of CO to CO2 and a CO boiler to recover sensible heat from the flue gas. The unit investment cost is usually lower for oxygen-lean regeneration due to the reduced regenerator, air blower, and flue gas system size. Steam production can be maximized by operating in an oxygen-lean mode of regeneration, due to combustion in the CO boiler. Regenerator heat removal systems (such as catalyst coolers) may be avoided in some cases if the unit is operated in an oxygen-lean mode of regeneration.

In some cases, complete CO combustion will allow the unit to operate with a lower coke yield and lower coke on the regenerated catalyst (CRC), thereby increasing the conversion. Different factors contributing to the production of coke in regenerators are described next.

2.4   DEFINITION OF DELTA COKE

Delta coke is defined as CSC – CRC = Coke yield/(C/O), where C is the coke yield in wt.%, C/O is the catalyst/oil ratio in wt/Wt. (Catalyst circulation rate in wt/Wt. feed flow rate). CSC is coke on spent catalyst expressed in wt.% of feed, and CRC is coke on regenerated catalyst expressed in wt.% on feed. Delta coke is the difference of coke on the spent catalyst (stripper outlet) and coke on the regenerated catalyst. Delta coke can also be seen as the amount of coke formed on the catalyst in a single pass in the reactor and is also equal to the amount of coke burned in the regenerator.

2.4.1   Types of Delta Coke

- *Catalytic*. The coke deposited on the catalyst when hydrocarbons are cracked by acid sites of the catalyst. Generally, the type of catalyst and the type of zeolite incorporated influence the catalytic delta coke. This is known to be typically around 65% wt. of total coke for distillate feeds.
- *Catalyst/oil coke or occluded/strippable coke.* Adsorbed or unstripped hydrocarbons from stripper contribute to this coke. A properly designed and

operated stripper reduces this type of coke. This is also influenced by catalyst matrix pore size. In some cases, while processing hydrotreated VGO
alone, a little hydrocarbon ingress to the regenerator is allowed along with
spent catalyst and contribute to coke to maintain regenerator temperature.

- *Contaminant.* Coke is produced by dehydrogenation activity of metals (Ni, V,
Cu, Fe). The coke produced through dehydrogenation by metals can be reduced
by use of nickel passivators (using antimony or bismuth) and vanadium by tin is
known to reduce the coke formed due to dehydrogenation and reduce hydrogen
production (indicated by reduction in hydrogen in sponge gas).
- *Additive or Concarbon coke.* Coke formed due to feed Conradson carbon
residue (CCR) is mainly the feed asphaltenes that deposit as coke. This is
low for distillate feed (CCR <1%), but for residual feed stock this type of
coke formed can be much higher (21%+).

2.5   POWER RECOVERY FROM REGENERATOR HOT FLUE GAS

Typically, regenerator off-gas pressure is reduced in the double-disc slide valve and
orifice chamber before letting the hot flue gas to the CO incinerator/CO boiler. It
is possible to expand the gas in a turbine to extract power to drive the air blower
and generate a considerable amount of electric power. The flue gas in the preceding
example (Table 2.5) is used and shown in simulation (Figure. 2.7). The expander can
produce more power than required for the drive of the air blower (Table 2.7). This

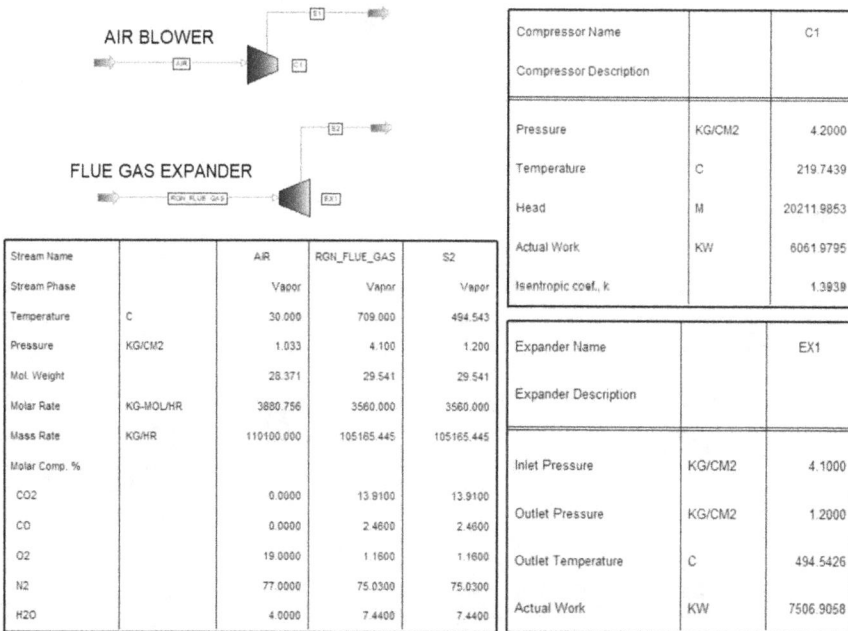

Compressor Name		C1
Compressor Description		
Pressure	KG/CM2	4.2000
Temperature	C	219.7439
Head	M	20211.9853
Actual Work	KW	6061.9795
Isentropic coef., k		1.3939

Expander Name		EX1
Expander Description		
Inlet Pressure	KG/CM2	4.1000
Outlet Pressure	KG/CM2	1.2000
Outlet Temperature	C	494.5426
Actual Work	KW	7506.9058

Stream Name		AIR	RGN_FLUE_GAS	S2
Stream Phase		Vapor	Vapor	Vapor
Temperature	C	30.000	709.000	494.543
Pressure	KG/CM2	1.033	4.100	1.200
Mol. Weight		28.371	29.541	29.541
Molar Rate	KG-MOL/HR	3880.756	3560.000	3560.000
Mass Rate	KG/HR	110100.000	105165.445	105165.445
Molar Comp. %				
CO2		0.0000	13.9100	13.9100
CO		0.0000	2.4600	2.4600
O2		19.0000	1.1600	1.1600
N2		77.0000	75.0300	75.0300
H2O		4.0000	7.4400	7.4400

FIGURE 2.7   Simulation showing power requirement air blower and expander.

TABLE 2.7

Power Requirement for Blower and Generation from Expander

Attributes	Expander (EX1)	Air Blower (C1)
Inlet pressure, (Kg/cm2) (a)	4.1	Atmosphere
Outlet pressure, (Kg/cm2) (a)	1.2	4.2
Inlet temperature, °C	709	30
Outlet temperature, °C	494.5	219.7
Power, KW	7506.9	6061.97

assumes an adiabatic efficiency of 87% for the expander and a polytrophic efficiency of 81% for the air blower and air temperature at a blower suction of 30°C (humidity of air not considered for simplicity).

Power generation from the regenerator flue gas expander required for the blower is presented for the problem in Table 2.7. An integrated scheme of flue gas system and power generation system is shown in Figure 2.8.

In a traditional system design, a steam turbine is integrated to the system to supplement the startup power requirement for the blower of the power recovery train (PRT). Once the unit is started and flue gas flow is maximized to the expander, the steam flow to the turbine is reduced to the minimum only to keep the turbine in operation. Alternatively, a simpler system having a separate turbo generator can be used to generate power from the flue gas, and the air blower can be driven by a steam

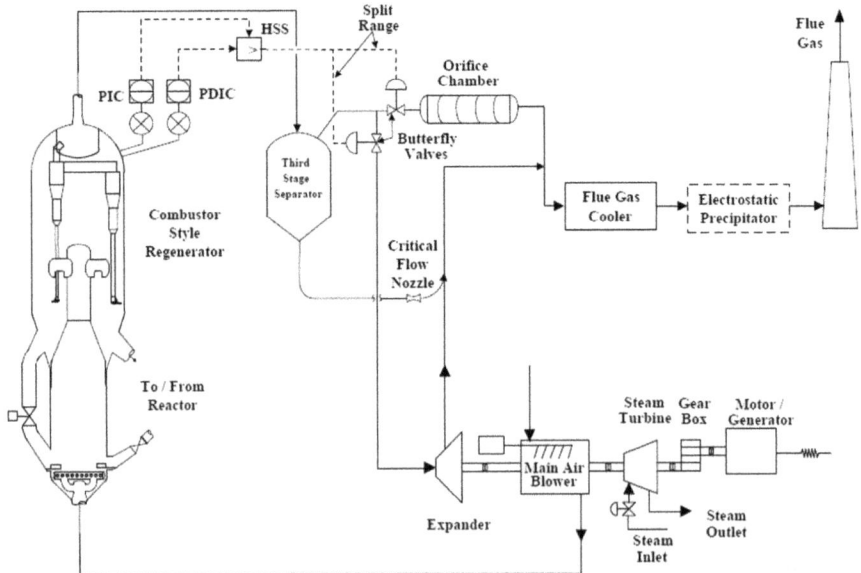

FIGURE 2.8 Facilities in the regenerator flue gas circuit for power generation.

turbine. The installation of the power recovery system is economically justifiable only for bigger FCCU (above 4 MMTPA capacity) as known. The system needs a tertiary stage separator (TSS) to arrest the particles above 20 microns going to the expander and to reduce the concentration of particles to typically less than 50 mg/m3. The TSS is provided with a bottom blowdown facility with a critical flow nozzle to route the blowdown with catalyst particles to the CO incinerator/CO boiler, as shown in Figure 2.8.

2.6 OPERATING VARIABLES

Important independent operating variables of the unit are presented as follows:

- Riser temperature
- Recycle rate
- Feed preheat temperature
- Fresh feed rate
- Catalyst makeup rate
- Gasoline endpoint

2.6.1 RISER TEMPERATURE

Increasing the riser temperature set point will signal the regenerated catalyst valve (RCSV) to open and increase the hot catalyst flow as required to achieve the desired riser outlet temperature (at same feed preheat). The regenerator temperature will rise due to the higher temperature of the catalyst entering the regenerator (from the reactor) and also due to an increase in coke laydown on the catalyst and liberation of more heat in the regenerator. At steady state, both the catalyst circulation and the regenerator temperature will be higher than they were at the lower riser temperature. The increased riser temperature and increased catalyst circulation (cat/oil) will result in increased conversion.

An increase in riser temperature produces the largest increase in dry gas production but results in less increase in coke yield. This makes increasing riser temperature an attractive option for increasing conversion when the unit is close to a regenerator air limit but has some spare gas-handling capacity in the fractionator and gas concentration unit (GCU). Increasing the riser temperature also significantly improves octane. The octane effect of increased reactor temperature is about 1× (R + M)/2 per 15°C increase of riser temperature. However, beyond a certain temperature, gasoline yield will be adversely affected. The octane effect may often support the economics and favor high riser-temperature operation.

2.6.2 RECYCLE RATE

The HCO and slurry injection facility from the main fractionator to the riser is typically provided to increase overall conversion and increase regenerator temperature when spare coke-burning capacity is available. Separate nozzles are normally provided in the riser for the same, typically at an upper elevation than the fresh feed nozzle. Coke and gas yield will be higher from cracking of HCO or slurry than from cracking incremental fresh feed, so the regenerator temperature and gas yield will increase significantly when recycling HCO or slurry to the riser. As such, recycle of slurry to the riser is an effective means to increase regenerator temperature, if required. Operation with HCO or slurry recycle together with a lower riser temperature is sometimes practiced when LCO maximization is desired. This maximizes LCO yield because the low riser temperature minimizes cracking.

2.6.3   Feed Preheat Temperature

Decreasing the temperature of the feed to the riser increases the catalyst circulation rate required for heating the feed to the specified riser outlet temperature. The alternative of lowering feed preheat temperature for increasing conversion produces a larger increase in coke yield but smaller increases in dry gas yield compared to raising the riser outlet temperature. So, when the FCCU is near the limit for production dry gas [wet gas compressor (WGC) and Gascon section reaching the limit], but has some spare coke burning capacity in the regenerator, reducing the preheat temperature may be a better way to increase conversion. Conversely, if in the FCCU the air quantity has attained its limit (in blower capacity or superficial velocity in the regenerator or in downstream flue gas circuit) but has excess light ends capacity, high preheat (and thereby increased riser temperature) is often the preferred mode of operation.

In the majority of cases, reducing preheat will lead to a lower regenerator temperature (because of higher outflow of hot catalyst from the regenerator and with the inflow of catalyst at the same reactor temperature). The increase in coke yield from the higher catalyst circulation (cat/oil) is typically not sufficient to meet the increased regenerator heat demand. In some cases, reducing the feed preheat temperature can result in an increased regenerator temperature. This can happen only if the feed preheat temperature is reduced to a very low value that affects atomization of feed in the feed nozzles (due to higher viscosity of feed at lower temperature) and lowers the feed atomization in the riser. The regenerator temperature can also increase with lowering of feed preheat, if catalyst stripping efficiency falls due to higher catalyst circulation rate and results in entry of additional oil (along with spent catalyst) to the regenerator. Both can increase coke laydown to a great extent and can result in higher regenerator temperature. Minimum preheat temperature is 190°C to 200°C, which does not increase the feed viscosity or cause problem of feed atomization and increased coke laydown for distillate feeds.

2.6.4   Fresh Feed Rate

Increasing fresh feed rate will need adjustment of so many variables, like reactor temperature, to contain the air blower capacity, catalyst circulation capacity (indicated by pressure drop across SCSV and RCSV), WGC capacity, and entire fractionation section capability.

2.6.5 CATALYST MAKEUP RATE

The conversion will usually increase with increasing activity due to the effect of higher catalyst activity. Some amount of catalyst addition is necessary to make up the loss of catalyst through the reactor and regenerator. Often some equilibrium catalyst is withdrawn from the regenerator and fresh catalyst is added to maintain the catalyst activity. A higher catalyst withdrawal and makeup is typical in petrochemical FCCU for high olefin yields. The riser outlet temperature or feed preheat temperatures can be adjusted to keep conversion constant as activity is increased and the same may result in decrease of coke and dry gas yields. This makes increasing catalyst makeup an attractive option in cases where the air blower or wet gas compressor is limiting, particularly when some increase in regenerator temperature can be tolerated.

2.6.6 GASOLINE END POINT

The gasoline/LCO cut point can be altered to shift product yield between gasoline and LCO while maintaining both products within acceptable specifications. Changing the cut point can significantly alter the gasoline octane and sulfur content. A lower cut point of gasoline results in lower sulfur content and generally higher octane, but the gasoline yield is reduced.

2.7 FLUIDIZATION AND RELATED EQUIPMENT IN FCCU

2.7.1 FLUIDIZATION

In the FCCU, catalyst particles behave as fluids when they are fluidized. In principle, the FCC standpipe is expected to behave analogous to a pipe full of water when fluidized. The FCC catalyst differs from water in the way that it is actually a fluidized solid. The gas flow supports the weight of catalyst. The pressure measured at any depth in the standpipe should be roughly proportional to the density of the fluidized catalyst and the height of the catalyst above the point where the pressure is being measured. Non-fluidized powders can support their own weight against the walls of their container unlike a liquid. If the catalyst loses its fluidization, it can also start supporting a portion of its weight against the standpipe walls and will see a reduced pressure buildup with respect to depth.

2.7.2 TYPICAL FCCU MAJOR EQUIPMENT/COMPONENTS

The entire FCCU along with the flue gas circuit having a CO incinerator and flue gas cooler and FGD is shown in Figure 2.3. The LPG recovery or gas concentration unit is also installed at the downstream of fractionator for recovery of LPG. The entire R-R section is depicted in Figure 2.9, showing by and large all the components, including the reactor (with internals), regenerator (with internals), and catalyst standpipes.

A different arrangement of R-R section is shown in Figure 2.10 with two risers. The second riser is deployed for cracking of gasoline components to increase

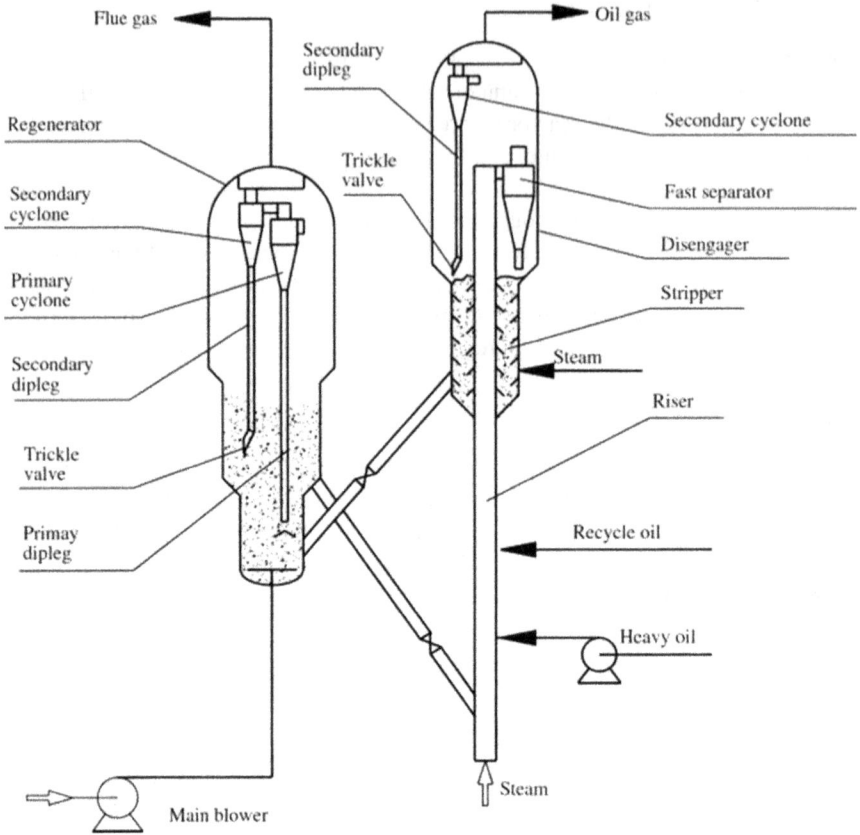

FIGURE 2.9   A simplified sketch showing components in R-R section.

LPG and light olefin yield and typically operates at a higher temperature than the main riser.

The Figure 2.11 shows the typical control in a single-stage regenerator (R-R section) where the reactor outlet temperature control acting on the regenerated catalyst slide valve and the stripper level acting on the spent catalyst slide valve.

2.7.3   REACTOR ARRANGEMENTS

Typically, there can be two stages of cyclones or reactor termination device (RTD) and one stage of cyclones as the disengagement device in reactors. There can be reactor cyclones or coupled cyclones (rough cyclones). Different reactor arrangements are shown in Figure 2.12 and Figure 2.13.

The stripper located at the bottom of the reactor vessel strips the hydrocarbons from catalysts and diverts them to fractionator. Trays or grids are mostly deployed in

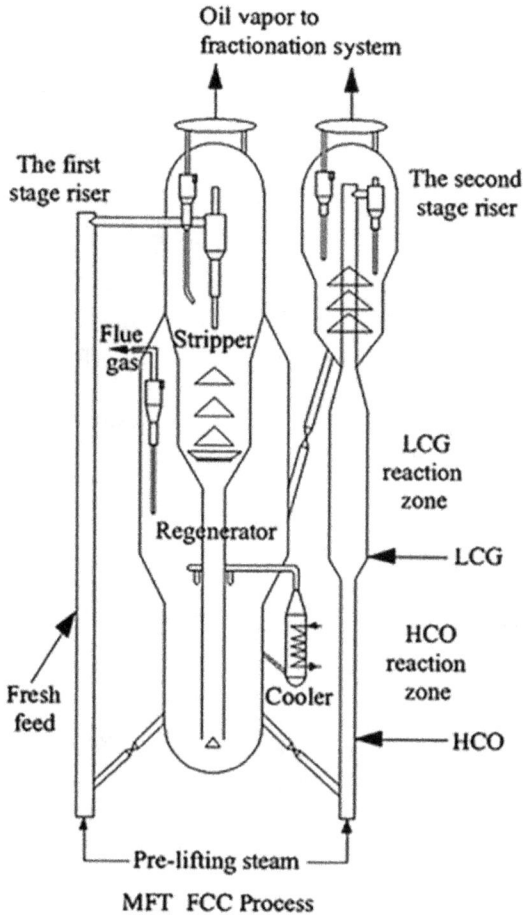

FIGURE 2.10   Multifunctional two- riser (MFT) in FCCU.

the section. The design of the stripper is continuously improved to reduce oil going to regenerator along with spent catalysts and contributing to additional coke in the regenerator.

2.7.4   DIFFERENT REGENERATOR CONFIGURATIONS AND INTERNALS

A typical FCCU regenerator (single regenerator) is shown with two stage cyclones in Figure 2.14, and a typical air grid is shown in Figure 2.15.

Different configurations of regenerators are shown in Figure 2.14 (conventional), Figure 2.16 (with external cyclones), and Figure 2.17 (with two-stage regenerator and catalyst cooler).

The regenerator drip legs are shown with flapper valves in Figure 2.18.

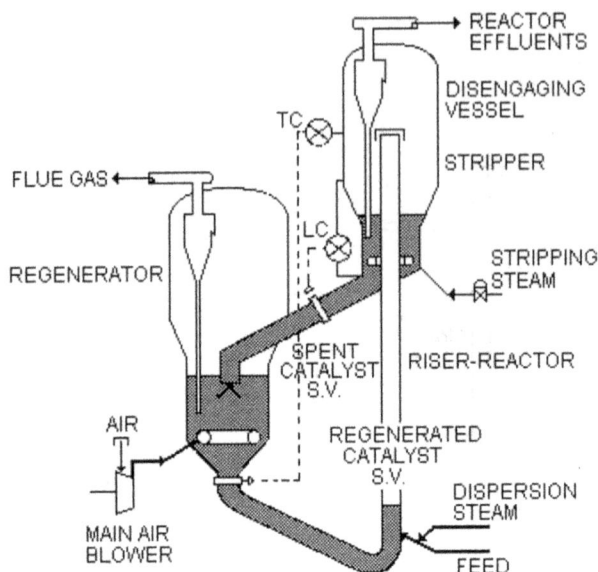

FIGURE 2.11   Control in R-R section.

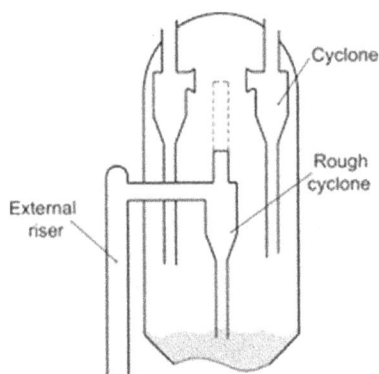

FIGURE 2.12   Riser (external riser) and reactor cyclones.

2.8   PHYSICAL PROPERTIES OF CATALYSTS RELATED TO FLUIDIZATION

A few important terms related to fluidization and catalyst are introduced here(source literatures on fluidization).

- *Apparent bulk density* (ABD). This is the density of the catalyst at which it is shipped in bulk volume or bags. It is the density at minimum fluidization velocity. ABD can be used for troubleshooting catalyst flow problems. Too

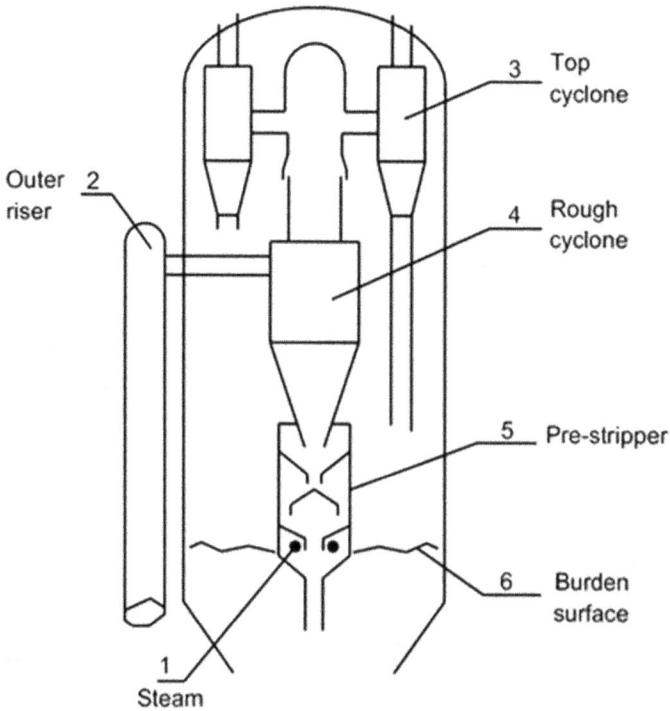

FIGURE 2.13   Reactor riser (external riser) and coupled cyclones.

high ABD can restrict fluidization and too low ABD can result in higher catalyst losses. Typically, ABD of equilibrium catalyst is found higher than fresh catalyst as a result of change in pore structure due to thermal and hydrothermal effects in prolonged operation in the unit.

- *Bed density* (ρ_{bed}). This is the average density of the fluidized bed of solid particles and gas. Bed density is mainly a function of gas velocity for an FCCU catalyst.
- *Minimum bubbling velocity* (U_{mb}). This is the velocity of gas that flows through the bed of catalyst at which discreate bubbles begin to form. The typical bubbling velocity for an FCCU catalyst is 0.03 ft/sec.
- *Minimum fluidization velocity* (U_{mf}). This is the lowest velocity of gas that flows through the bed of catalyst at which the full weight of catalyst is supported by the gas used for fluidization. It is the minimum gas velocity at which the packed bed of solid particles will begin to expand and behave as a fluid. For a FCCU catalyst, the minimum fluidization velocity is about 0.02 ft/sec.
- *Ratio of minimum bubbling velocity to minimum fluidization velocity* (U_{mb}/U_{mf}). This is a very important parameter. The higher the ratio, the easier to fluidize the catalyst. The expression to compute the ratio is presented later in Section 2.14.

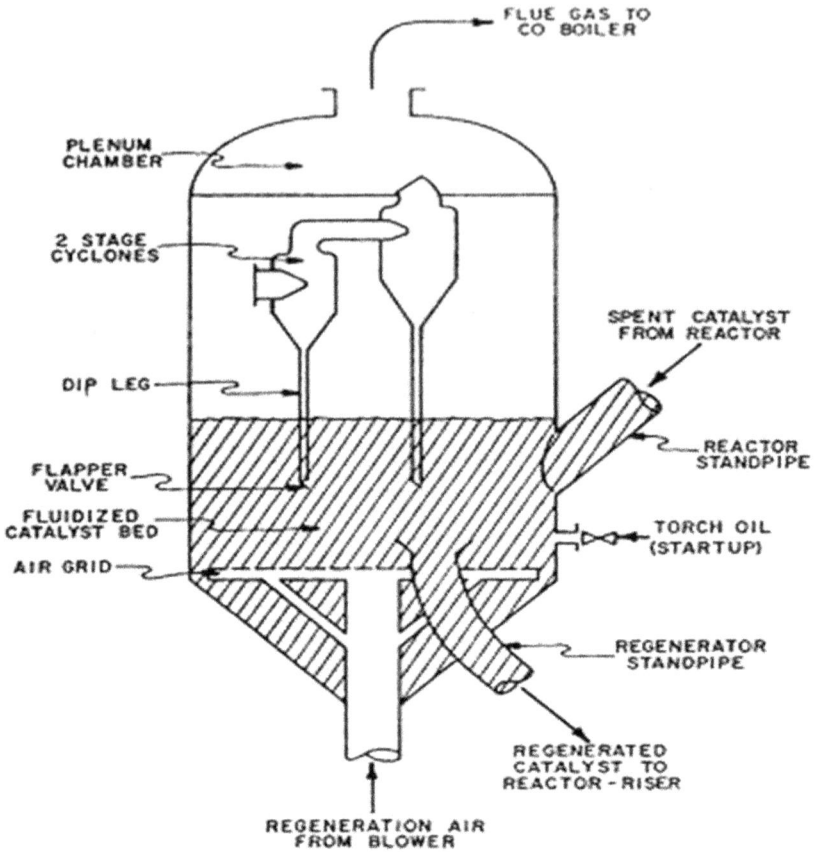

FIGURE 2.14   Typical regenerator arrangement.

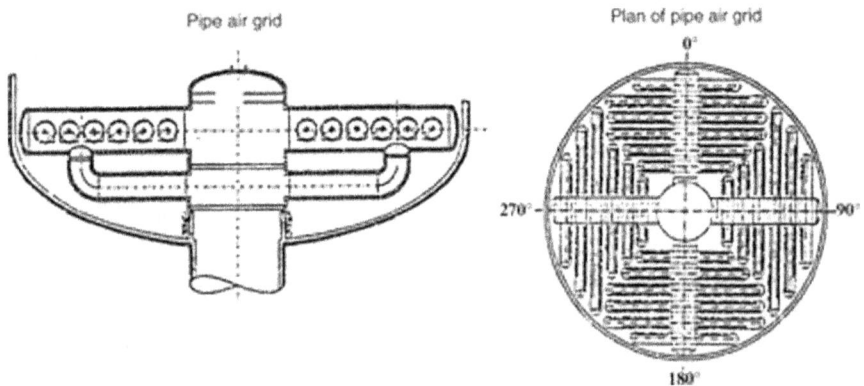

FIGURE 2.15   Different types of air grids.

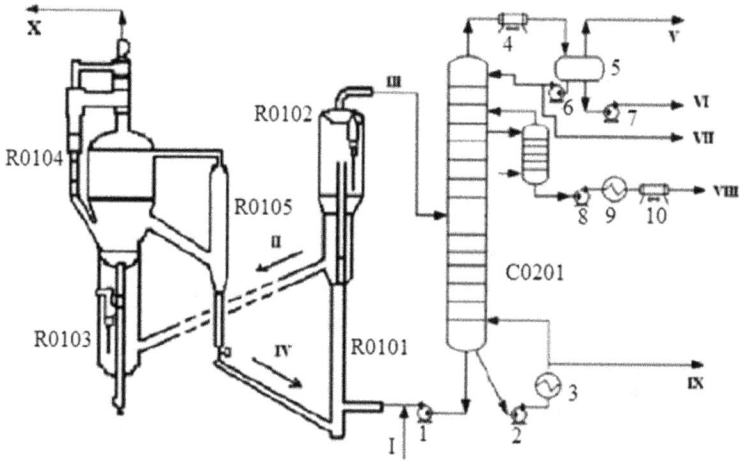

FIGURE 2.16 Regenerator with external cyclones.

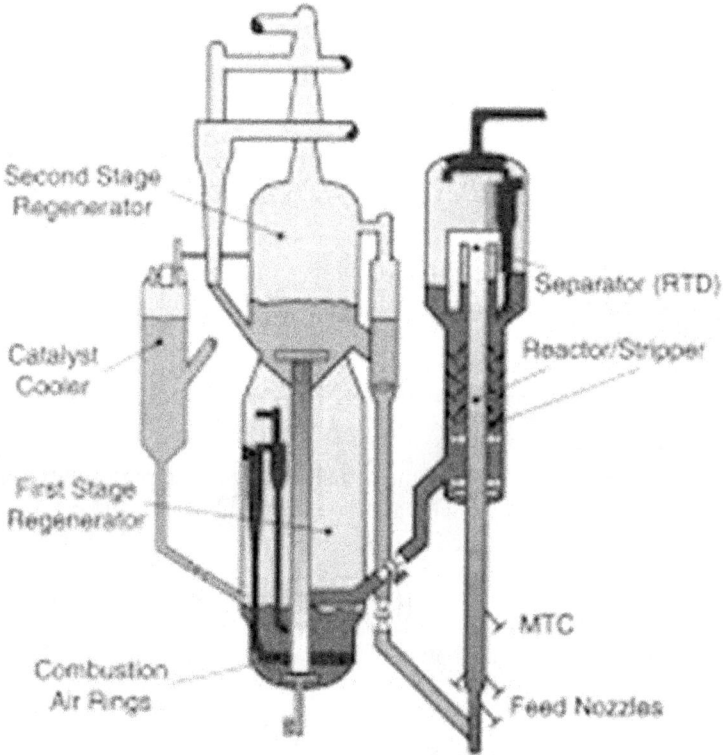

FIGURE 2.17 Two-stage regenerator (R2R) with catalyst cooler.

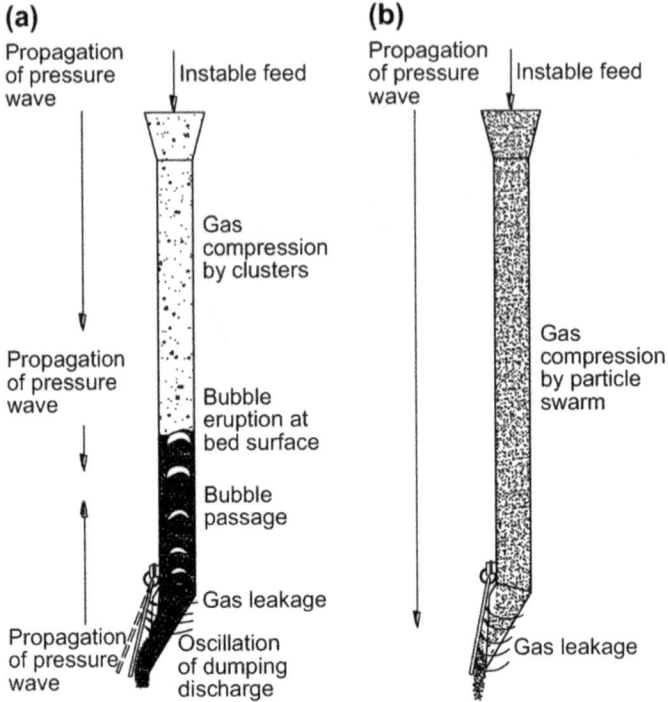

FIGURE 2.18   Regenerator with flapper valves in diplegs.

- *Particle density* (ρ_p). The particle density is the actual density of the solid particles taking into account any volumes due to voids (pore) within the structure of the solid particles. Particle density is calculated from skeletal density and pore volume:

$$\rho_p = \text{skeletal density}/((\text{skeletal density} \times PV) - 1))$$

  where PV is the pore volume.
- *Skeletal density* (SD). This is the actual density of the pure solid material that constitute the individual catalyst particles. For an FCC catalyst, the skeletal density is calculated as follows:

$$SD = 100/((Al/3.4) + (Si/2.1))$$

  where Al is the alumina content of the catalyst in wt.% and Si is the silica content of the catalyst in wt.%
- *Pore volume* (PV). Pore volume, in cc/gram, provides an indication of the volume of void in the catalyst particle and can provide a clue for the cause of catalyst deactivation in an operating unit. Hydrothermal deactivation is known to have little effect on change in pore volume, whereas thermal deactivation leads to reduction in pore volume.

- *Superficial velocity.* Superficial velocity is the velocity of the gas through the vessel or pipe without any catalyst present. It is the volumetric flow rate of the fluidization gas divided by the cross-sectional area.

2.9 GUIDELINES FOR DESIGN OF CYCLONES

2.9.1 BASICS

Cyclones use centrifugal force to separate catalyst particles from the gas. The particles are forced (by centrifugal action) towards the walls of the cyclone and fall into the dipleg. The gas exits through the outlet tube at the top of the cyclone. The recovery efficiency of a conventional two-stage cyclone system is very high at over 99.99%. Different sections of the cyclone are shown in Figure 2.19 and Figure 2.20, including the nomenclature of each part.

Velocity is a key operating parameter for cyclone performance. Collection efficiency increases with velocity and then drops off due to catalyst re-entrainment. Catalyst attrition to microfines occurs within cyclones and increases with velocity. The overall cyclone collection efficiency depends on numerous factors including the following:

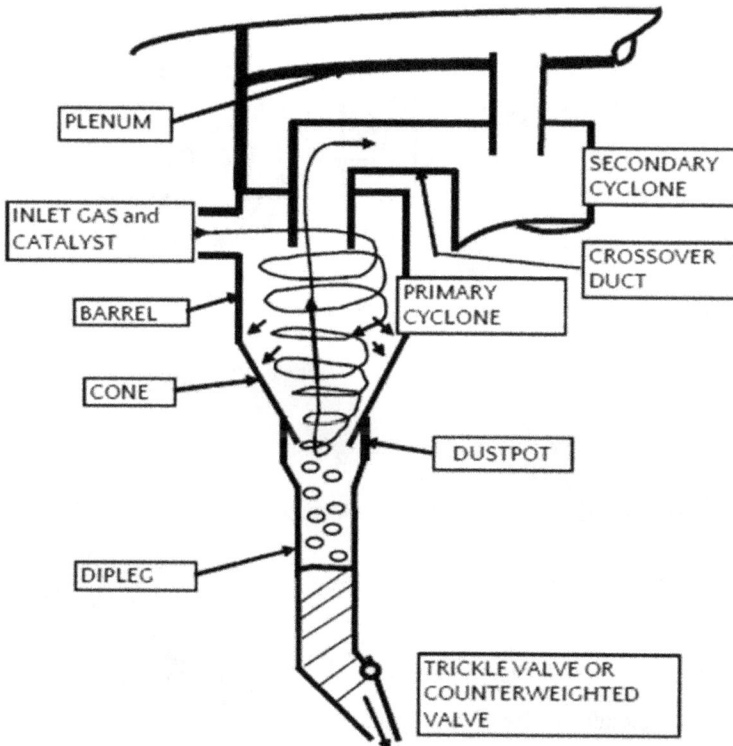

FIGURE 2.19 Cyclone with trickle valve.

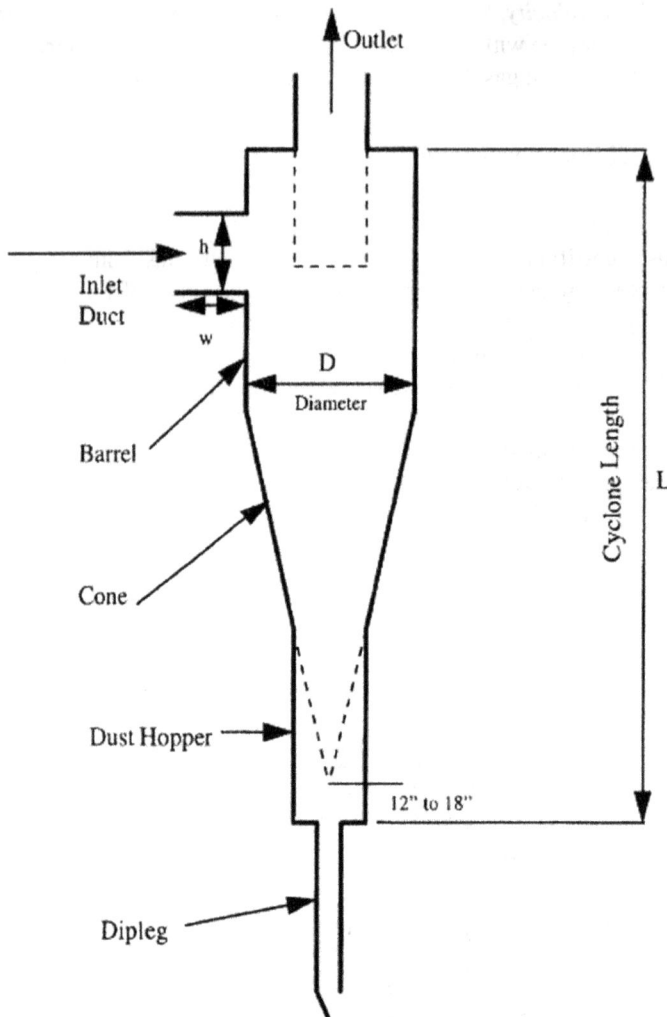

FIGURE 2.20   Nomenclature of different parts of a cyclone.

- Number of spirals within the barrel and cone
- Inlet velocity
- Particle density and size
- Catalyst loading

2.9.2   ACCEPTED DESIGN LIMITS FOR CYCLONES

For primary cyclones the design limits are as follows:

- Maximum inlet velocity, 65 ft/s
- Outlet velocity, 150 ft/s

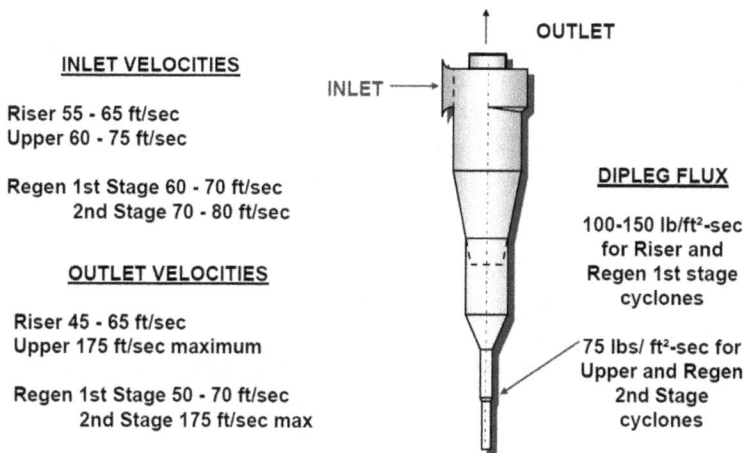

INLET VELOCITIES

Riser 55 - 65 ft/sec
Upper 60 - 75 ft/sec

Regen 1st Stage 60 - 70 ft/sec
2nd Stage 70 - 80 ft/sec

OUTLET VELOCITIES

Riser 45 - 65 ft/sec
Upper 175 ft/sec maximum

Regen 1st Stage 50 - 70 ft/sec
2nd Stage 175 ft/sec max

OUTLET

INLET

DIPLEG FLUX

100-150 lb/ft²-sec
for Riser and
Regen 1st stage
cyclones

75 lbs/ ft²-sec for
Upper and Regen
2nd Stage
cyclones

FIGURE 2.21   Typical design parameters of cyclones.

- Operating temperature (for 304H SS), 1400°F
- Dipleg mass flux, 150 lb/ft²·s

For secondary cyclones the design limits are as under:

- Maximum inlet velocity, 75 ft/s
- Outlet velocity, 175 ft/s
- Dipleg mass flux, 40 lb/ft²·s

High velocities can increase catalyst attrition. Dipleg choke can be a concern for primary cyclones. Sustained high temperatures will reduce the life of the cyclones.

Typical design parameters of a cyclone are also presented in Figure 2.21.

2.10   PROBLEMS RELATED TO CYCLONES

Common operational problems in the FCCU cyclones are described next.

2.10.1   HIGH CATALYST LOSSES

The typical tolerable FCCU catalyst loss is 0.1 lb/bbl of feed processed. A higher one can be 0.15 lb/bbl. The presence of a lower proportion of particles of 0–40 micron (expressed in percentage) (lower than the start) and a higher average particle size (APS) may also indicate higher loss from the R-R system.

High loss from the reactor is typically reflected in the following:

- Increase in ash content in slurry.
- Slurry fines and/or equilibrium catalyst (ECAT) particle size distribution (PSD) shift.
- Increase in pressure drop in slurry pumparound circuit.

Similarly, higher catalyst losses on the regenerator side is reflected in the wt.% of solids in the purge water and rate of purge from the scrubber or flue gas desulfurization unit (FGDU) and the total amount in tons collected for disposal with electro static precipitator (ESP) or other dry separation devices and is also immediately indicated by the opacity of the stack discharge. Higher loss from the regenerator is also reflected by reduction of the regenerator level.

The following data needs thorough analysis to estimate catalyst loss:

- Ecat fines and PSD old and present.
- Regenerator off-gas scrubber purge water or third-stage separator fines analysis historical and current chemical and PSD analysis.
- Slurry analysis for current ash content and PSD (with chemical analysis if available)), along with slurry product flow rates (present and old results for comparison).

High catalyst loss can be caused by the following:

1. Flooded diplegs of the cyclone leading to catalyst carryover. The probable reasons for flooded diplegs are:
   - High catalyst loading/higher cyclone ΔP/high catalyst bed level
   - Plugging of diplegs by refractory, debris, coke, or catalyst deposit
   - Trickle valve, flapper plate, or counterweight plate movement restricted
2. Unsealed diplegs leading to excessive gas leakage and catalyst entrainment
   Catalyst entrainment through cyclone can happen due to the following:
   - Low bed level (at startup)
   - Loss of sealing plate of a dipleg when the dipleg is terminated in dilute phase
   - Flapper valve plate stuck in an open position
   - Holes in the cyclone system can be due to catalyst erosion caused by refractory/metal from high velocities. The erosion rate is known as proportional to the velocity to the power of 3 to 5. Holes can result in gas leakage and disruption of cyclone operation
     - Often a hole will lead to a gradual increase in losses as the hole gradually enlarges due to erosion
     - Higher losses from first-stage cyclones may be partly handled by second-stage cyclones
     - Mechanical problems will require unit shutdown and entry to repair

If the catalyst loss increases with time along with a reduction in (0–40) micron percentage of particles in the equilibrium catalyst, it may be indicative of a hole or crack in the plenum chamber and may call for a unit shutdown. If losses are steady at a higher level and fines decrease in the equilibrium catalyst, it may indicate that something has broken inside and further may point at flooded or plugged diplegs. The action can be lowering of the cyclone velocities by any other action or by increasing the operating pressure to reduce velocity.

TABLE 2.8
Problems and Potential Causes

Problem	Potential Causes
Higher catalyst losses from regenerator	Mechanical problem or operation issues like higher superficial velocity and too high velocity in RG cyclone. Steady loss indicates more catalyst entry to cyclones, flooded/partial blocking in dipleg or improper dipleg sealing. Increasing loss indicates a hole or crack in the plenum or cyclone.
Higher catalyst losses from reactor	Mechanical problem or operation issues like higher superficial velocity and too high velocity in RX cyclone. Steady loss indicates partial blocking in dipleg. Increasing loss indicates a hole or crack.
Higher losses from both	Lower system pressure and resultant higher superficial velocities or catalyst issues.
Fine fraction (0–40) micron reducing in equilibrium catalyst	Poor fine retention, Low fines make up along with fresh catalyst.
Fine fraction (0–40) micron increasing in equilibrium catalyst	High stream velocities in dense phase and attrition producing more fines. Catalyst may be more prone to attrition. A check for orifice plates in purge points, particularly in standpipes, can be conducted, orifices can be installed, and bypasses can be closed, and stripping steam can be reduced.

Equilibrium catalyst average particle size can be an indicator of regenerator problems. A hole or plenum crack will often present as a gradual increase in losses from either the reactor or regenerator.

Table 2.8 summarizes the problems and potential causes.

Likely areas of failure in the cyclone system are shown in Figure 2.22.

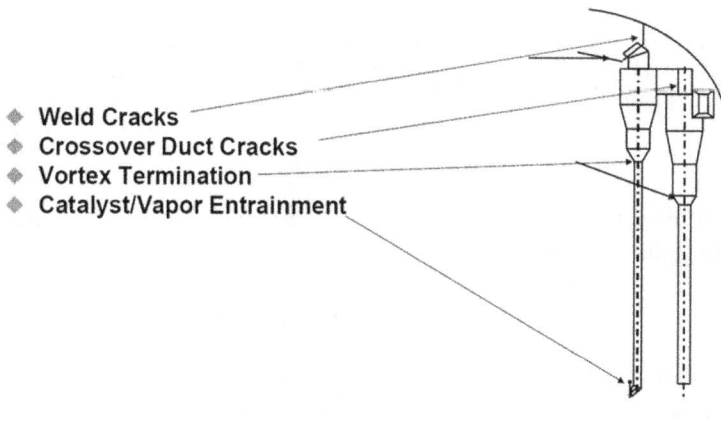

◆ **Weld Cracks**
◆ **Crossover Duct Cracks**
◆ **Vortex Termination**
◆ **Catalyst/Vapor Entrainment**

FIGURE 2.22 Likely areas of failures in cyclone system.

2.10.2   FINE GENERATION

1. Excessive velocities in most of the cases causes fine generation and can result due to the following:
   - Missing restriction orifice on steam or air nozzles used for aeration, torch oil nozzles, etc.
   - Eroded or lost stripping steam distributor and/or air grid nozzles
   - Feed injectors, air grid, or cyclones operating above design guidelines
2. Higher catalyst loading to cyclones
3. Higher catalyst circulation rate
4. Catalyst properties/unsuitable fresh catalyst attrition properties
5. Excessive air rate for pneumatic conveying during catalyst loading to and from hopper

Fine generation can be controlled by checking all flow distributors. It should be noted that jet velocities greater than 300 ft/s can cause catalyst attrition. Fluidization taps should be checked for normal flow of aeration agent and use of dry steam if used, for aeration. Torch oil steam flow valves should be checked to eliminate likelihood of passing/leaking. PSD may be verified by sending a sample of catalyst to the catalyst supplier or checked in the refinery.

2.10.3   REACTOR CYCLONE COKING

Coke will accumulate and grow in cracks and crevices in the refractory, pushing the refractory away from the metal. Hex mesh anchors in the reactor should be fully welded along each seam, and any cracks should be properly repaired during turn-arounds. Coke will often deposit on the outside reactor cyclone gas tubes (extended inside the cyclone). In the event of an upset/thermal shocks, *this coke may spall and block the cyclone dipleg.* As a preventative measure, installation of V anchors are suggested by experts to prevent coke from spalling. Coke formation is usually due to condensation of heavy hydrocarbons in the "dead" area behind the cyclone inlet horn and outside of the gas outlet tube. Coke formation can be minimized with the actions like proper functioning of the feed injection device, higher riser stream, introduction of feed when unit is hot.

2.10.4   POOR CATALYST CIRCULATION

Symptoms of catalyst circulation problems that commonly occur in standpipes include the following:

- Low slide valve (or plug valve) differential pressure.
- An inability to circulate additional catalyst despite higher opening of the slide valves. This is often associated with difficulty to control/increase reactor outlet temperature.
- Erratic slide valve differential pressure.
- Physical bouncing or hopping of catalyst standpipes.

2.10.5   POOR PRODUCT YIELDS

- Product yield loss can be due to poor circulation.
- Lower catalyst activity on account of improper regeneration of catalyst, activity loss due to metal poisoning, catalyst of inherently low activity, and refractory nature of feedstock.

2.10.6   CONCLUSION ON CYCLONES: REASONS FOR FAILURES

Cyclones are listed as one of the top three reasons why the run lengths FCCU end early.

- Many people run higher inlet velocities than designed. It has inherent risk of ending a run early due to mechanical damage. Further, higher velocity leads to higher pressure drops in the cyclone and can increase the height of catalyst in diplegs and result in entrainment and loss.
- Erosion to a cyclone is a function of velocity to at least the third power (may be also higher).

Units can run for an extended time period with cyclone damage but need to be aware of its effect on fluidization due to loss of small particles (smaller than 40 microns). Locations prone to formation of holes in cyclones is shown in Figure 2.23.

2.10.7   REMEDIES

Cyclone lifespan can be increased by minimization of erosion by (suggested by experts)

- Increasing cyclone length (minimum L/D of 4)
- Designing to avoid excessive velocities
- Adding a vortex stabilizer to the secondary cyclones as shown in Figure 2.25
- Ensuring good inspection and maintenance of refractory during each turnaround
- Controlling afterburn with use of CO promoter and ensuring even rise in temperature in regenerators
- Distributing air and spent catalyst to minimize creep and sigma phase embrittlement due to local high temperatures

It is learnt that a recent PSRI program studied different FCC cyclone technologies, as the erosion of cyclone and its reliability were highlighted by users as the major concerns of FCC operation. The majority of problems were identified as the erosion in the secondary cyclones, particularly in the lower cone and in the transition to the dipleg. Experts observed that the highly loaded first-stage cyclones normally experience little to no cone erosion, whereas lightly loaded second-stage cyclones can exhibit severe cone erosion. This appears little difficult to apprehend at the first

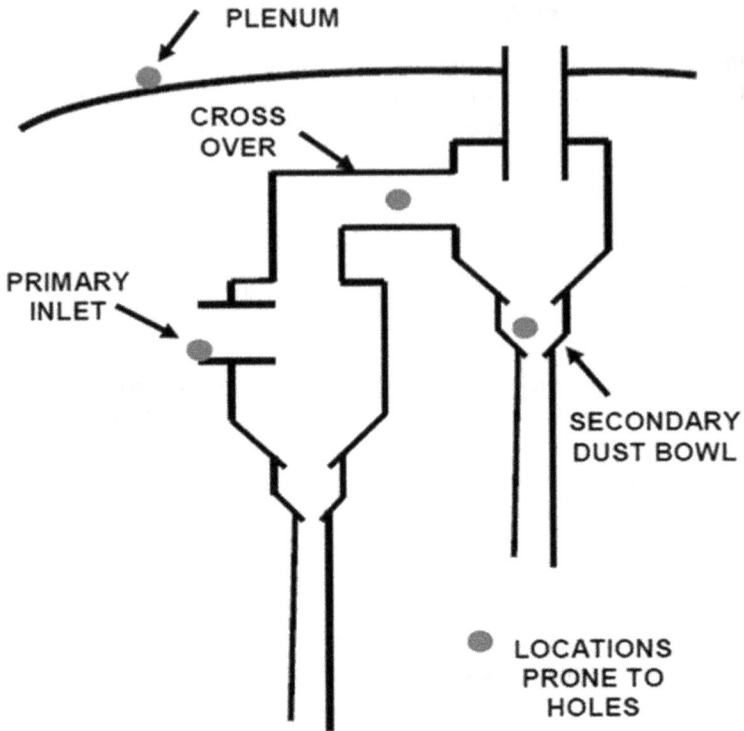

FIGURE 2.23   Locations prone to formation of holes in cyclones.

instance. However, the key difference in erosion pattern can be attributable to the differences in the solids flow patterns and vortex formation in the first and second stage cyclones.

It had been opined by experts that high solids loading and low gas velocity in a typical FCC primary cyclone, gravitational force takes the key role.

The solids are likely to fall rapidly down into the cyclone cone and dipleg, as shown above in Fig. 2.19 taking only one to two full turns to exit through the bottom of the cyclone.

The vortex length in the highly loaded primary cyclone is much shorter because the high solids loading dampens the formation of a robust vortex. Therefore, the vortex does not whip the solids at a high velocity around the cone in the primary cyclone.

Whereas in a FCC second-stage cyclone, the solids loading is approximately 1/1000 to 1/10,000 of the loading in the first-stage cyclone. Due to lower solids loading and high gas velocity, the vortex can be relatively long, energetic, and, more importantly, unstable, can move asymmetrically about its axis. Experts further observe that the swirling solids in the outer vortex approach the cone in a second-stage cyclone. The long, rapidly rotating vortex accelerates the solids stream and causes it to intensify its rotation. The outer vortex in a second-stage cyclone typically takes from four to six turns before exiting the bottom cone, and the spinning

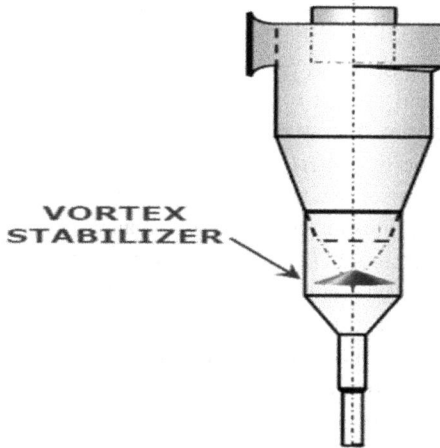

FIGURE 2.24   A vortex stabilizer in secondary cyclone.

continues into the top portion of the dipleg below the cone. The concentrated solids stream rotates at a high velocity, and the unstable, continuous movement of the vortex causes the significant erosion observed in the cone and the top of the dipleg of second-stage cyclones.

The PSRI study proposed three different solutions to mitigate the damaging erosion occurring in FCC second-stage cyclones:

- Increasing the cyclone length (L/D) of a conventional cyclone
- Increasing the angle of the cone of a conventional cyclone
- Adding a vortex stabilizer as shown in Figure 2.24 to a conventional cyclone simulating the improved cyclone technology

2.11   RISER SEPARATION SYSTEM AND POSITIVE/ NEGATIVE PRESSURE CYCLONES

2.11.1   DESCRIPTION OF RISER SEPARATION SYSTEMS

Post-riser residence times for hydrocarbons in the reactor dilute phase were typically 15–45 seconds. This is significantly longer than the contact time in the riser (typically, 1–5 seconds) and is of great concern. The hydrocarbons can be thermally degraded if they are exposed to high temperatures for a long time in the presence of significant amounts of catalyst in the dilute phase and is known to result in overcracking and with eventual production of more gas. In order to increase olefin production or conversion, the reaction temperature can increase to 550°C or more. In those conditions, thermal degradation will be more pronounced.

Thermal degradation and overcracking lead to higher volumetric flow rates of gas (due to increase in number of moles and lower molecular weight of the gas).

A reduction in production of gases can be achieved by limitatng the post-riser cracking introduces the flexibility to increase the overall unit capacity without changing this costly equipment, particularly the reactor cyclones, fractionator, wet gas compressor, and the entire gas concentration section. Thus, the ways to reduce post riser thermal degradation and overcracking were investigated and solutions were proposed by experts.

In order to limit post-riser reactions, the disengagement devices for gas and catalyst were improved, and introduced immediately at the riser outlet, leading to the evolution of new riser separators (cyclones) or riser terminating devices (RTD). This device separates the catalyst from hot oil vapors right at the riser outlet and tries to limit the post-riser contact time and resultant over cracking. The RTD forces the catalyst into the stripping zone and hydrocarbon vapors to the upper portion of the dilute phase close to reactor cyclones. Figure 2.25 shows three different arrangements inside a reactor. A simple arrangement is shown in scheme A with single-stage cyclones. A riser termination device with reactor cyclones (not connected) is shown in (B). A riser termination device with connected reactor cyclones is shown in (C).

A reactor with a vortex separation system (VSS) as the riser termination device and reactor cyclone is shown in Figure 2.26. This system is known to effectively reduce post-riser cracking and reduce gas production from the unit.

Another option to further suppress thermal degradation is the injection of recycle liquids into the vapor after primary catalyst separation to quench the reaction. The modification of the riser termination can induce dramatic changes in unit performance. The improvement can lead to about a 25% decrease in dry gas flow rates

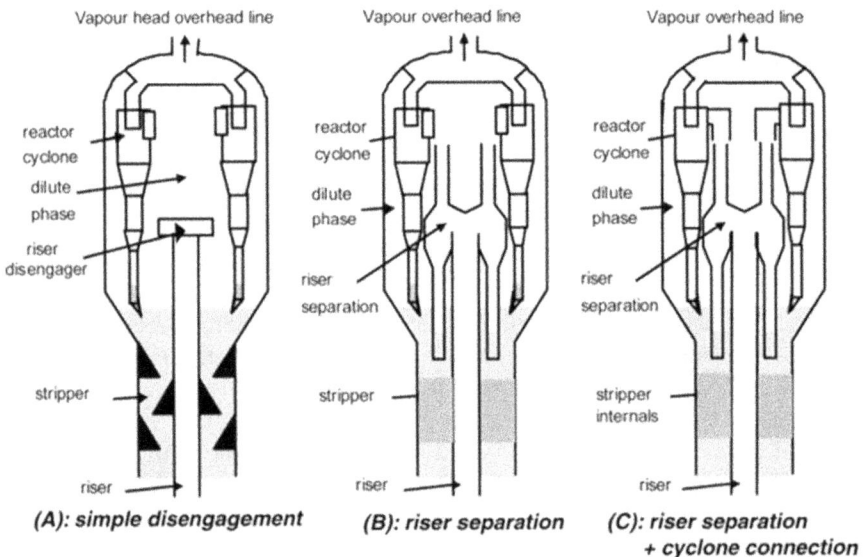

FIGURE 2.25 Different reactor cyclone arrangements.

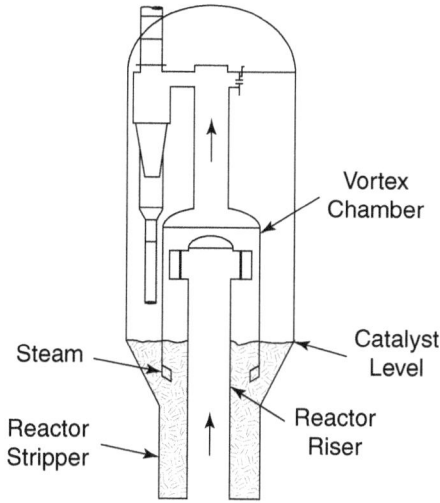

FIGURE 2.26 Vortex separation system.

Therefore, it is likely that the refiner could increase the unit throughput significantly with the mentioned benefits.

The author had the opportunity to operate a unit with a reactor with two-stage cyclones converted to an assembly with a riser termination device (VSS as RTD) followed by two-stage reactor cyclones (existing originally) system. There was a marked reduction in dry gas yield and some reduction in catalyst loss from the reactor was also observed. The unit capacity was increased by about 50%. The main modifications implemented in the gas concentration section were installation of parallel WGC and change of column internals in the fractionator and Gascon section.

One of the main difference between riser separation systems and conventional reactor cyclones is that, for most applications, the riser separation system is at a higher pressure than the surrounding reactor. In Figure 2.27, two identical cyclones are represented in a fluidized bed vessel. The cyclone on the right is fed by an external transport line (coupled with a riser). It discharges gas in the dilute phase through the gas exit pipe. The cyclone barrel is therefore at a higher pressure than the dilute phase (even after the pressure drop inside the cyclone). If a dense phase forms in the dipleg, its height should establish below the bed level, as shown in Figure 2.27. The left cyclone is fed from the dilute phase. Due to inlet acceleration and cyclone pressure drop, the cyclone barrel is at a pressure lower than the dilute phase. As a consequence, a rise of catalyst occurs (above catalyst bed level) and a dense phase establishes in the dipleg. The left cyclone is called a *negative pressure separator* and the right cyclone a *positive pressure separator*. Reactor cyclones and regenerator cyclones behave like negative pressure cyclones. Most of the riser separation systems are positive pressure separators.

The height of catalyst in the dipleg of the negative pressure cyclone will be much higher than the positive pressure cyclone and will be above the level of dense phase

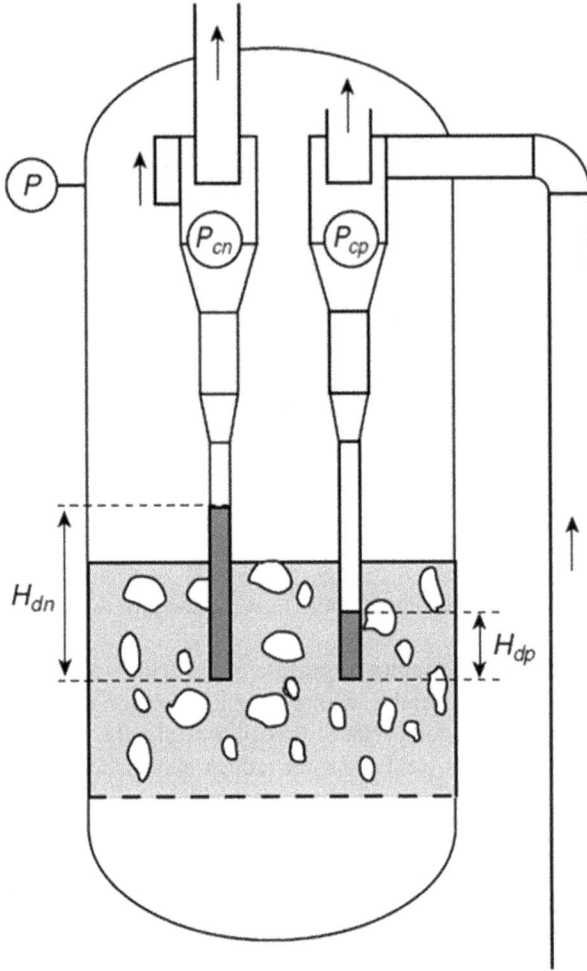

FIGURE 2.27   Reactor cyclones with dipleg sealing.

as shown in Figure 2.27. It can be shown that the level difference of dense phases in both cyclones is proportional to the pressure drop ΔP_c of the cyclones (pressure difference from gas inlet to gas outlet) without consideration of the entry and exit losses at the inlet and outlet of the cyclones:

$$\rho_d g \left(H_{dn} - H_{dp} \right) = \Delta P_c$$

The pressure drop in the cyclone can be considered proportional to the square of the volumetric flow rate through the cyclone (neglecting the inlet and outlet losses) or $\Delta P_c = K c Q^2$. So, if velocity increases, the pressure drop will also increase and will increase the catalyst height in diplegs and may increase entrainment.

A dense phase buildup is a good way to limit gas flow (underflow) in diplegs for positive pressure separators. Only interstitial gas in the dense phase can then migrate to the fluidized bed. However, to get a dense phase, it is necessary to immerse the dipleg in the bed deep enough. If the dipleg immersion in the bed is not sufficient, no dense phase will build up and diluted catalyst flow with a lot of gas will flow through the dipleg to the fluidized bed. Cyclone failures or obstructions can occur due to poor fluidization around dipleg discharge. Coke or refractory dislodged from top, contaminant deposits in the dipleg, or mechanical failure during a thermal cycle can also result in obstruction in diplegs. Contaminant deposits forming in the diplegs can completely block the dipleg and can completely stop flow through the dipleg.

2.12 ESTIMATION OF TRANSPORT DISENGAGING HEIGHT IN REGENERATORS (SOURCE PUBLICATIONS)

2.12.1 ESTIMATION OF BED LEVEL IN REGENERATOR

The bed level should be estimated in absolute terms (i.e., height above reference point) using pressure survey data and/or level instrument reading. Fluidized FCC catalyst behave like any other fluid and generate a static pressure head. However, FCC catalyst does not have a fixed fluidized density. The density (ρ_{bed}) depends on the fluidization conditions, so the density needs to be measured inside the bed to calculate bed level as $\rho_{bed} = P_{bed}/h_{bed}$, where ρ_{bed} is the regenerator bed density, and P_{bed} is the pressure differential inside the bed over a known vertical height of h_{bed} as illustrated in Figure 2.28. Automatic bed level indicators are common and follow the calculation methodology mentioned earlier.

The transport disengaging height (TDH) is the height above the regenerator bed where the catalyst density (entrainment) is constant (relatively low). Catalyst entrainment is much lower above the TDH, as illustrated in Figure 2.28. TDH should terminate below the primary cyclone inlet in order to minimize catalyst entrainment and losses. Experts suggest the following equation to calculate TDH:

$$\text{Log}_{10}\text{TDH}_{20} = \text{Log}_{10}20.5 + 0.07(V - 3), \text{and}$$

$$\text{TDH} = \text{TDH}_{20} + 0.1(D - 20)$$

where D is the regenerator diameter (ft) and V is superficial velocity (ft/s) in the regenerator (maximum known to be 3.3 ft/sec or around 1 meter/sec to avoid overloading of cyclones). TDH calculation is demonstrated next for a vessel with a diameter of D = 30 ft and regenerator superficial of V = 3 ft/s:

$$\text{Log}_{10}\text{TDH}_{20} = 1.3117 \text{ or } \text{TDH}_{20} = 20.5$$

$$\text{TDH} = \text{TDH}_{20} + 0.1(D - 20) \text{ or}$$

$$\text{TDH} = 19.86 + 0.1(30 - 20) = 20.5 + 1.1 = 21.6 \text{ ft}$$

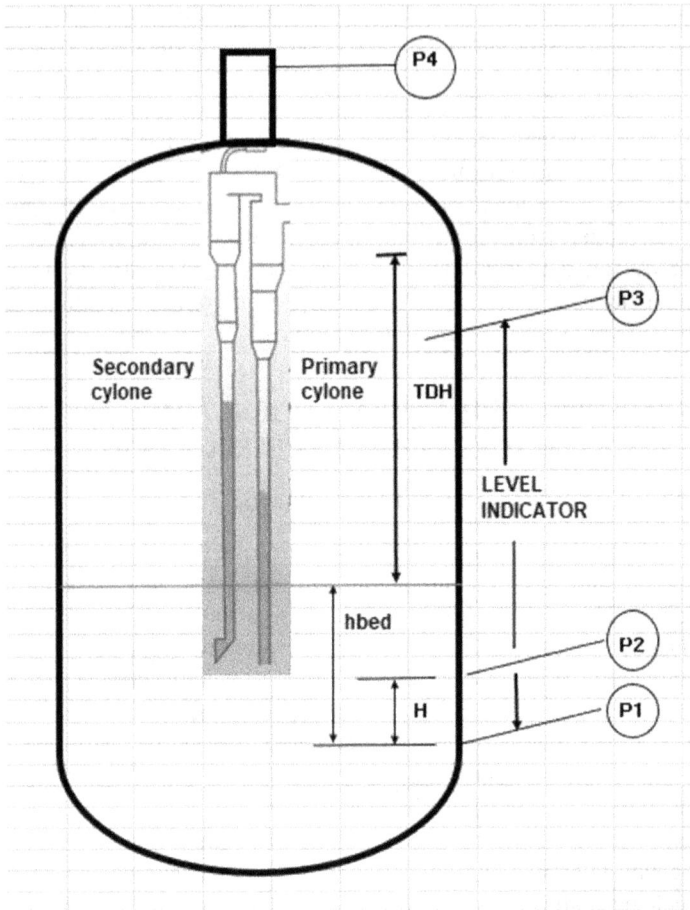

FIGURE 2.28   Regenerator bed level measurement and TDH in a regenerator.

Catalyst levels inside the cyclone diplegs can be estimated from pressure survey data of the cyclones or can be computed. Dipleg levels should terminate below the cyclone body. The necessary actions may include the following if the cyclone dipleg level is high:

- Reduce air rate
- Increase pressure
- Lower bed level

The following guidelines need to be followed regarding catalyst level in diplegs relative to the bed level:

- The diplegs should not be submerged if this was not intended in the design.
- The diplegs should be properly submerged if they were intended to be submerged (too high submergence leads to high level in cyclones).

- Inadequate submergence leads to gas flow up the dipleg (for negative pressure cyclones).

The primary cyclone dipleg level for a negative pressure cyclone can be calculated as ΔP in the primary cyclone divided by density of the catalyst in primary cyclone plus the dipleg submergence multiplied by the ratio of bed density to cyclone dipleg density. Primary cyclone dipleg height is normally not a concern.

Similarly, the secondary cyclone dipleg level can be calculated as ΔP in cyclones (ΔP primary and secondary summed for coupled cyclones) divided by density of the catalyst in the secondary cyclone (ρ) plus the dipleg submergence multiplied by the ratio of bed density to cyclone dipleg density). ΔP cyclone can be estimated as $\Delta P = KG \, (Q)^2$. ρ can be taken as 20–25 lb/ft^3 for primary cyclones.

2.13   DESIGN GUIDELINES FOR CYCLONE DIPLEGS

The dipleg is typically installed at the cyclone bottom and terminates in a fluidized bed or in the dilute region of a fluidized bed (or dilute phase) r). For smooth dipleg operation, the downward-flowing particles must be fluidized The best way to ensure that the downward-flowing particles are fluidized is to correctly design the dipleg. The design of a cyclone dipleg, to achieve the desired particle flow behavior, depends on whether the cyclone is the primary, secondary, or tertiary.

2.13.1   DIPLEG DIAMETER

Design of a dipleg for a primary cyclone (or for a lone cyclone) uses the solid flux in the dipleg as a guiding parameter. In this dipleg, the *solids flux* (the mass flowrate of particles divided by the cross-sectional area of the leg) typically must be between 75 and 150 lb/ft^2-s. A more common design criterion for the solids flux in a primary cyclone dipleg is simply 100 lb/ft^2-s; bigger isn't always better. With a correctly designed primary cyclone dipleg, an added benefit is that little or no gas from the bottom flows into the cyclone (for negative pressure cyclones).

The dipleg diameter for a secondary cyclone (and a tertiary cyclone, if there) typically is 8 inches or more, mostly to allow larger pieces of debris, such as dislodged refractory, to flow through the dipleg. This solids flux criterion as primary, if used would indicate the requirement of a much smaller dipleg diameter of the secondary than advisable based on the practical considerations. So, gas may tend to flow upward in the dipleg of secondary cyclones.

2.13.2   DIPLEG TERMINATION

To prevent excessive upward gas flow, especially during startup, the dipleg of a secondary cyclone is often terminated with a trickle valve or flapper valve, as shown in Figure 2.29. The trickle valve (Figure 2.29 a) consists of a movable plate mounted on hanger rings over the dipleg outlet. A mechanical stop limits the degree of plate opening. The flapper valve (Figure 2.29 b) consists of a movable plate (or flapper)

Dipleg outlet valves for preventing excessive upward gas flow

a. Trickle valve

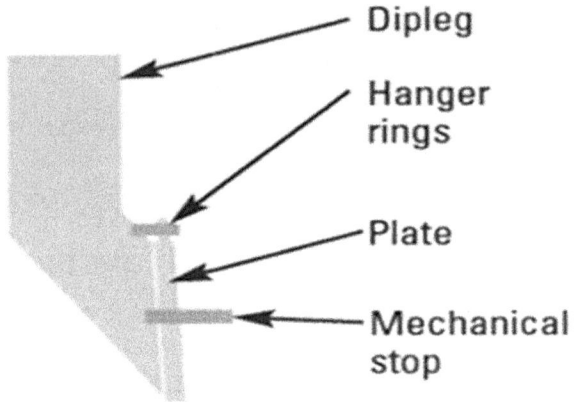

Dipleg

Hanger rings

Plate

Mechanical stop

b. Flapper valve

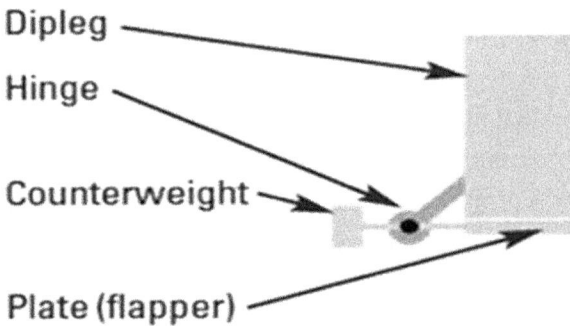

Dipleg

Hinge

Counterweight

Plate (flapper)

FIGURE 2.29   Dipleg termination with trickle valve and flapper valve.

mounted on a hinge at the dipleg outlet. In this case, a counterweight limits the degree of plate opening.

Unfortunately, both of these valves can have problems. *The trickle valve can also function when submerged in a fluidized bed, but the flapper valve works only in a dilute region/phase.* Neither valve works when immersed in a packed bed. Both valves can also periodically stop the particle flow and cause defluidization. This may plug the dipleg, which can make the secondary cyclone ineffective. To prevent this, fluidization can be maintained by adding aeration to the lower portion of dipleg, if mounted outside.

It is apparent that *horizontal counterweight (CWT) flapper valves do not allow catalyst and spalled refractory to slide off as the valves do not get wide open while in operation and a buildup of rubble is not unlikely (if refractory is spalled) and can lead to blockage in the flow or can also cause the valve flapper to get stuck open.*

It must be ensured that the trickle or flapper valve has a tight seal because even a small gas leak can cause entrained particles to significantly erode the seal. Maintaining a sufficiently tight seal is difficult. It needs to be regularly inspected and the valves serviced during shutdowns.

2.14 DESIGN AND OPERATION GUIDELINES FOR FCC CATALYST STANDPIPES

2.14.1 PRESSURE PROFILES IN THE FCC STANDPIPE

One way to assess the workings of the FCC standpipe is to conduct a pressure survey along its length. Usually, the most readily available way to do this is to conduct a single-gauge survey. This essentially measures pressure at various locations along the stand-pipes. The pressure measured at any depth in the standpipe should be nearly proportional to the density of the fluidized catalyst and the height of the catalyst above the point where the pressure is being measured. Figure 2.30 shows the ideal pressure profile that would be observed in a perfectly behaved FCC standpipe. The pressure is seen to be linearly increasing with the depth in the standpipe. If the catalyst loses its fluidization, it can start supporting a part of its weight against the standpipe walls and will deliver a reduced pressure buildup with respect to depth at the point of measurement.

A pressure survey can indicate local problems in fluidization along the height, if any. Figure 2.30 shows the ideal pressure profile and Figure 2.31 shows a pressure

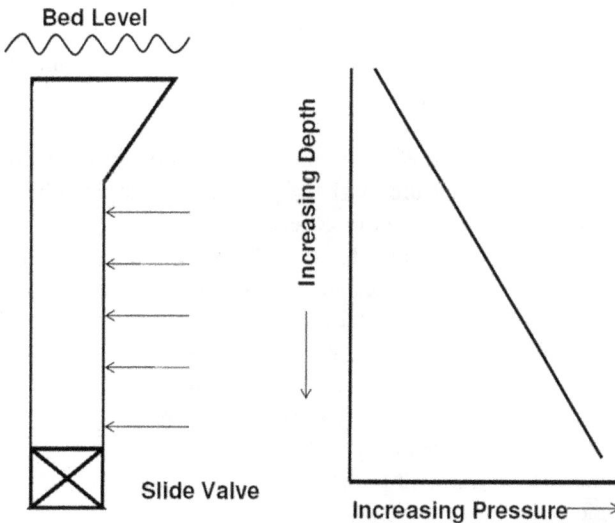

FIGURE 2.30   Ideal pressure profile of a standpipe.

Bed Level

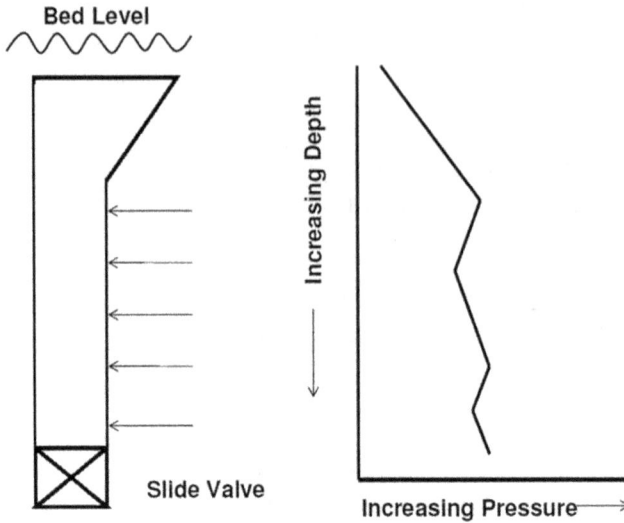

FIGURE 2.31   Pressure profile of a standpipe with circulation problem.

profile for a standpipe that is experiencing problems with pressure buildup and have moderate circulation difficulties.

2.14.2   Catalyst Behavior and Operating Range

FCC catalyst has a narrow operating range within which it will flow well in stand-pipes. The operating range lies between point of *incipient fluidization and to the point of incipient bubbling.* Incipient fluidization represents the beginning of the well-behaved operating range. If the gas present is lower at this point, in the mixture of catalyst and gas, the fluidization of the catalyst will be lost and the bed would revert back to a state of packed bed.

On the other hand, if any higher amount gas were present or is admitted in the fluidized catalyst and gas mixture, over the point of incipient bubbling, then there would be a tendency for the system to generate bubbles. In standpipes, bubble for-mation tends to impede catalyst flow because the bubbles act as obstacles. A severe circulation problem can be experienced as shown in the profile in Figure 2.31 where loss of standpipe pressure is substantial.

These two extremes of fluidization are often measured in terms of velocity, and many articles have been written on measuring the superficial gas velocity at incipi-ent bubbling (U_{ib}) and the superficial gas velocity at incipient fluidization (U_{if}) in beds of FCC catalyst. In an operating FCC standpipe, however, gas velocity cannot be clearly defined as it is difficult to determine gas velocity within flowing catalyst in the stand pipes.

Thus, the range of densities between the density of the emulsion (ρ_{if}), at the point of incipient fluidization and the density (ρ_{ib}), at its point of incipient bubbling have been considered by experts over which a fluidized FCC catalyst is expected to remain well fludized in a standpipe. Thus, as the bubbling bed of catalyst begins to enter an FCC standpipe, it needs to shed some of the excess gas bubbles and avoid dragging them down into the standpipe. A standpipe inlet configuration needs to be properly designed to allow this initial removal of excess gas in an orderly fashion. Often catalyst withdrawl wells are provided in regenerators to shed the excess gas/ air before the catalyst enters the regenerated catalyst standpipe.

After the fluidized FCC catalyst emulsion enters the top of the standpipe and begins its descent, it starts undergoing a form of compression / compaction as a result of increase in pressure down the standpipe due to the catalyst head. This increasing pressure compresses the interstitial gas that is surrounding the catalyst particles and in the catalyst pores. The net result of compression of this gas is a decrease of the volume of the catalyst emulsion and consequent increase in the density of the emulsion.

If the standpipe is long enough, and if no aeration is provided along the length of the standpipe, then this process of compression will continue as the catalyst travels deeper into the standpipe. This can cause the density of the catalyst emulsion to continue to increase and the catalyst emulsion can reach its incipient fluidization density resulting in a shift of behavior from a fluidized catalyst to a packed one and may give rise to problems in circulation. Thus, aeration of standpipe is required along the length of stand pipe to maintain density in the desired range. The purpose of the standpipe aeration is to supply just enough additional gas to the catalyst to restore the catalyst emulsion to its original emulsion density, as in Figure 2.32.

2.14.3 STANDPIPE DESIGN GUIDELINES

The ratio of densities ρ_{if}/ρ_{Ib} is termed by experts as the "stable expansion ratio" (MSER) for a fluidized catalyst. The higher this ratio, the catalyst is expected to tolerate a larger density change due to compression, before loosing its fluidization, while descending down the standpipe or the catalyst is more flexibile to density changes and makes the catalyst circulation easier.

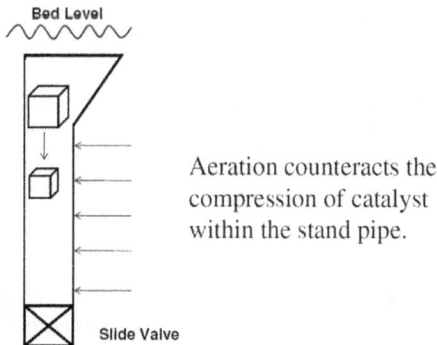

FIGURE 2.32 Aeration of a standpipe.

Abrahamsen and Geldart have shown that the ratio of the superficial gas velocity at incipient bubbling to the superficial gas velocity at incipient fluidization is a function of the physical properties of the catalyst:

$$U_{mb}/U_{mf} = \left(2300\rho_G^{0.126} \times \mu^{0.523} \times e^{0.716F}\right)/d_p^{0.8} \times g^{0.934} \times \left(\rho_p - \rho_G\right)^{0.934}$$

where

U_{mb} = superficial gas velocity at incipient bubbling, m/sec
U_{mf} = superficial gas velocity at incipient fluidization, m/sec
ρ_G = gas density, kg/m3
ρ_p = particle density, kg/m3
μ = gas viscosity, kg/m sec
F = 0–45-micron fines fraction in catalyst
d_p = mean particle diameter, meters
g = gravitational constant, 9.81 m/sec2

They also show that the MSER can be estimated from

$$MSER = \rho_{if}/\rho_{ib} = U_{mf}/U_{mb}$$

It is apparent from the preceding equation that very low fines content in the equilibrium catalyst greatly reduces the MSER. Thus, standpipes that normally operate well may fail when the cyclone performance deteriorates and the 0–40-micron fines content of the equilibrium catalyst drops. Catalyst with a very high equilibrium apparent bulk density (ABD) also can aggravate standpipe circulation problems because the maximum stable expansion ratio decreases with an increase in density of the catalyst particles.

2.14.4 CATALYST FLUX

High catalyst flux can also lead to standpipe problems. Many FCC standpipes operate with a flux as high as 980–1220 kg/(m2) (second) (or 200–250 lb/ft2 sec). Some standpipes have been known to operate as high as 1465 kg/m2 second (300 lb/ft2 sec) without any problem. Factors like defluidization, bubble formation and a actual obstruction due to presence of dislodged refractory or a foreign material can also contribute to problems of circulation.

2.14.5 STANDPIPE AERATION PRACTICE

It is very important to calculate the aeration required by each individual aeration tap location along the length of the standpipe. The theoretical aeration is calculated with the help of skeletal density of the Ecat s suggested by experts.

$$V_{gas} = V_{emulsion}\left(1 \quad \left(\rho_{emulsion}/\rho_{skeletal}\right)\right)$$

where $V_{emulsion}$ is the 1000 ($Q/\rho_{emulsion}$), $\rho_{emulsion}$ is the assumed standpipe density and is typically 560.65 Kg/m3 or around 35 lb/ft3, and Q is catalyst flow rate in metric ton per minute. V_{gas} is the volume of gas circulated with catalyst in M3/min. During the flow through catalyst standpipe the volume reduces with increased pressure. The reduction in gas volume ΔV_{gas} needs to be compensated by fresh aeration in the first aeration point and can be evaluated as

$$\Delta V_{gas} = V_{gas} (1 - P \text{ in } /P \text{ tap1})$$

where P_{in} is the pressure at the standpipe inlet, P tap1 is the pressure at the first tap or tap1 from the inlet. The quantity of aeration required is calculated from the aeration gas molecular weight, pressure, and temperature at the tap point to compensate the loss of V_{gas} at the point. This evaluates the theoretical aeration rate required. Experts observe that a 60%–70% of theoretical aeration flow is maintained in actual practice, however, in some cases the theoretical amount is also required for aeration.

Overaeration can lead to a dramatic loss of standpipe pressure build up and supports the usual practice of use of less than theoretical quantities while starting standpipe aeration and may be continued in all the taps along the length of standpipe. The change in catalyst emulsion density from the inlet to first tap or from tap to tap $\Delta \rho_{emulsion}\% = 100 \times (\Delta V_{gas}/V_{gas})$. This signifies the required density increase from tap to tap, in other words the compression requirement from tap to tap.

A lower density increase between taps indicates lower likelihood of circulation issues during run with change of catalyst size distribution. A typical aeration point arrangement is shown in Figure 2.33.

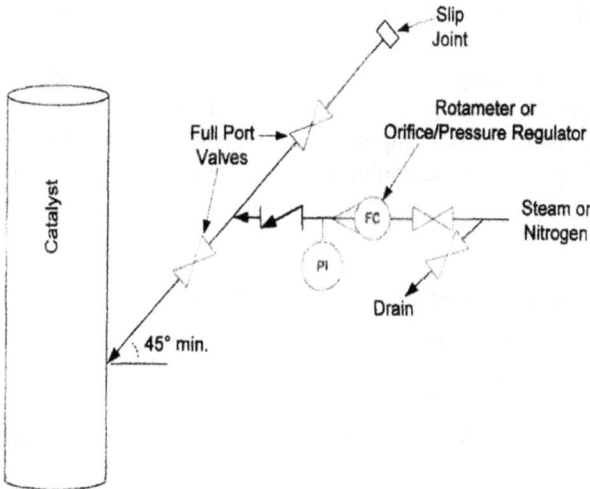

FIGURE 2.33 A typical aeration arrangement in standpipes.

2.14.6 Choice of Aeration Gas

Aeration media can be air, steam, fuel gas, or hydrogen. A comparative analysis is presented next:

- Air or nitrogen is the best aeration medium. They have higher viscosity and density than steam.
- Steam is not as effective as air. Condensation complicates the application.
- The low gas viscosity of hydrocarbons/fuel gas make it ineffective as an aeration medium.
- Low gas viscosity and very low density makes hydrogen a very ineffective aeration medium.

2.14.7 Practice of Aeration in Regenerated Catalyst Standpipe (RCSP)

Traditionally, steam is used for aeration. Air can be also used. But there can be apprehension about use of air, as that can give rise to oxygenated product in gasoline and affect stability of gasoline. Some use air upstream of the slide valve (RCSV) and steam downstream in RCSP. Air can be used upstream and nitrogen downstream to avoid likelihood of formation of oxygenated compounds and to avoid a higher amount of air entering the reactor in case of a missing restricted orifice or orifice with enlarged hole size in the aeration taps. However, for ΔP measurement across the slide valves, nitrogen (or same aeration agent) may be used for both upstream and downstream pressure taps.

2.14.8 Practice of Aeration in Spent Catalyst Standpipe (SCSP)

Traditionally, steam is used at the upstream slide valve (SCSV) as the aeration medium and air is used downstream. The upstream aeration agent can be nitrogen and air can be used downstream to improve performance and conserve nitrogen.

Improvement in standpipe pressure buildup and stability were immediate and marked in many FCCU as a result of changes made to air/nitrogen for standpipe aeration. Performance of standpipes can be improved by proper fluidization before entry to standpipe, proper standpipe geometry, proper aeration and right catalyst properties.

2.15 REACTIONS IN FCCU AND CATALYSTS FEATURES AND ADDITIVES

2.15.1 Reaction Mechanism

The first step in the reaction is formation of a positive-charged carbon atom and the phenomenon is called carbocation. Carbocation can occur by carbenium and carbonium and ion mechanism. Carbonium ions are not stable, thus, primarily, carbenium ions initiate the catalytic cracking reactions and govern the reactions in catalytic cracking. Once formed, the carbenium ions can lead to various reactions based on the nature and strength of catalyst acid sites. The main reactions that take place in catalytic cracking are as follows:

- *Scission of carbon–carbon bond.* The cracking or beta (β) scission is the main feature of catalytic cracking. β This is scission of the C–C bond two bonds away from the positively charged carbon atom. (This is different from the free radical mechanism as in thermal cracking.)
- *Isomerization* of hydrocarbons occurs to a considerable extent and shows a higher branched (ISO/Neo) to normal molecule distribution in the products.
- *Hydrogen transfer* to terminate the reactions shown:
  Olefin + Olefin → Aromatic + Paraffin
  Olefin + Paraffin → Aromatic + Paraffin
  Olefin + Naphthene → Aromatic + Paraffin

Other reactions are dehydrogenation initiated by metals, particularly nickel and vanadium, and formation of coke on catalyst and also corroborated by increase in H2/ CH4 ratio in sponge gas.

The catalysts used in FCCU is essentially zeolites. The zeolite, or more popularly faujasite, is the key ingredient of FCCU catalyst. Zeolites are crystalline alumina silicates having a regular pore structure. The basic building blocks are silica and alumina tetrahedra. The performance of the catalyst depends largely on the nature and quality of zeolite. The right property of zeolite is chosen for the desired product yield. Compared to previous amorphous silica–alumina catalysts, the zeolite catalyst exhibit much higher activity and selectivity. The catalyst normally have some rare earth component in them (known typically up to a maximum of 4% wt. in catalyst). The rare earths contribute primarily to enhancing the hydrothermal stability of the catalyst. Rare earth is a generic name for the 14 elements of the lanthanide series, which contain atomic numbers from 57 through 71 plus scandium and yttrium. They are highly similar in chemical properties. The similarity complicates the separation of one from another. Thus, they are often supplied in the mixture of oxides as extracted from ores. The typical rare earth consists of 46% cerium oxide, 20% lanthanum oxide, and 15% neodymium oxide as well as other oxides in the series. The main purpose of adding rare earth to improve the hydrothermal stability of cracking catalyst exposed to high regenerator temperature is to arrest the likelihood of sintering of X-type zeolites and becoming amorphous, resulting in loss of activity.

The subsequent development of Y-type zeolite improved the inherent stability of catalysts without rare earth and can withstand similar regenerator temperatures as earlier varietiy with rare earths. These Y-type zeolites are known to withstand a temperature as high as 1400°F. Rare earths are known to be added to zeolite catalyst through ion exchange. During this process a portion of the acidic protons and sodium located in zeolite are known to be exchanged with rare earth ions. Rare earths inhibit dealumination of a zeolite and thus a higher concentration of acid sites will be found in rare-earth-exchanged catalysts.

This leads to an increase in activity and hydrothermal stability of catalyst. As a result, the cracking activity as well as the hydrogen transfer activity are enhanced. Primary cracking reactions involve initial scission of the carbon-to-carbon bond to form lighter products. Primary hydrogen transfer occurs between cracked products to terminate the cracking reaction in the gasoline range, thus reducing overcracking of gasoline range components to C3s and C4s.

The hydrogen transfer reactions are greatly enhanced with addition of rare earths to zeolite and increases gasoline yield. Adding rare earths decreases the octane of gasoline and also lowers the cetane index of LCO, as shown in the following reaction:

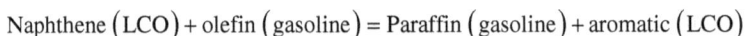

$$\text{Naphthene} \left(\text{LCO}\right) + \text{olefin} \left(\text{gasoline}\right) = \text{Paraffin} \left(\text{gasoline}\right) + \text{aromatic} \left(\text{LCO}\right)$$

It also reduces olefins found in the products. and reduces olefin proportion in LPG. The addition of rare earth reduces the wet gas generation to give relief to the wet gas compressor and to the gas concentration section.

2.15.2 Additives in Catalyst

Traditionally, some additives are used in FCCU to enhance the performance of the unit and are mentioned as follows:

- *CO promoter.* CO promoter is added to facilitate CO combustion in the regenerator dense phase to minimize afterburn and consequent temperature excursion in the dilute phase. This helps in reducing coke on regenerated catalyst (CRC) and thus ensures high activity of the regenerated catalyst. It is mostly used in full or partial burn types of units mostly having a single regenerator. The active ingredients of the prompter used is platinum group metals. The platinum concentration is typically 300 ppm to 700 ppm dispersed on a support. CO promoter is frequently added to regenerator catalyst.
- *Nickel passivators.* The common metal compounds in FCCU feed are nickel and vanadium. They poison the catalyst and reduce the activity of regenerated catalyst. Nickel, particularly, is known to increase the dehydrogenation reaction and typically results in higher sponge gas production and increases the hydrogen-to-methane ratio of the sponge gas. Refiners usually increase fresh catalyst addition or increase low metal equilibrium catalyst makeup, or use nickel passivators to reduce the effect of the metals. In case of high nickel, antimony or bismuth-based compounds are known to be most effective. Use of nickel passivator is economical when the nickel on Ecat is greater than 1200 ppm or the passivator can help restrict higher production wet gas and enable a higher feed rate.
- *Vanadium traps.* High level of vanadium in Ecat is converted to V2O5 by reacting with balance oxygen of air in regenerator and further likely to react with water (formed from burning of hydrogen of coke in the regenerator or carry over of steam from reactor) producing vanadic acid and that may attack

the zeolite catalyst and lead to collapse of zeolite crystal lattice leading to loss of activity. Vanadium traps are typically used when the nickel + vanadium increases beyond than 1600 ppm to retain activity of the catalyst.

- *ZSM-5.* This is a type of shape-selective zeolite added to the unit to increase gasoline octane and yield of light olefins. ZSM-5 achieves this by converting low octane components of gasoline into light olefins of the range of C3, C4, and C5. The units processing paraffinic feedstocks will achieve the greatest benefit by using ZSM-5, and gain will be lesser in the units using a higher reactor temperature and naphthenic feedstocks. Typically blended with catalyst in proportion of 10%–15% to enhance light olefin yield, the addition of ZSM-5 increases the load of wet gas compressors and Gascon section.
- *SO_X additives.* The sulfur in coke burns to produce primarily SO2 and some SO3, commonly referred to as SO_X. They are discharged in the atmosphere and add to pollution. The extent of SO_X emission largely depends on coke yield, coke sulfur content, and regenerator conditions. There are few common methods to reduce SO_X emission:
  1. Flue gas scrubbing
  2. Feedstock desulfurization
  3. Use of SO_X additive

  The SO_X additive is usually metal oxides added directly to the catalyst inventory. The additive functions by combining chemically with the SO_X in the regenerator. The stable sulfates formed in the regenerator are taken to reactors along with the catalyst where sulfates are converted to H2S and metal oxides by reacting with hydrogen produced in the reactor. The H2S goes to the fractionator along with the reactor effluent, and metal oxide is recirculated back in the regenerator.

2.16 CORROSION IN FCCU AND ITS PREVENTION/MITIGATION

Ammonium chloride (NH$_4$Cl)-based salt deposition in the FCCU, particularly in the main fractionators (MF), is known to be a recurring problem. This is due to potentially high chloride levels in feeds processed in some refineries. The increasing proportion of residue processed in FCCU results in higher chloride content of feed. In addition, atmospheric residue feedstock containing higher chloride is often processed (either neat or in admixture) and results in a higher chloride level in feed.

For many of these refiners, withdrawal of a gasoline side-cut from the main fractionator for subsequent hydrotreatment to near-zero sulfur levels has resulted in low main fractionator top temperatures (as low as 100°C), compared to previous temperatures in the range of 125°C–130°C. The lower temperatures have led to higher salt deposition at the top of the main fractionator. Other likely causes of NH$_4$Cl salt deposition include:

- Poor crude desalter operation
- Reprocessing of slops in MF

- Leaking overhead condenser (using sea water)
- Overhead receiver boot water entry along with reflux
- Improper distribution of cold reflux stream (cold spot)
- Feedstocks containing organic chloride from additives used to increase oil recovery

Catalyst vendors rule out chlorine ingress from catalysts. The majority of chlorides used in catalyst formulation are known to get removed in the manufacturing process (in calcination). The balance on fresh FCC catalyst is removed in the regenerator before the catalyst makes an entry to the FCCU's reactor section.

As experienced, NH4Cl deposition primarily occurs at the top of the main fractionator and is often accompanied by an increase in pressure drop. It can also be encountered to a lesser extent in the overhead line, where the gas is passed through the air and water coolers, or the downstream FCCU gas plant. Other important symptoms of NH4Cl deposition may include:

- Flooding of MF top section
- Plugging of top product draws
- Loss of separation efficiency between gasoline and LCO as a result of flooding in the section of column.
- Increase in reactor pressure and consequently regenerator pressure

In addition, corrosion may also be an issue, especially for packed columns. A summary of the consequences of salt deposition may include:

- Corrosion of trays and packings
- Increased unit delta coke (higher reactor pressure due to higher ΔP in column)
- Reduced air blower capacity (higher regenerator pressure due to higher pressure in reactor as a result of higher ΔP in column)

2.16.1 PREVENTION OF AMMONIA CHLORIDE DEPOSITION

To prevent NH4Cl deposition in the overhead line, water is usually added, with typical quantities in the range of 6–7 vol.% water on a fresh feed basis. Addition of an anti-fouling additive in the reflux stream can prevent the formation of NH4Cl deposits on the trays and packing. Another recommendation (by experts)is to water wash the main fractionator. Water is injected either periodically in the reflux stream, and the main fractionator top temperature is reduced to approximately 85°C to allow water to condense inside the column to dissolve the salt. This method is practiced in many FCC units. The methods to remove and reduce deposition of NH4Cl is described next.

- The water is preferably removed on a dedicated tray, where it is separated from the heavy cracked naphtha. This procedure has been known to be successfully practiced by Saudi Aramco and elsewhere. Alternatively, the main fractionator top temperature can be increased [for instance, to above

135°C (275°F)] for a given period of time to enable dissociation of the salt. But, this may result in a higher end point of gasoline during the time period. Alternatively, the water draw-off facility may be provided in the main fractionator tray that will allow water draw along with the dissolved salts to be removed without contaminating the heavy cracked naphtha.

- The installation of a two-stage desalter is also considered to improve performance of desalter operation in the crude unit and reduce chlorides in feed.
- Another option includes installation of a gasoline splitter and collecting the full range naphtha through overhead of the main fractionator (maintaining a higher top temperature) and subsequently split in the splitter.
- Hydrotreating FCC feed resolves the problem, as this removes most of the feed chloride and significantly improves the yield.

2.16.2   CALCULATION OF NH4CL DEPOSITION TEMPERATURE IN THE MAIN FRACTIONATOR

The calculation is illustrated in the following example.

Computation of the NH4Cl deposition temperature in an FCC unit main fractionator (MF) processing residue feedstock includes the following conditions:

- Feed rate = 500 MT/h (500,000kg/h)
- Feed nitrogen content = 1500 ppmw
- Feed chloride content = 1.5 ppmw
- MF top pressure = 1.8 bara (27.8 psig)
- Total steam to MF = 60 MT/h = 3333 kmol/h
- Dry gas production = 2500 kmol/h
- LPG product = 2000 kmol/h
- LCN product + Reflux = 4000 kmol/h
- Total moles of hydrocarbon + Steam to the MF top = (4000 + 2500 + 2000) + 3333 = 11833 kmol/h

The example assumes 20% wt. of the feed nitrogen goes to form NH3:

- Nitrogen (in the feed) = 500,000 × 0.15 wt.% = 750 kg/h
- Production of ammonia = 750 × 0.2 = 8.82 kmol/h
- Partial pressure of NH3 = 1.80 × (8.82/11833) = 1.34×10^{-3} bara
- Production of HCl (from the feed) = 500,000 × 0.00015 × 10^{-2} = 0.75 kg/h = 0.75/36.5 = 2.05×10^{-2} kmol/h
- The partial pressure of HCl = 1.80 × (2.05 × 10^{-2}/11833) = 3.11×10^{-6} bara
- P.P$_{NH3}$ × P.P$_{HCl}$ = 1.34 × 3.11 × 10^{-9} = 4.17×10^{-9} bara

Deposition temperature is evaluated by using the ammonium chloride deposition charts in Chapter 4 (Figure 4.9A and 4.9B). Alternatively, the following equation can be also used (suggested by experts):

$$ln\left(K_P\right) = -21183.4/T + 34.17$$

where $K_P = P.P_{NH3} \times P.P_{HCl}$, and T is the minimum main fractionator top temperature required to avoid salt deposition (measured in K). The minimum top temperature evaluated by the above formula is found to be almost the same (123.2°C) as earlier computed by the chart. A fractionator top temperature higher by at least 5°C–7°C over the estimated deposition temperature is preferred.

2.16.3   OTHER TYPES OF SALT DEPOSITION

The other main types of salt deposition are from ammonium hydrosulfide ($NH_4(HS)$) and iron sulfide (FeS). The deposition of (NH_4) HS is controlled by the equilibrium reaction:

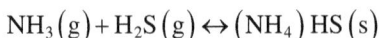

$$NH_3(g) + H_2S(g) \leftrightarrow (NH_4) HS(s)$$

The reaction takes place at a lower temperature than for ammonium chloride. Deposition of this salt is most likely to occur in the overhead line, coolers, and sometimes in the wet gas compressor itself, particularly, when processing feeds with high nitrogen and sulfur content. The deposition of NH_4HS is best reduced by the use of a water wash.

Iron sulfide is usually a corrosion product, which is mostly found on main fractionator trays and packings. Since the salt is pyrophoric, its accumulation is a potential hazard during the opening of the column. Proper procedures for shutdown of the FCCU, and the application of a chemical wash to oxidize the FeS prior to vessel entry, should be followed.

2.16.4   USE OF WATER WASH

An effective water wash dilutes and scrubs corrosive species, such as hydrogen sulfide, ammonia, chlorides, and cyanides, from the FCC off-gas. It may be necessary to inject wash water at several points, because of the possible limitations imposed by flow regimes. The best location for a water wash is downstream of each stage of gas compression, ahead of the water coolers as corrosives gases are more soluble in water at higher pressures.

A water wash in the main fractionator overhead vapor line can also be useful as a first step in removing volatile acid gases. The traditionally recommended water wash rate into the gas compression stages is 1.0–1.5 gpm/1000 bbl of oil feed to the FCCU. Or an addition of water of 6–7% vol. on fresh feed is also practiced in the overhead line.

A measured quantity of water should be injected on a continuous basis through a spray nozzle. Common wash water sources include boiler feed water, steam condensate, and stripped process condensate. Recycling from high to low pressure, often referred to as a backward cascading water wash, is not preferred as it may aggravate corrosion. It is often necessary to supplement the water wash with chemical additives, such as ammonium or sodium polysulfide and filming inhibitors.

2.16.5 USE OF POLYSULFIDES

Polysulfide is injected into a water wash to react with hydrogen cyanide, producing thiocyanates, as the thiocyanates do not react with the protective sulfide film on metal surfaces as cyanide does, and hydrogen penetration is reduced. Either sodium or ammonium polysulfide can be used, but ammonium polysulfide is most commonly used. Based on their susceptibility for oxidation, care should be taken to prevent polysulfides from being contaminated with oxygen. The polysulfide solution should be blanketed with hydrocarbon both to prevent odor problems and to reduce oxygen contamination and degradation while in storage.

Concentrated ammonium polysulfide is soluble in alkaline sour water. If the pH drops below 8, the polysulfide will decompose into ammonia, hydrogen sulfide, and sulfur, and thus will be ineffective. The formation of sulfur may foul the equipment.

Excess polysulfide can increase corrosion at elevated temperatures and, because of its instability, can foul process equipment.

2.16.6 USE OF CORROSION INHIBITORS

In most FCCU, corrosion inhibitors used are filmers. These provide a thin barrier of organic material on the inside surface of the equipment at risk. Dosages are only a few parts per million, based on the mass flow rate of the treated process. This thin barrier can prevent an aqueous phase from reaching the surface, thereby preventing the corrosion that can cause hydrogen permeation or carbonate cracking.

Some inhibitors are passivators, which are known to interact chemically with the surface to anodically or cathodically inhibit the electrochemistry of corrosion.

2.17 SIMULATION GUIDELINES

The simulation for design and rating of the fractionation section can be conveniently done using a steady state simulator. The feed to the fractionator can be constituted by blending of all the product streams in a mixer, heating it to the ROT and feeding to the bottom of the fractionator. For a new design or a new product slate, pilot plant data (or data from mathematical models) is usually made available as a blend of pure components and normal boiling point (NBP) cuts. Alternatively, it can be also available as distillation of the cuts along with their densities and few other properties. The column will have partial condenser and typically have three pump arounds and two or three side cuts with side strippers. FCCU feed is superheated vapor, thus, needs to be cooled down to its saturation temperature for further rectification. The cooling of the superheated vapor is traditionally achieved via the main column bottom pump around (MCB PA). Other pump arounds are provided as per requirement. The main fractionator column profile for an FCCU is presented from a simulation in Table 2.9. In the simulation for revamp design of a unit, the feed composition was entered as a combination of pure components and NBP cuts and taken from pilot plant data. The simulation considers two side draws LCO and HCO with individual strippers and three pump-around refluxes.

The simulation of feed preheat circuit is straight forward and can start with the feed distillation, density to define the feed and followed by heating to the required

TABLE 2.9

An FCCU Main Fractionator Vapor-Liquid Profile from Simulation

Tray	Temp	Pressure	Liquid	Vapor	Feed	Product	Heater Duties
	Deg C	Kg/CM2			KG/Hr		M*KCAL/HR
1C*	43.0	1.38	36059.5			75379.4V	−22.4602
						60000.0L	
						15651.1W	
2	115.0	1.98	41777.8	187090.0			
3	124.7	2.00	81824.9	192808.3	30000.0L		
4	131.9	2.02	45270.5	202855.3		30763.0L	
5	139.7	2.03	30952.6	197064.0			
6	153.8	2.05	424233.3	182746.0	5242.1V		−13.0000
					328563.8P		
7	183.6	2.07	20702.0	242220.8		328563.8P	
						67582.4L	
8	242.5	2.09	13796.4	234835.8			
9	251.5	2.10	12002.2	227930.2			
10	253.0	2.12	10810.5	226136.0			
11	253.8	2.14	377538.7	224944.3	2009.8V		−16.0000
					294838.1P		
12	288.6	2.16	109260.6	294824.5		294838.1P	
						5183.8L	
13	318.0	2.17	94042.3	326568.4			
14	333.5	2.19		311350.1	230000.0V	270237.1P	−22.0000
					270237.1P	12692.2L	

* Trays with free water.

The GASCON section simulation starts with the off gases from fractionator overhead condenser and
        overhead naphtha as feed (not shown).

preheat temperature. Heat exchange with MCB PA is normally considered to preheat the feed.

The simulation of flue gas circuit starts with evaluation of flue gas quantity, composition and temperature from each regenerator based on quantities of coke burnt in each (for a two stage regenerator) and composition of flue gas can be derived for each (by configuring conversion reactors or by manual computations). The flue gas generated from regenerator 1 (RG1) is further admitted to CO incinerator to burn the CO present with excess air in CO incinerator, supported by auxiliary fuel gas firing (by configuring a conversion reactor (R2) for reaction of CO with oxygen of air for full conversion of CO) and arrive at the flow rate, temperature and composition of flue

FLUE GAS CIRCUIT

Flue gas from RG1
129 MT/HR, 676 C   Flue gas from RG2

70 MT/HR, 742 C   Flue gas Cooler

Fuel gas   941C
                                                      To Stack
0.634 MT/HR

                                    1018 C   200 C

AIR for combustion   CO INCINERATOR   DUTY 51.03 GCAL/HR
45.3 MT/HR

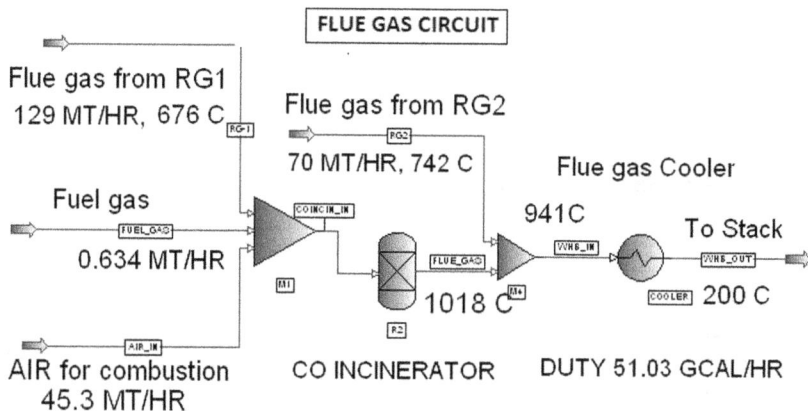

FIGURE 2.34   Simulation flow diagram of flue gas circuit of a FCCU with R2R.

gas at the outlet of CO incinerator. The flue gas generated is admitted into flue gas cooler along with flue gas from regenerator 2 (RG2) to cool to the desired flue gas temperature and routed to the stack. A simulation flow diagram of flue gas circuit of an operating FCC unit with R2R configuration is presented in Figure 2.34.

BIBLIOGRAPHY

A. R. Abrahamsen and D. Geldart. Behaviour of gas-fluidized beds of fine powders part I. Homogeneous expansion. *Powder Technology*, 26, 1980, pp. 35–46.

Ashis Nag. *Distillation and Hydrocarbon Processing Practices*. Tulsa, OK: Pennwell, 2015.

Grace Davison. Answers to the 2011 NPRA, Q&A FCC questions. no. 110, 2011. pp. 28–32.

J. Wilson. *Fluid Catalytic Cracking Technology and Operation*. Tulsa, OK: Penwell, 1997, pp. 188–189.

Jason Goodson and Joe Mclean. *Catalyst Losses and Trouble Shooting*. Galveston, TX: REFCOM, 2019.

Jeff Koebel. Troubleshooting FCC catalyst losses. *Refinery Operations*, 2, no. 1, January 05, 2011, Vol 2, ISS:1.

M. Melin, C. Baillie, and G. McElhiney. Salt deposition in FCC gas concentration units. *Catalgram*, no. 107, Spring 2010, p. 36.

Michel Melin, Collin Baille, and Gordon McElhiney. Salt deposition in FCC gas concentration units. Grace Davison Refining Technologies. no. 107, 2010, pp. 34–39.

Ray Cocco. *How Standpipes and Dip Legs Aid Fluidized-Bed Processes, Question-Answering*. Particulate Solid Research Inc.

Raymond W. Mott. Troubleshooting FCC standpipe flow problems. *Catalagram*, 106, Fall 2009, p. 11.

Rebecca Kuo. *Troubleshooting Catalyst Losses in the FCC Unit*. Galveston, TX: RefComm, 2016.

Reza Sadeghbeigi. *Fluid Catalytic Cracking Handbook*, 2nd ed. Houston, TX: Gulf Professional Publishing, 2000.

Robert A. Mayer. *Handbook of Petroleum Refining Processes*. 2nd ed. New York: McGraw-Hill, 1996.

Russell C. Strong, Veronica K. Majestic, and S. Mark Wilhelm. FCC corrosion-conclusion
basic corrosion control methods solve varied problems. *Oil and Gas Journal*, 1991.

T. Gauthier, J. Bayle, and P. Leroy. FCC: Fluidization phenomena and technologies. *Oil &
Gas Science and Technology – Rev. IFP*, 55, no. 2, 2000, pp. 187–207.

ABBREVIATIONS

APS	average particle size
BFW	boiler feed water
BS&W	basic sediment and water
CLO	clarified oil
CRC	coke on regenerated catalyst
DDSV	double-disc slide valve
DO	decanted oil
FGDU	flue gas desulfurization unit
HCO	heavy cycle oil
MCB	main column bottom
MCB PA	main column bottom pumparound
PSRI	Particulate Solid Research, Inc.
ptb	pounds per thousand barrels

3 Coker

3.1 INTRODUCTION

Delayed coking is a thermal cracking process converting residual oils to liquid products. Byproducts are off gas and petroleum coke. In delayed coking the heated feedstock is routed to the coke drum, allowing adequate time for cracking and coking.

Feedstock characteristics vary widely. Some feedstocks pose challenges that affect operation of the heater and the run length of the unit. It is important to maintain the delayed coker online for as long as possible and at maximum throughput. If a refinery does not continue to have the delayed coker in operation, it would have to cut back on the crude unit capacity and shut down or cut back various other units that take products of coker as feedstock. Typical products and their yields (of a low recycle vacuum residue coker) are given in Table 3.1.

The yield of coke is a function of feed Conradson carbon residue (CCR) and other operating variables like drum pressure, temperature, and recycle ratio. Typical yield of coke is evaluated as $1.35 \times CCR$ for low recycle coker units.

3.2 PROCESS DESCRIPTION

3.2.1 GENERAL PROCESS SCHEME

The feed to the coker unit can be reduced crude oil (RCO) or vacuum residue (VR) or a mixture. The feed is typically preheated by heat exchange with hot products and pumparounds from the main fractionator and fed to the fractionator bottom spout. In the spout the feed combines with the recycle from the upper section of the fractionator. The combined feed is picked up by a high-pressure feed pump and pumped through the tube passes of the furnace. Typically, it is heated to 498°C–505°C. The furnace outlet is routed through the four-way switch valve to a coke drum ready for feed in. The other drum is in some stage of decoking operation. The vapors from the drum outlet are quenched mostly by heavy coker gas oil (HCGO) product from the column to around 425°C before it is fed to the column through ring valve. HCGO quench is introduced through a spray nozzle to evenly spray the liquid in the line. Often blowdown scrubber (BDS) bottom liquid is also fed as quench fluid. A few coker unit flow schemes are presented in Figures 3.1 and 3.2. There are several sections in the unit:

- Fresh feed preheating section
- Coker furnace
- Coke drums with accessories and valves
- Coke cutting system
- Blowdown scrubber and accessories

DOI: 10.1201/9781003268246-3

TABLE 3.1

Typical Yields of Products [Vacuum Residue (VR) as Feed]

Product	Wt.% Yield
H2S	—
Gas	3.72
LPG	3.1
LT naphtha	6.56
HVY naphtha	5.36
LCGO	23.34
HCGO	25.43
Coke	31.28
Total	**100**

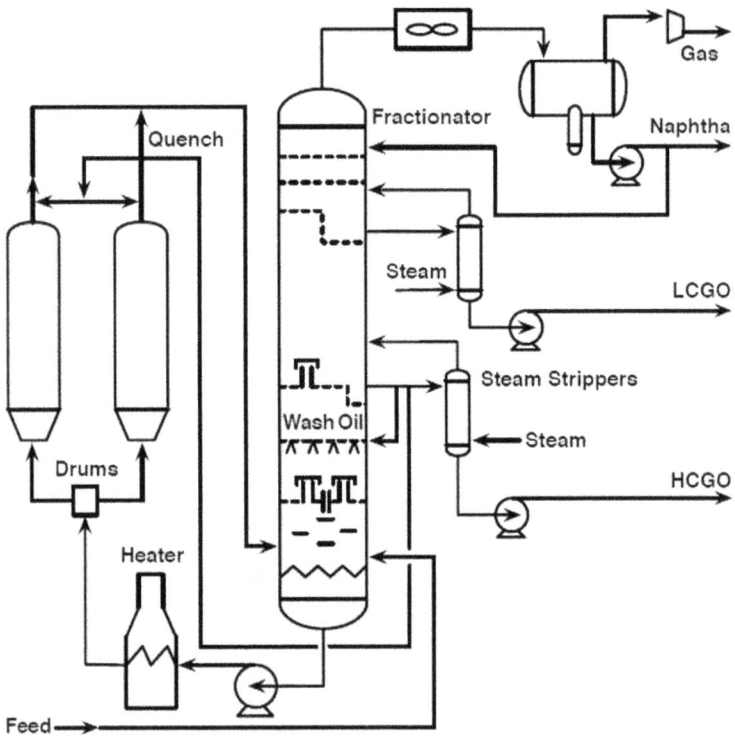

FIGURE 3.1   Flow scheme of a coker (main sections).

- Coker fractionator with side strippers
- Vapor recovery/LPG recovery section
- Water handling system for cooling and coke cutting

Normally, two coke drums are fed by a single furnace. For bigger units, the number of coke drums is increased (beyond a maximum diameter of 32 ft for a coke drum). Accordingly, a four-drum coker has two coking furnaces and a six-drum has three coking furnaces. For bigger coker units deploying more drums, additional BDS may be required.

Sufficient height above the maximum coke level needs to be kept in the coke drum to avoid carryover of coke or foam to the fractionator. Measurement of coke height is very important. Accordingly, instruments (gamma ray detection) for coke height measurement are provided (with adequate redundancy). The typical height above the maximum coke level is 12 meters and higher. Coke drum overhead vapor velocity is typically maintained at 0.5–0.6 ft/sec. A facility for injection of antifoaming agent at the top of drum is provided to arrest foaming and carryover. Sometimes, the coke drum pressure is marginally increased to maintain specified superficial velocity (and arrest carryover) even at the cost of production of more gas and coke.

A coker fractionator is provided with a cleanup circuit at the bottom, comprised of a pump (high flow and low head) with twin/duplex strainers at the suction to arrest the coke fines in the strainer, from building up at the bottom, and their inadvertent entry to furnace feed pumps and downstream. Coker flow schemes are shown in Figures 3.1 and 3.2.

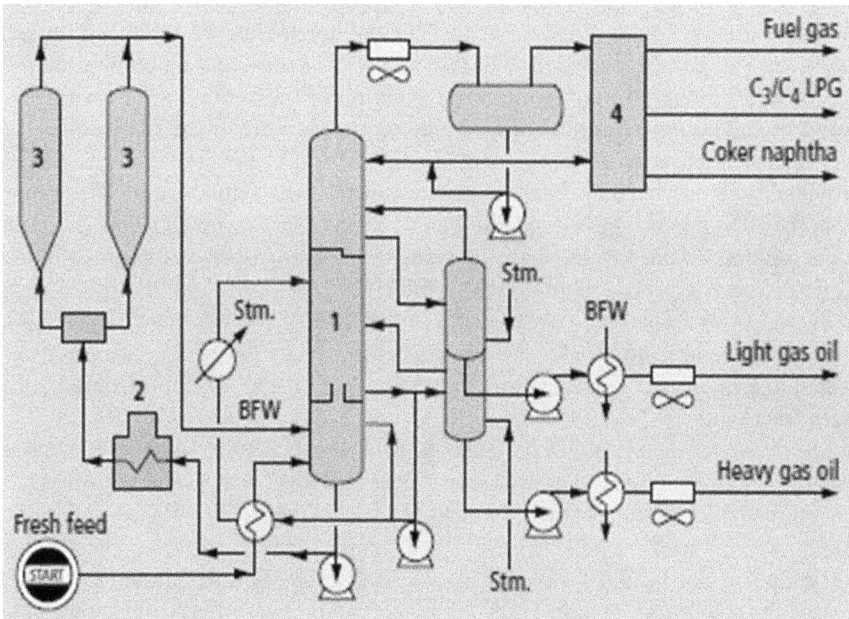

FIGURE 3.2 Generalized flow scheme of a coker.

In Figure 3.2, the number 3 indicates coke drums, 2 the furnace, 1 the column with side strippers, and 4 the vapor recovery section. Quench is introduced at the top of a drum immediately at the vapor outlet of the drum to cool the hot vapor (at 440°C–460°C to around 425°C) and also to try to keep the line wet to avoid coking in the line. Usually BDS bottom liquid or HCGO is used for quenching the hot vapor. *Some designers additionally allow some atmospheric cooling of the line from the drum to column (sometimes for generating some condensate to keep the line wet* by providing less insulation on the line or part of the line).

Typically, the quench flow plus wash oil flow (above the wash bed in the column) is reduced to half or less (often referred to as the drying ratio) when contacted with hot vapors from the drum and hot vapor in the column. A heavy oil is normally preferred as the quench oil to the overhead line to raise the dew point of the vapor and to increase the likelihood for generation of some condensate. But in low pressure coking units, no condensate is normally generated in the drum overhead line primarily due to lower pressure and presence of large amount of steam (coil velocity steam, valve gland steam and purge steam).

The dried quench oil plus dried wash oil join the fresh feed at the column bottom, constituting the recycle. Typically, wash oil is introduced through two nozzles at two elevations above the wash zone in the column. Each has provision of medium-pressure (MP) steam to flush the header and the spray nozzles.

During decoking operation, like final steaming, water cooling, and vapor heating, the BDS is used (shown in Figure 3.3). The steam formed during cooling of coke is admitted to the BDS. In vapor heating, part of the vapor from the drum in operation is introduced from top of the drum to be heated. The vapor leaves the drum through the bottom, flows to the coke condensate drum (CCD), then to the BDS. The BDS is provided with typically 10–14 trays and a pumping facility is provided to pump the bottom liquid to the top of the column after cooling the liquid in an air cooler for maintaining the top temperature of the BDS well above the water condensation point (around 140°C–170°C). The overhead of the BDS is provided with overhead air cooler and blowdown drum (reflux drum). The vapors from the BDS overhead are condensed and collected in blowdown drum. The non-condensable is routed to the flare or main fractionator overhead receiver. Often a liquid ring compressor is provided to route the gases from the blowdown drum to the main fractionator overhead drum at the inlet of the wet gas compressor to recover the stream and avoid loss of hydrocarbon to flare. Once the coke drum gets heated to around 200°C (water free) the vapors from the CCD can be routed to the main fractionator.

During water quenching of hot coke, a lot of steam is generated and the vapors (mostly steam) from the top of drum are routed to BDS. The hydrocarbons from the vapor are condensed by the pump around reflux provided and collected in the BDS bottom and pumped out to slop (or to the drum overhead as quench). The steam is condensed in the BDS overhead air cooler and collected in the blowdown drum, joining the sour water stream. Thus, the BDS column gets engaged during different activities in the decoking cycle. Once the coke drum is cooled/filled, the water from the drum is drained in the coke pit from where it flows through the iron maze and

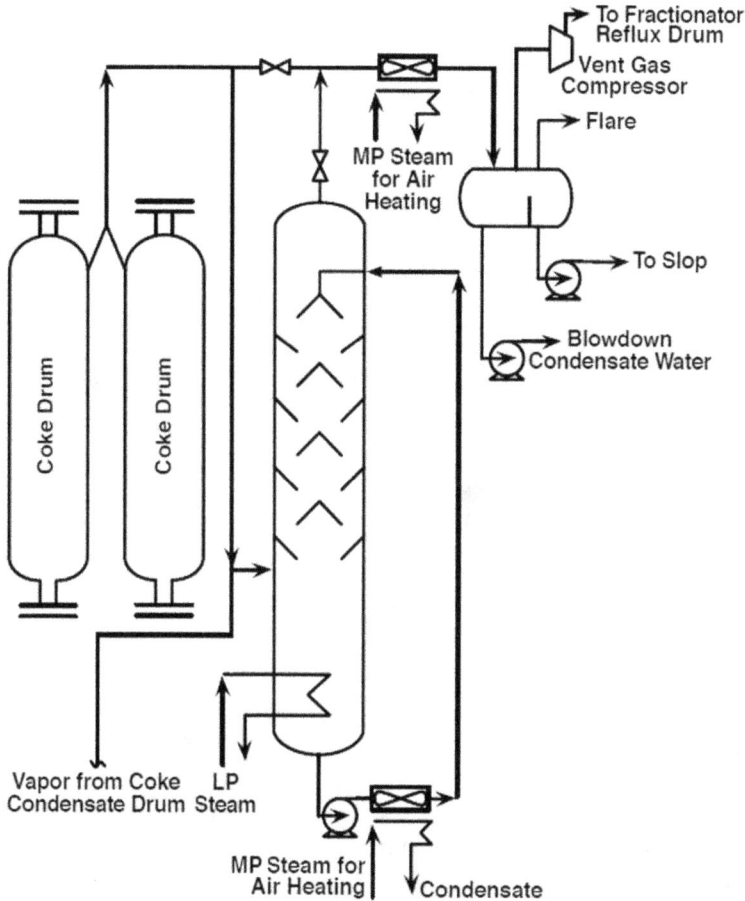

FIGURE 3.3 Flow scheme of a coker (scheme of BDS/drum).

coke settling channel to the water settling tank, then stored for further reuse. The BDS column with its accessories are shown in Figure 3.3.

Figure 3.4 shows the ball valves and ring valve around the coke drums. There will be always one transfer line from a set of two drums leading to the column with a ring valve (Figure 3.4). For a six-drum coking unit there will be three transfer lines going to the column, each with a ring valve. The purpose of the ring valve is to enable its partial closure and pressurization of the drum under operation to the extent to divert part of the hot vapor to the idle coke drum for vapor heating.

After the coke is cooled, the coke drum is then depressurized and unheaded. The coke is drilled and cut with a hydraulic jet, and then coke and cutting water flows through a chute (fixed and telescopic chutes are common) to the coke pit. Water flows out again through maze and settling channels, then stored in water tanks. Coke is removed from the coke pit normally by electric overhead traveling (EOT) cranes to receive coke from the next drum. The coke drum overhead and feed lines are

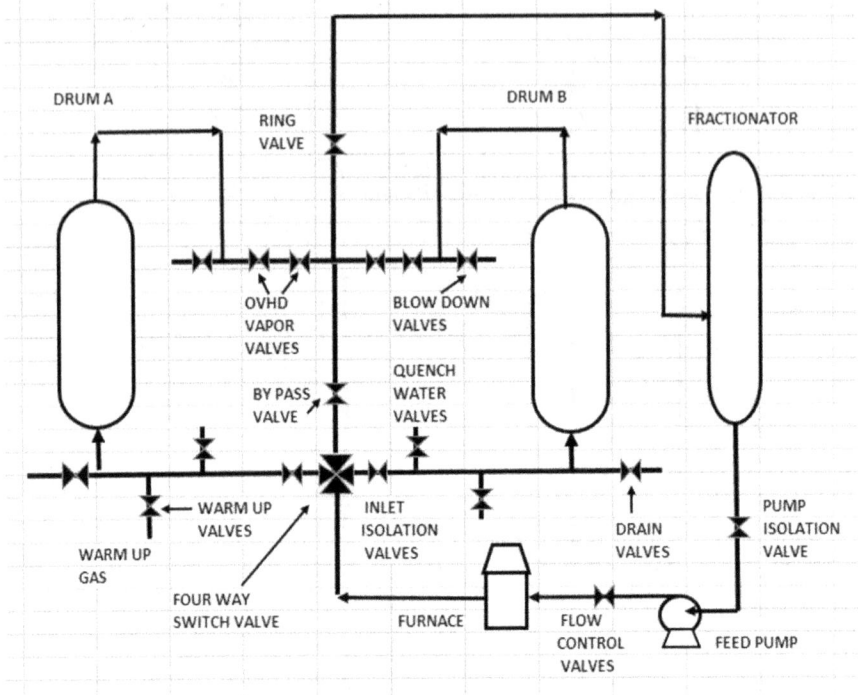

FIGURE 3.4 Valves of a coker (ball valves) and ring valve.

provided with ball valves. These valves are mostly remotely operated. The different activities are controlled by operation of the ball valves.

Figure 3.5 shows the top and bottom slide valves for drum heading and unheading, along with the isolation valves (ball valves) marked with the number 10.

The side products of the coker unit are light coker gas oil (LCGO) and heavy coker gas oil (HCGO/HHCGO) are drawn through side strippers with steam as the stripping agent. The top naphtha is normally split into two cuts: light coker naphtha and heavy coker naphtha (mostly in a separate splitter). The light coker naphtha (LCN) is treated in a unit with silicone guard for desulfurization, diolefin saturation, and can be blended in motor spirit or can be sold as naphtha.

Heavy coker naphtha (HCN) and LCGO are typically treated in the diesel hydrodesulfurization (DHDS) unit to lower the sulfur content and improve the cetane number, then can be sold as diesel. The HCGO/HHCGO is typically fed to VGO hydrotreaters/hydrocrackers in combination with straight run vacuum gas oil (VGO) for further desulfurization and cracking. The typical HCGO quality desired from a coker are

- CCR 0.48 wt.%
- Asphaltenes <100 ppm w
- Ni+V 1.82 ppm w

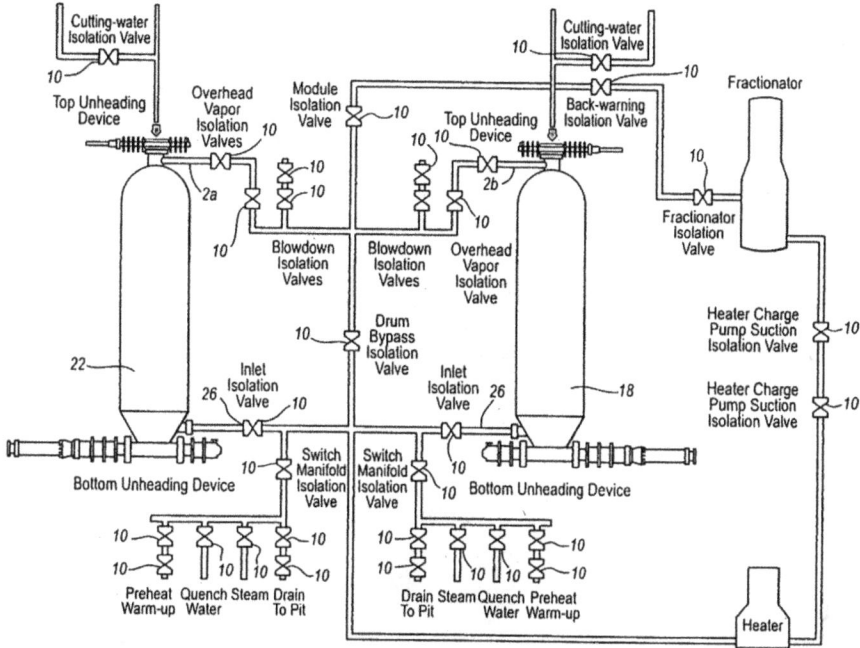

FIGURE 3.5 Heading and unheading slides; valves at top and bottom of coke drum.

3.2.2 FEED DEVICE

Traditional coke drum designs had a central or bottom feed entry arrangement producing uniform upward flow of feed in the coke drum Figure 3.6. But with the arrangement, bottom unheading was difficult, time consuming, and involved manpower intensive high-risk activities. Later, semiautomated bottom unheading facility evolved and is in practice. It was possible to reduce time for unheading but risk (of blowout) continued. Finally, the slide valve for bottom unheading was developed, introduced, and mostly in practice (Figure 3.5). It is fully automatic and safe. However, this introduced a new challenge for the introduction of feed to coke drums. With the induction of automatic slide valves for unheading, it became obvious to introduce feed through one side (side feed arrangement, Figure 3.7) from the earlier practice of center feeding. The issues with single-side feeding systems apprehended/observed include:

- Non-uniform flow in the coke drum
- Non-uniform temperature profile in the coke drum leading to enhanced drum leaning/bending to one side (banana movement) and enhanced thermal fatigue
- Hot spots
- Blowout
- Vibration

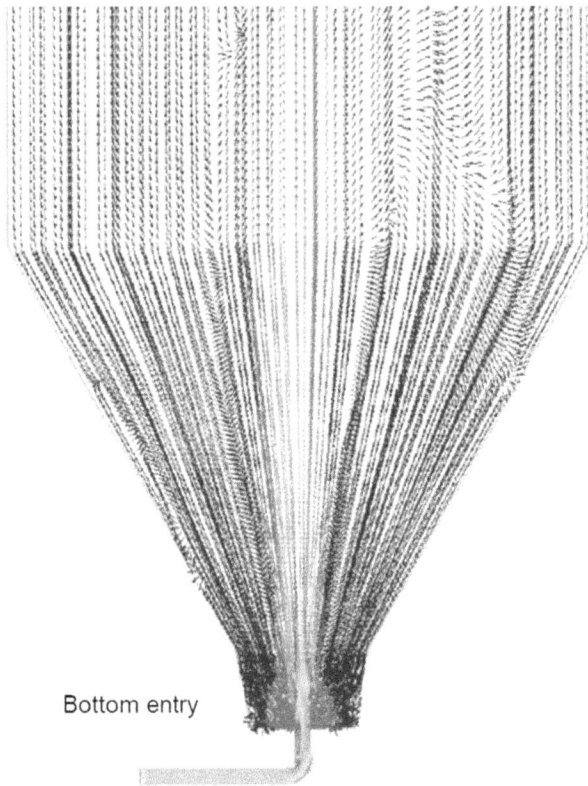

Bottom entry

FIGURE 3.6   Bottom or central feed entry device. (From Amec Foster Wheeler catalog.)

The double-side entry feeding arrangement was developed (as an improvement) to reduce the adverse effects of the single-side entry feed nozzle. But some issues, like complicated piping, reported leaks, and balancing of the flows, continued to remain a concern. Retractable center feeding devices have been recently developed, simulating traditional bottom feed entry systems and are also being offered by valve manufacturers. However, their functioning may need to be further established.

3.2.3   Wash Zone

The wash zone in the fractionator is located below the HCGO tray and primarily performs the following:

- Controls the heavy tail of HCGO
- Minimizes the entrainment of coke fines in the products primarily in HCGO
- Sets the recycle cut point and recycle ratio

The wash liquid is supplied by the HCGO internal reflux introduced through the pump. The zone is prone to fouling with deposition of coke and may lose efficiency

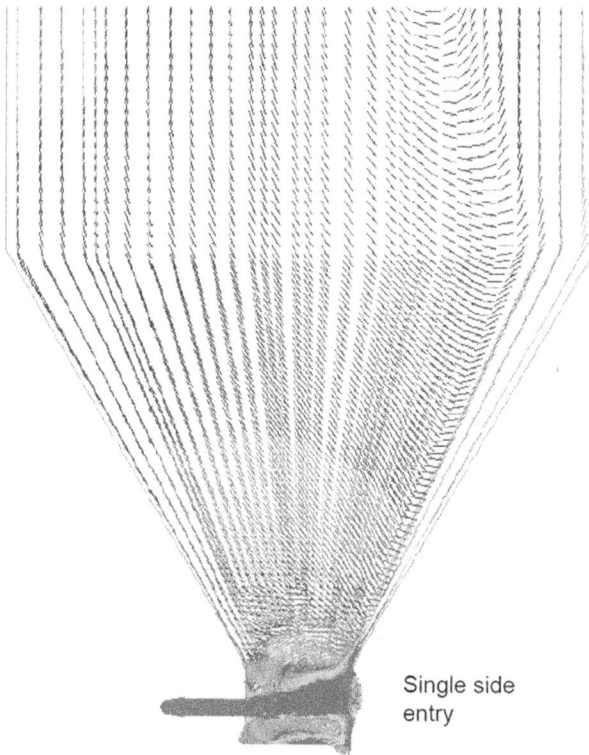

Single side
entry

FIGURE 3.7   Single-side entry feed device. (From Amec Foster Wheeler catalog.)

for cleanup and quality control of the HCGO product. The design of the wash section involves a tradeoff between HCGO quality and fouling of the zone or run length. The wash zone controls the quality of the HCGO. Different kinds of wash section arrangements are (figure 3.8)

1. Conventional cross-flow trays
2. Combination of grids and packing (counterflow device)
3. Several layers of shed decks
4. Open spray chamber

The efficiency of washing may be highest in trays and progressively declines as we go down the from trays to grids, shed decks, and spray chambers. But the fouling resistance follows a reverse trend, viz., is highest in the open spray chamber and reduces progressively going up the ladder. Conventional cross-flow trays pose problems due to accumulation of coke and they normally flood after prolonged operation. The author has changed the section of column to grid packing assembly in a coker unit. *Modern coker units are usually provided with either open spray chambers or shed decks to work as wash zones.* For a recycle ratio near 5% a lower open spray

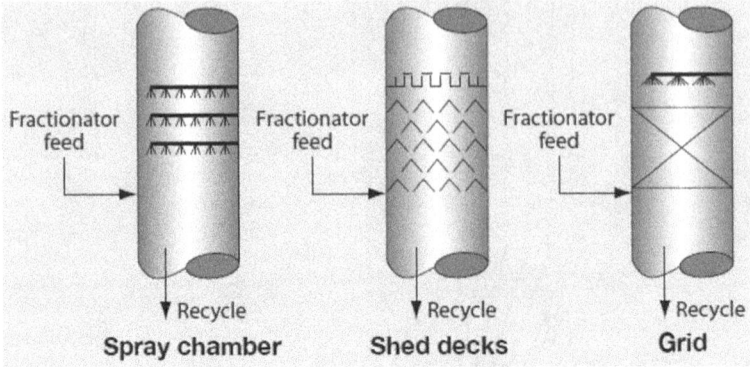

FIGURE 3.8   Typical wash zones of coker fractionator. (From *PTQ*, Q2, 2009.)

chamber is normally recommended by experts. When an open spray chamber is used, very large height is provided for contact of vapor and liquid in the wash section. For simulation of the coker main fractionator, typically one stage is considered in the wash bed. When an open spray chamber is used as a wash zone *the maximum C_{SB} recommended by experts is 0.25–0.28 ft/sec to contain the carryover of coke fines to the HCGO. In revamp applications, C_{SB} can be as high as 0.3 ft/sec.* The location of the wash zone in a coker fractionator is shown in Figure 3.9.

Typically, wash oil is introduced through two sets of nozzles at two elevations above the wash bed. Each has provision of MP steam to flush the header and nozzles.

Typically, an additional amount of steam gets introduced in the system through the purge steam, valve seat, and valve gland purge steam over and above the furnace coil purging steam. This purge steam quantity is best evaluated through sour water generation from the reflux drum by deducting the quantity of coil steam and wash water (from the quantity of sour water generation).

3.3   OPERATING VARIABLES

3.3.1   COIL OUTLET TEMPERATURE

Typical coil outlet temperature (COT) is 498°C to 505°C. If the temperature is increased, volatile combustible matter (VCM) reduces and coke yield reduces. Too high a temperature may make the coke hard and difficult to drill and cut through with a water jet. Typical VCM in coke is 10.5% to 11.5% wt.

3.3.2   RECYCLE RATE

Depending on the configuration of the unit, a recycle stream is specified to meet the desired HCGO product specifications based on the further processing plan of the HCGO. The typical recycle ratio is 5% to 10%. If the recycle ratio is increased the coke and gas yield increases and fresh feed reduces at a constant secondary feed rate. The quench to drum overhead line and wash oil in the column dry up in contact with hot

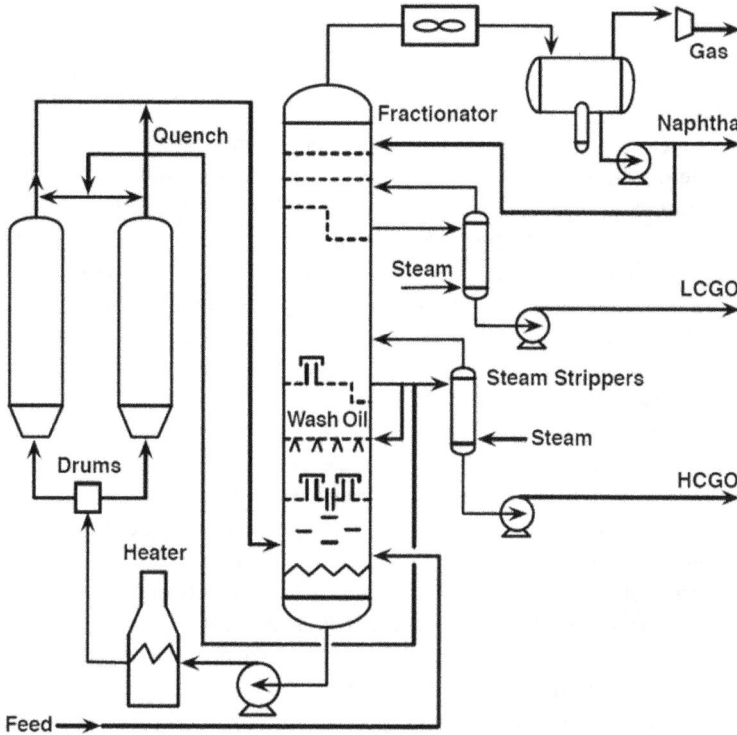

FIGURE 3.9 A coker unit showing the location of the wash section.

vapor and provides the recycle. If wash oil in the column is reduced, the recycle reduces. If the performance wash section drops, the recycle increases with the same wash oil flow. A typical wash section is shown in Figure 3.8 and Figure 3.9.

3.3.3 DRUM PRESSURE

Coke drum pressure is maintained by fractionator reflux drum pressure or the suction pressure of the wet gas compressor (WGC) provided to pick up vapors from the reflux drum, and compress and direct the vapor to the vapor recovery or gas concentration section. If the drum pressure is increased, more liquid will be retained in the coke drum and crack. Coke and gas yield will increase with an increase in drum pressure.

3.4 ACTIVITIES IN DECOKING OPERATION

The major activities involved in decoking cycle are as described next:

- *Steaming/wax tailing.* Soon after switching of feed to another drum, the drum full with coke is steamed to recover the residual oils and to remove the volatile matter content from the coke and initially routed to the fractionator to recover the same. It is subsequently routed to the BDS. Some cooling also takes place during steaming.

- *Cooling.* The coke bed is cooled by filling water in the drum, initially at a slow rate then ramped up. A lot of superheated steam is generated and continues to go to the BDS. Too fast quenching may lead to thermal shock and deformities in the drum. The quenching rate is monitored by skin temperature indications in the coke drum and the skirt support. The coke drum bottom cools first during water filling and the bottom section diameter shrinks. This produces a lot of stress in the welding joint between the coke drum and skirt support. Sometimes, the top pressure of the BDS increases due to resistance in the flow path of steam through the BDS overhead air cooler. Often the sour water produced from the BDS overhead reflux drum (blow down drum) is compared with the inlet water flow rate to estimate the extent of vaporization. When the extent of vaporization reduces in the drum, filling starts. The portion of coke that comes in contact with water gets cooled but may leave some hot areas inside the coke bed.
- *Water draining.* Water is drained from the drum once the drum outlet is cooled to around 120°C–160°C with water. This is also corroborated by reduction in sour water generation from the blowdown drum and lower skin temperatures of the coke drum. The drum vent is opened through the silencer and draining is started. After draining appears complete, there can be still some water locked up inside the drum.
- *Unheading.* The unheading is done with opening the top and bottom covers of the drum, under decoking. There may be a considerable amount of trapped water that drains after the bottom cover is opened. This is drained through the chute to the coke pit.

  Shot cokes are produced during coker operation, steam, cooling, etc. Under certain circumstances, these tar balls (shot coke) can be rapidly ejected from the bottom head opening creating an avalanche of shot coke.

  Although infrequent, if the coke within the drum is improperly drained and the coke bed shifts or partially collapses, residual water can come in contact with the isolated pockets of hot coke, resulting in a geyser of steam, hot water, coke particles, and hydrocarbon from either or both drum openings after the heads, have been removed. These are the hazards associated with the coke drilling cutting operation. In view of this, operators engaged in coke cutting are housed inside a closed cabin and provided with remote switches and joysticks to control the operations (with viewing glass panels of the closed chamber).
- *Coke drilling/cutting.* After the water is drained, coke drilling/cutting is performed with high-pressure water. The high-pressure water is supplied by a high-pressure multistage pump provided with a spill back connection to the decoking water storage tank. A sketch of the coke cutting system is given in Figure 3.10 and the water circuit is presented in Figure 3.11. The pressure of the coke cutting water typically ranges from 150 Kg/Cm^2 to 350 Kg/Cm^2. This depends on the coke drum diameter and nature of the coke.

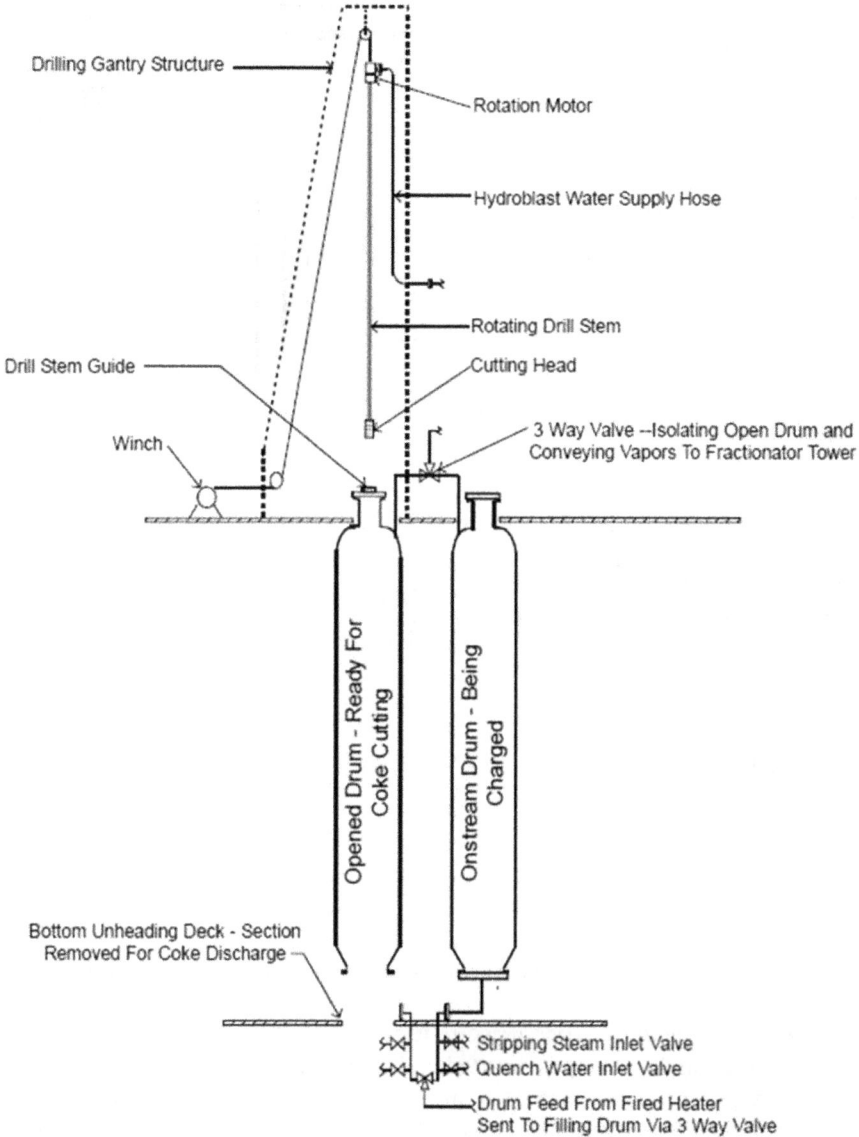

Drilling Gantry Structure

Rotation Motor

Hydroblast Water Supply Hose

Rotating Drill Stem

Drill Stem Guide

Cutting Head

Winch

3 Way Valve --Isolating Open Drum and Conveying Vapors To Fractionator Tower

Opened Drum - Ready For Coke Cutting

Onstream Drum - Being Charged

Bottom Unheading Deck - Section Removed For Coke Discharge

Stripping Steam Inlet Valve

Quench Water Inlet Valve

Drum Feed From Fired Heater Sent To Filling Drum Via 3 Way Valve

FIGURE 3.10 A sketch of coke cutting system. (From OSHA, "Hazards of Delayed Coker Unit (DCU) Operations.")

A correlation to estimate the flow and pressure of the coke cutting water is given next:

- Coke cutting water flow can be estimated as 50 gpm/ft diameter of coke drum or by some other correlations.
- Similarly, coke cutting water pressure can be estimated as 150 psi/ft diameter of coke drum or by some other correlations.

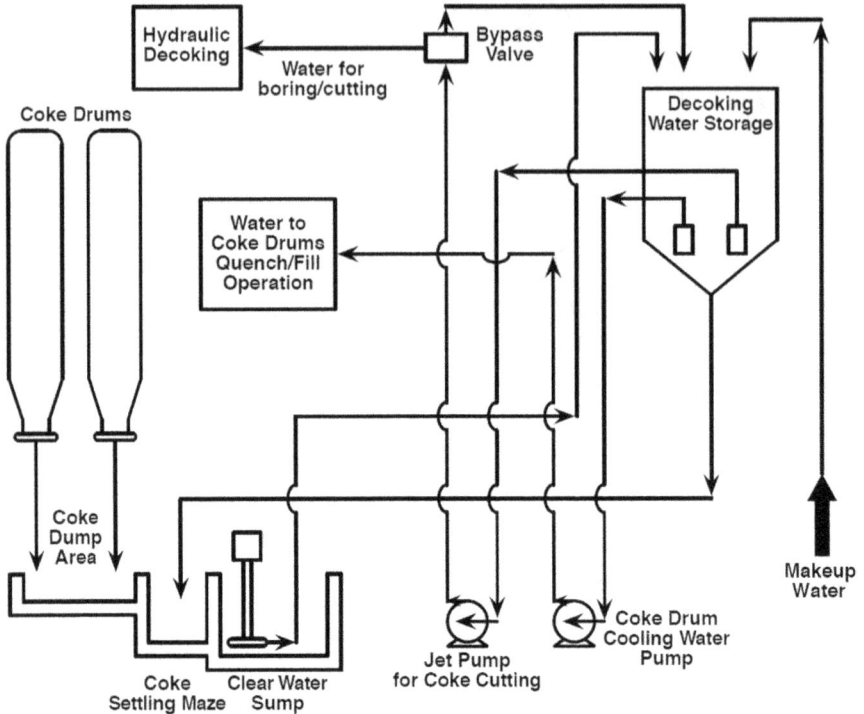

FIGURE 3.11   A sketch of coke cutting water circuit.

During coke cutting very high-pressure water is supplied through a hose involving joints, coupling, etc. A leak in the system may be hazardous.

- *Heading.* Once the coke cutting is completed the coke drum under decoking is headed.
- *Pressure testing.* After heading the drum is pressure-tested with MP steam.
- *Drum heating.* After pressure-testing, a part of the vapor from the drum under coking is diverted to the top of the drum to be heated. The vapor condensate leaves from the bottom and is routed to the CCD. From there the vapor is typically routed to the BDS and condensate is pumped to the BDS. The facility sometimes provided in design to route the vapor to the fractionator as well. Some people continue to route the CCD vapor and liquid to the BDS. Some route to the fractionator after attaining a temperature of the coke drum around 150°C. Finally, the coke drum is heated to a temperature of 180°C–220°C before switching (feed in). Switching at a lower temperature causes stress on the coke drum, which may cause deformity of the drum.
- *Feed switch.* Feed switching is done very slowly in steps to avoid any shock. Steam for wax tailing is started with the drum being closed.

TABLE 3.2

Decoking Activities in Different Coking Cycles

Activity	48-Hour Cycle (Time in Hours)	36-Hour Cycle (Time in Hours)	28-Hour Cycle (Time in Hours)
Coking	24	18	14
Switch + steaming	1.0	1.0	0.5
Steam to blowdown	1.0	1.0	1.0
Water quenching	5.0	4.0	4.0
Draining	3.0	2.0	2.0
Cover removal	1.0	0.5	0.5
Boring/cutting	4.0	4.0	2.5
Reheading/pressure testing	2.0	1.0	0.5
Drum heating	7.0	4.5	3.0
Total	**48**	**36**	**28**

> Once switching is completed, the new drum under operation takes time to liberate vapors, but the vapor continues to the column from the drum under wax tailing.

The coke cutting water circuit is shown in Figure 3.11.

The activities involved in the decoking cycle and their duration are presented again in Table 3.2.

3.5 EFFECT OF FEEDSTOCK

Crude oil is a complex mixture of hydrocarbons that for simplicity are classified into several categories:

- *Saturates.* Can be either paraffinic or naphthenic with long alkyl chains.
- *Paraffinic feeds.* Crack easily and make more gas, lighter distillates, and less coke. High sodium in feed could result in accelerated heater fouling. Stability of the feed when blended with asphaltenic residue could be an issue. May reject asphaltenes due to incompatibility.
- *Aromatics.* Made up of many unsaturated rings with short alkyl chains.
- *Aromatic feeds (decant oil, DO).* High aromatic content gives high HCGO yield. CLO or DO in feed have potential to change coke morphology from shot to sponge.
- Resins. About 40%–53% aromatic with intermediate paraffin chain on naphthenic structure. Also contain aromatic rings.
- *Asphaltenes* (C7 insoluble). Normally stabilized, highly condensed naphthenic and aromatic molecules with small side chains. Almost all crude oils are comprised of different quantities of each of these classes. Typical

coker feed CCR-to-asphaltene ratio is 2 or higher. As the ratio reduces, i.e., asphaltene increases, fouling of furnace increases and coke becomes hard and difficult to cut.

3.6   COKER FURNACE AND FACTORS AFFECTING FAST FURNACE FOULING AND FREQUENT DECOKING OPERATION

3.6.1   Brief Description

The coker furnace is the most critical equipment of the unit. The heart of the coking unit is its furnace where the feed is heated to a typical temperature of 498°C–505°C. The entire heat duty for the unit is provided by the coking heater. It includes the total heat of the endothermic reaction for conversion of feed to products. It also includes the sensible heat and heat of vaporization of the products, including the recycle, and the heat loss from the coke drums and transfer lines.

Paraffinic feed cracks more and needs more heat. During operation, coke lay down takes place inside the coils and the tube skin temperature increases. Skin thermocouples are provided on tubes to monitor the skin temperature of the tubes. Periodic scanning is also practiced to measure the skin temperatures of the tubes and cross-check the temperature indications shown by the fixed skin thermocouples. Once the maximum allowable skin temperature is approached, the furnace is taken out for steam air decoking, spalling, or pigging.

Normally, two coke drums are fed by a single furnace. Thus, a four-drum unit will have two furnaces. Bigger units may have six drums with three furnaces. Earlier furnaces were of the single-fired type. Modern furnaces are double fired to increase the average heat flux by about 150% while maintaining the same maximum heat flux as a single-fired furnace. This reduces the length of coil required after the cracking temperature is attained. This in turn reduces the residence time and reduces coke lay down inside the furnace tubes. Turbulizing or velocity steam is further used to increase velocity and lower coke laydown. The furnaces are normally fuel gas fired. The preheated feed from the column bottom is introduced in several passes inside a furnace convection zone. Different sections in the firebox are constituted by bridge walls inside. The tubes are vertical hairpin types. All the radiant tubes of a pass are placed between two bridge walls or between the end wall and bridge wall in a vertical plane between two rows of burners. Gas burners are normally inclined toward the walls to avoid impingement of flames on the tubes. Each pass (section) has an independent firing system and temperature control for the pass placed in the section. Two types of fireboxes are shown in Figure 3.12 and Figure 3.13.

Tube sizes usually are 4–4.5-inch OD. The diameter of the tubes is selected (in the range of 4–4.5 inches) to optimize the pressure drop as well as maintain high velocities in the tubes. High in-tube velocities (6 fps minimum) are maintained in the tubes. The convection section preheats the feed before entry in the radiant section. *Generally speaking, a higher mass velocity results in a longer run length.* However, if the mass velocity is too high, the pressure drop will be high, and the run will be terminated by the high pressure drop rather than high tube metal temperature. There

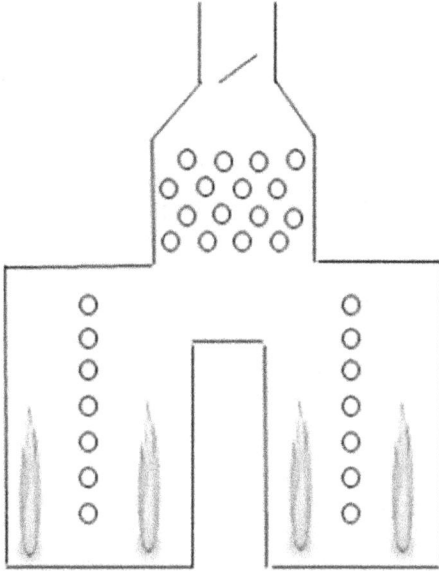

FIGURE 3.12 A double-fired coker furnace.

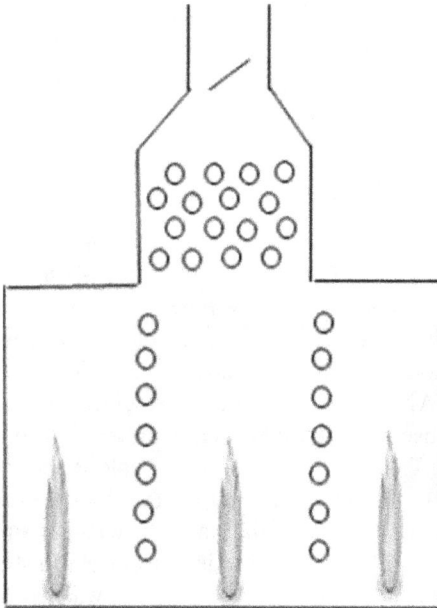

FIGURE 3.13 Variation of a double-fired coker furnace.

are an even number of coils per heater or per pair of coke drums. The cold oil velocity should be about 1.83 meters per second (6 feet per second).

The fired heater run length is the key component affecting the delayed coking unit *onstream factor*. The run length of the coker is primarily governed by the run length of the coker furnace. In general, the refiners decide in advance the desired decoking method(s) so the licensor of the delayed coker unit can make provisions for the desired method in the heater design and heater piping layout. The options are

- Online spalling and pigging
- Offline spalling and pigging

It may be necessary to note that organic fouling can be effectively removed by spalling and pigging. Inorganic Fouling can be effectively cleaned by pigging only. Out of the means of coke removal pigging is considered to be most effective.

3.6.2   THE COIL OUTLET TEMPERATURE

The coil outlet temperature is required to meet the required conversion and product specifications as well as the volatile combustible matter (VCM) of the coke specified. Too high a heat rate will cause excessive coke laydown on the tube wall. Steam injection is used to increase the linear velocity to decrease the residence time in the coil. Quantity is calculated to increase the cold oil velocity to 3.05 meters per second (10 feet per second) at the point where reaction begins. The amount of steam required is calculated by the heater designer and provided in a graphic form or table as a function of the process fluid (feed) flow. At turndown the quantity of steam required is higher to maintain the desired cold oil velocity.

3.6.3   METALLURGY OF TUBES

The usual metallurgy of tubes is 9 Cr–1 Mo. This has a limiting design temperature, according to API 530, of 704°C (1300°F) but is subject to oxidation above 650°C (1200°F). This indicates that the tubes can scale at higher temperatures, which will reduce tube life. The metallurgy has been proven to be successful in operation on various feed tocks including high S, high minerals, and high total acid number (TAN). Type 347 stainless steel (SS) metallurgy is sometimes used/suggested because it has a higher temperature limit of 1500°F. If pressure drop is a limitation, TP347H can be used to replace the coil material. Due to the higher strength at elevated temperatures, the thickness required will be less (and the internal diameter can be higher) and the end-of-run temperature will be higher. TP347H is a stabilized alloy but is susceptible to stress corrosion cracking. During downtime, it will need to be alkaline washed to get rid of polythionic acid. SS tubes are known to have better spalling ability on organic fouling due to thermal expansion properties. But SS tubes are more prone to erosion in return bends, and chlorides must be checked in fuels and feeds to avoid chloride-induced stress corrosion cracking of tubes.

3.6.4   Reasons for Rapid Fouling and Additional Coke Buildup in Furnace Tubes

Additional coke laydown is primarily caused by condensation and precipitation of asphaltenes. When a mixture of crudes or different feedstocks are used, it should be ensured that they are compatible and do not lead to asphaltene precipitation and foul the furnace tubes. A well-designed heater with proper mass velocity and steam injection will have turbulent flow, but the area at the tube wall will be slow moving and possibly stagnant. The temperature at the tube wall (film) is higher and residence time is longer. Material at the tube wall will undergo the cracking reactions, but as the velocities are low there, more time is available for the condensation reactions to occur and together with the precipitated asphaltenes layer of coke will form.

3.6.4.1   Low Mass Velocity

Depending on the mass velocity, some of the precipitated asphaltenes and other deposits can be sheared and pushed downstream. If the mass velocity is low, more will collect on the tube wall. The flow meter (both for feed and velocity steam) can be out or reading incorrectly, which will cause differences in the rate of fouling. Ultrasonic heater pass meters are considered more reliable and widely used. Design values of turbulizing steam are between 1 and 2 wt.% of the heater feed, but this is highly dependent on the flow rates to each pass. More steam is always better for the heater, limited by the maximum superficial velocity in the drum.

- Too much steam can cause solids carryover from the drum and hit wash zone velocity limits resulting in poor quality HCGO.
- Some sections of the column may experience too much vapor traffic and result in jet flooding.
- Can put extra load on column overhead air cooler/water coolers.
- Much increase in heater velocity steam will add to the sour water production from the coker.The first elbow in the heater convection outlet is often considered the best location for injection of steam.

Long residence time—The coking reactions are a function of temperature and time. Longer residence time favors the condensation reactions. The longer the material is in the tube, the more likely that coke will deposit on the tube walls.

High coil outlet temperature—Requires higher heat rates, which increases the local tube metal temperature. Low COT pushes the cracking initiation temperature farther back in the coil. This will result in longer residence time above the cracking temperature of around 425°C.

3.6.4.2   Effect of Feed Quality on Rapid Fouling

Shifting to lower quality heavier crudes results in higher levels of asphaltenes, metals, sulfur, and naphthenic acids. Coking units no longer operate on a fixed feed slate. A variety of available crudes are selected from the market. Some feeds are not compatible. Further, refinery expansion often requires higher severity operation.

All have contributed to shorter run lengths. A *good indicator of expected run length is CCR and asphaltene content of feed. Higher CCR and asphaltene content leads to shorter run length.* The addition of aromatic stream will improve the solubility of asphaltenes. Adding aromatic-rich clarified oil from the fluid catalytic cracking unit (FCCU) at a rate of 3%–5% is often practiced to lower fouling of tubes . This will marginally increase the heater duty and heat rate but it will have a net effect of increasing the run length.

Feed impurities like sodium and calcium also adversely affect the run length of the furnace. The deposits formed due to sodium and calcium can be effectively removed by pigging and not by steam air decoking, online spalling, etc. The sodium gets in the feedstock from the salts present in crude. Sodium also gets in the feedstock from the caustic injected in crude after the desalter to reduce chlorides in the atmospheric tower overhead and arrest corrosion. Total sodium and calcium needs to be limited to 15 ppm in the feed stock. A good g performance of desalter and efficient caustic injection system may help to control the sodium content of the coker feed.

3.6.5 Factors That Can Act As Catalyst for Rapid Fouling of Furnace

Metals such as iron, nickel, and their oxides, chromium and titanium, catalyze the coking reaction. In other words, the tube itself acts as catalyst initiated at 355°C.

A quick increase in temperature is suggested by experts until the metal surface is covered with a thin layer of coke.

3.6.5.1 Feed Interruptions

Feed interruptions is one of the most common causes of shortened run lengths. Unplanned trips result in high tube metal temperatures due to loss of feed flow, which results in coke formation. Heater refractory walls, floor, and roof have a certain degree of heat storage. If feed is lost, even if the fuel valves close and the firing is stopped, the coils will remain hot. Hydrocarbon content remaining in the coil will continue to be heated and coke will lay down in the tubes. Upon loss of feed flow, the coil should be immediately purged with a sufficient quantity of steam to force the oil out of the coil. Availability of a sufficient quantity of emergency steam at all times is of paramount importance.

3.6.5.2 Feed Preheating

Coker fresh feed is typically heated by hot products, and hot pumparound refluxes from the column are then introduced in the column bottom spout and mixes with recycle at the column bottom, and then are subsequently pumped to the convection section of the furnace for further preheat. The feed from the convection section is thereafter introduced in the radiant section for further heating to the coil outlet temperature of 498°C to 505°C.

A typical fresh feed preheat of 280°C to 315°C is common. The temperature of secondary feed is typically 10°C higher than fresh feed after mixing with recycle. Often feed preheat is further increased with a preheat furnace to around 330°C or higher to debottleneck the coking furnace duty constraint during capacity revamps. The preheat increase may be taken up with lot of care and caution, and thorough analysis of the temperature profile along the length of the coil. Under no circumstances should the cracking temperature of around 425°C be reached anywhere in the convection section. The cracking temperature may be only attained in the radiation section where heat flux is higher and coil steam is present to further increase mass velocity and reduce coke lay down. The opinion of the furnace designer or licensor may be sought before deciding on increasing the preheat. The extent of preheat increase needs to be examined and reviewed by an expert.

3.6.5.3 Burners

All burners should have the same fuel pressure at the burner and the same air amount. If the furnace is forced draft, all the burners should have the same damper or register opening. If burners are natural draft, all registers or dampers should have the same setting. It may not be advisable to attempt to operate at excess air levels too low. Especially in forced draft systems with multiple burners, operating at excess air levels too low will almost guarantee that some burners will have insufficient air.

The heater firebox should be monitored daily to ensure all burners are firing properly and all flames look uniform. Any burner whose flame does not looks alright should be taken offline and the burner tips cleaned. The heater should also be checked for the appearance of haze, which is an indication of secondary combustion.

3.6.5.4 Means to Monitor Tube Skin Temperatures

Heaters are typically equipped with tube skin thermocouples. These give an indication of the tube wall temperature at fixed locations in the coil. They may miss hot spots. They may read the wrong temperature if they go bad, which they frequently do. A scan of the coil is made periodically using a digital pyrometer to look for hot spots and to verify the readings from the tube skin thermocouples. Most of the recycle material will vaporize in the coil, increasing the velocity in the heater tube thus reducing the residence time. Decant oil (FCC slurry oil) is highly aromatic and tends to dissolve asphaltenes. Experts' views on design and operating parameters are listed next.

3.6.6 Gist of Actions to Improve Run Length (Suggested by Experts)

1. During a turnaround, the outlet piping from the heater to the coke drum inlet should be cleaned out.
2. Use sonic or ultrasonic meters for feed flow to avoid errors in flow measurement.
3. Coil outlet temperatures should always be the same; .
4. Inspect tubes (with an optical pyrometer or thermal scan) to balance heater firing. Measurement of temperature of the tube supports as well as the tubes may be necessary to verify the heat flux.
5. Ensure the flow meters (pass feed and steam) working well. This might require using ultrasonic meters.

3.6.6.1   Transfer Line Configuration

The transfer line affects the heater operation by the pressure drop associated with the transfer line hydraulics and the added reaction volume. The diameter usually is not that significant. It is the length and number of fittings or flow path in the outlet piping to the coke drums that causes the hydraulic issues and added reaction volume. Too many bends, valves, and fittings can significantly add to the pressure drop in the transfer line.

3.6.6.2   Locations in a Transfer Line that can Cause Problems (According to Experts)

1. Line size changes—asphaltene can settle out on the pipe when the velocity changes in the transfer line.
2. Changes in flow direction—flow patterns in the transfer piping can cause low velocity zones due to eddy current patterns.
3. Erosion at the outlet—spalling will cause erosion problems in not only return bends in the heater but also in the outlet of the heater, i.e., the first elbow of the outlet. The high erosion area needs reinforcing of the pipes. This extra wall thickness needs to be applied to the outside of the tube/pipe so that the line can be easily pigged.
4. Smart or few cleanout locations—cleaning the flanges is critical and should be provided as needed.

3.6.6.3   Ways to Fix Problems with a Problematic Transfer Line

1. Use more steam.
2. Simplify the piping/transfer line; a larger line may help but is not guaranteed.
3. There is some flexibility on the coil diameter but generally 3 to 4 NPS is used. Smaller tubes may not be the best solution for reducing fouling. If done correctly, enlarging the last few tubes can be a good solution but is highly dependent on the feed quality and operation condition. Changing of any of these factors could make this tube arrangement worse.

3.6.6.4   Steam-Air Decoking

1. Difficult and labor intensive; must watch air/steam ratio to prevent over-heating the tubes with accelerated combustion.
2. Not practiced as much.
3. Requires a heater/unit shutdown.
4. Can cause damage to the tubes if the tubes are overheated leading to carburization of tubes.
5. Requires some spalling to remove the bulk of the coke before the actual air burn begins.

3.6.6.5   Pigging or Mechanical Coke Removal

1. Very easy for operations and a the job is outsourced and done by expert agencies.
2. Requires heater/unit shutdown.
3. Can damage the tube if the pig metal studs are improperly used.

(Tungsten carbide studs have a Brinell hardness of 600–800. Most furnace tube materials will have a Brinell hardness of 150–225.)

3.6.6.6   Online Spalling

1. Can be difficult initially. Operation needs to carry out the process carefully.
2. Does not require unit shutdown.
3. Every effective in removing coke in the lower radiant section of the heater but not effective for removing inorganic solids in the convection section of the heater.
4. Risk of plugging the coil if the spall is done too aggressively and/or if there is too much coke in the tubes between 1/4 and 1/2 inch is a good maximum thickness.
5. The return bend in the heater and the 90° bend directly outside the heater need to be thicker to prevent erosion from spalling coke.
6. General practice is to online spall and decoke with pigging when the opportunity is available.

3.6.6.7   Design and Operating Parameters: Firebox

1. Flame impingement will rapidly foul the affected area.
2. Ultralow NOx burners have very small fuel orifices at the burner tip and may plug with time, resulting in an irregular flame pattern.
3. The fuel should be filtered with a fuel gas coalescer/filter.
4. The fuel gas line from the coalescer to the burners may be of SS to lower corrosion and avoid corrosion products getting in burner tips.
5. Steam trace the fuel gas line downstream of knock out drum.
6. In a retrofit the box height needs to be reviewed by experts.
7. Ultralow NOx burners extend the flame and can cause flame impingement.
8. O2 levels may not be controlled too closely (less than 3%). Operating higher O2 (greater than 5%) will help reduce fouling by lowering the tube wall temperature.
9. Higher O2 will shift heat to the convection section and reduce radiant flux. Increasing the O2 from ~3% to ~8% is reported to lower the tube wall temperature by ~75°F. Multiple O2 analyzers are needed in a typical firebox.

3.6.7   Decoking Operation

Decoking methods are spalling, steam air decoking, pigging, and online spalling. For spalling/steam air decoking one block, comprised of a furnace and two coke drums, is isolated. Spalling and steam air decoking is performed while the other blocks continue to operate.

Similarly, during pigging of a furnace, the furnace and associated coke drums are also taken out of service. Thus, one block becomes unavailable for processing of feed and results in loss of processing capacity. During pigging, polyethylene pigs (with cleaning studs on the outside surface) are introduced in the furnace tubes and pushed by high-pressure water. The pig travels through the tubes and scrapes the inside surface of the tubes and removes the deposits. The flow direction is reversed at intervals. This job is normally outsourced and done by agencies with specialization to carry out the task. Often smart pigs are used that display the flow path and position of the pig. The different types of pigs used are shown in Figure 3.14A–3.14B.

The typical setups used by experts are represented in Figures 3.15 and 3.16.

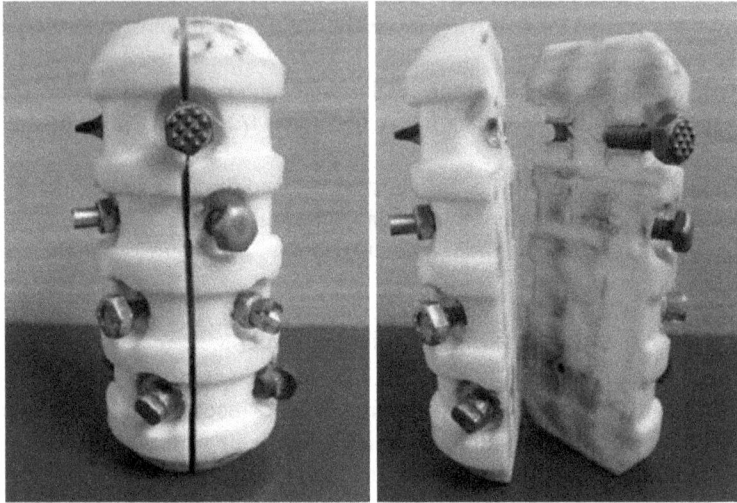

FIGURE 3.14A   Cleaning pig type 1.

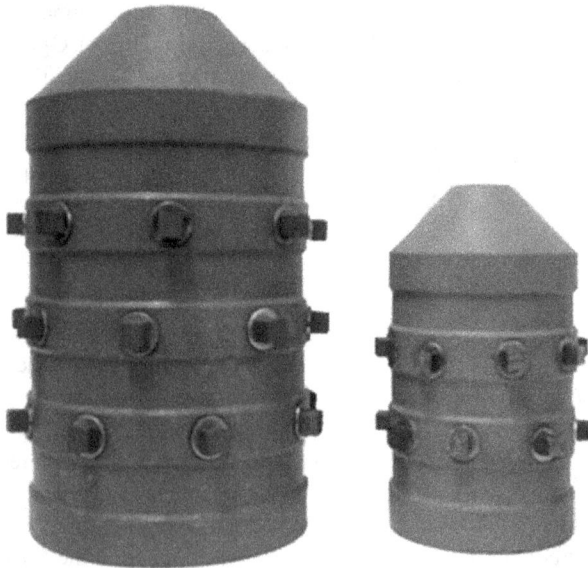

FIGURE 3.14B   Cleaning pig type 2.

FLOW DIAGRAM OF PIG DECOKING

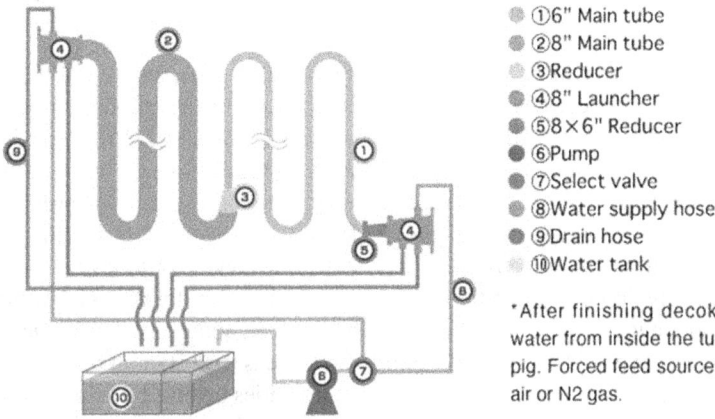

①6" Main tube
②8" Main tube
③Reducer
④8" Launcher
⑤8×6" Reducer
⑥Pump
⑦Select valve
⑧Water supply hose
⑨Drain hose
⑩Water tank

*After finishing decoking, remove water from inside the tube by using a pig. Forced feed source can be either air or N2 gas.

FIGURE 3.15 A flow diagram of decoking with pigging.

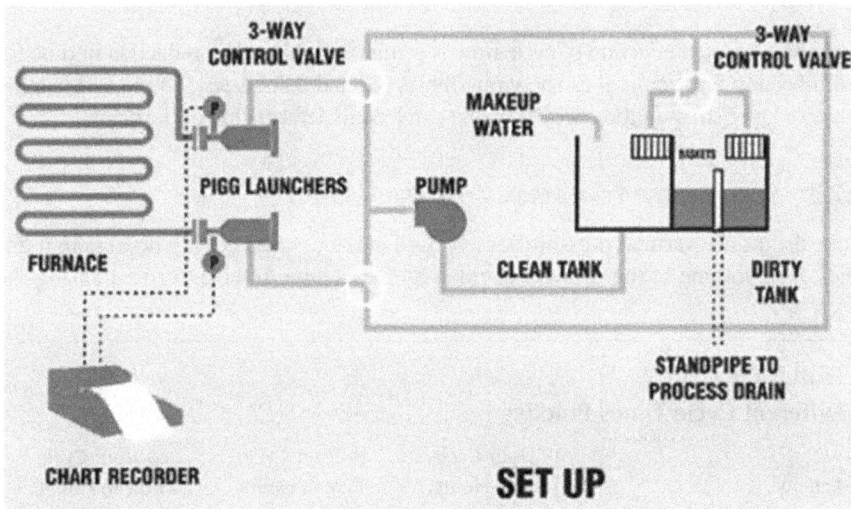

FIGURE 3.16 Sketch of a pigging arrangement.

Often online spalling is practiced to achieve some lowering of the skin temperature without stoppage of the furnace and significant loss of feed processing. During online spalling feed flow through one pass is stopped and a mixture of steam and boiler feed water is injected in the particular tube pass. Meanwhile, feed continues through other tube passes and the furnace is kept in operation. Special care is taken to monitor the skin temperature of the pass being spalled. Normally, the furnace designer performs the procedure.

3.7   COKE DRUM

3.7.1   EFFECT OF CYCLE TIME

In the past, delayed coker units were designed for 48-hour operation: 24 hours for coking and 24 hours for decoking. The heading and unheading operations were done manually or with semiautomatic devices. Feed to the coke drums was introduced centrally through bottom. Modern coker units operate with fully automatic heading and unheading devices (slide valves; delta valves are very common) and feeding provision is from one side (not bottom feeding). This results in a different pattern of flow feed (furnace outlet fluid) inside the coke drums. Further, the usual coking cycle time is reduced to typically 36 hours: 18 hours for coking and 18 hours for decoking. Some units operate with a lower cycle time to process higher feed quantities. The reduction in cycle time is achieved in heading/unheading, and there is a substantial reduction in the activities of vapor heating and cooling. A list of activities involved for the 36-hour and 48-hour cycle times is shown in the Table 3.3. The table shows that lowering of cycle time is primarily achieved by lowering of time for two main activities.

1. Water quenching
2. Drum heating

It can be seen that reduction of cycle time is primarily achieved by reduction in time for drum heating and cooling, as shown in Tables 3.2 and 3.3. Faster cooling and heating causes a lot of stress in the coke drum and can lead to deformation of the drum.

3.7.2   SOME SALIENT FEATURES IN COKE DRUM DESIGN

Coke drums are vertical pressure vessels used in the delayed coking process in refineries. The volume of the drum is decided by the volume of coke formed during the

TABLE 3.3
Different Cycle Times Practiced

Activity	48-Hour Cycle (Time in Hours)	36-Hour Cycle (Time in Hours)	28-Hour Cycle (Time in Hours)
Coking	24	18	14
Switch + steaming	1.0	1.0	0.5
Steam to blowdown	1.0	1.0	1.0
Water quenching	5.0	4.0	4.0
Draining	3.0	2.0	2.0
Cover removal (unheading)	1.0	0.5	0.5
Boring/cutting	4.0	4.0	2.5
Reheading/pressure testing	2.0	1.0	0.5
Drum heating	7.0	4.5	3.0
Total	**48**	**36**	**28**

coking cycle, leaving a top clearance height of 12–14 meters. The height above the highest coke level is provided to arrest carryover of coke fines and foam. The diameter of the drum is decided primarily to limit the vapor superficial velocity of 0.5 ft/sec in the drum cross section to avoid carryover. The conservative coke yield of a low recycle (low pressure) vacuum residue coker is considered as $1.35 \times CCR$ of the feedstock for estimating coke drum height. The actual flow rate of vapor should consider the distillate yield plus recycle, including the coil steam, valve gland, and seat steam. This mass flow is converted to actual cubic meter per hour (ACMH) of flow across the cross section of the drum and is used for computing the superficial velocity then arriving at the conservative diameter of the drum. A coke drum is made of 1.25 CR 0.5 Mo with 3 mm SS 410 lining inside.

3.7.3 Coke Drum and Drum Deformities

The author has seen a drum diameter maximum of 32 ft. Beyond this diameter, the number of coke drums is increased to contain the coke. Normally, two drums are fed by a single furnace. Higher capacity coker units have a higher number of coke drums along with the associated furnaces.

The coke drums undergo primarily three types of deformities:

- Cracks at the weld of the coke drum bottom and skirt support
- Bulging of the surface of the coke drum
- Bending of the coke drum to one side like a banana

3.7.4 Some Key Points of the Coking Cycle and Its Effect on Coke Drums (Expert Opinion)

- The drum grows larger and taller when it is hot.
- It is filled with a lot of hard material as the hydrocarbon cracks and releases vapor and hinders contraction when cooled.
- Some cokes will bind to the wall, and flow channels develop within the coke bed when hot oil is stopped (diverted to another drum).
- Steam is used to remove volatile vapor and flows mostly through channels.
- Water enters from the bottom to cool the coke bed, becomes steam, and flows up the center or outside along the walls.
- The coke drum shrinks in diameter and height as it cools.
- Eventually water can form and accumulate, and the drum becomes cooler.

3.7.5 Fast Quench Issues (As Explained by Experts)

Traditional analysis methods assume a uniform average flow of water upward to remove heat from the coke bed and shell at the same time.

- Coke bed formation determines the path of least resistance for water flow.
- Flow channel area and friction determines the flow.

- Plugging and channel collapse takes place.
- Permeability and porosity is affected.
- Collapse strength of coke matrix is affected.
- Temperature measurements suggest fast quench with flow near the wall being common.
- This creates greater stress in the shell/cladding bond and skirt weld.
- This increases the likelihood that hot zones remain in the coke bed after quenching.

Fast quench problems and damage caused in coke drums (permanent deformation pattern of vessels in cyclic service) include the following:

- The skirt is attached to the cylinder by welding. Welds fail due to low cycle fatigue.
- Cracking often initiates at edges of the weld cap interface to the clad and the crack grows through the base metal.

A fast quench can damage a coke drum in the following areas:

- Skirt attachment weld
- Shell circular seams
- Cone circular seams

A fast quench can harm a drum due to the following:

- Constraint created by components at different temperatures (i.e., thermal expansions)
- Different material properties (yield), expansion, conductivity, diffusivity

Typical drum deformations during steam and water quenching are shown in Figure 3.17.

3.7.6   GIST OF FINDINGS OF THE REASONS FOR DRUM DEFORMITIES (FROM THE EXPERTS)

The skirt-to-shell junction weld on coke drums is susceptible to fatigue failure due to severe thermal cyclic stresses. One method to decrease junction stress is to add slots near the top of the skirt, thereby reducing the local stiffness close to the weld. The most common skirt slot design is thin relative to its circumferential spacing. A new slot design proposes wider slots. Thermal-mechanical elastoplastic 3-D finite element models of coke drums are created by experts to analyze the effect of different skirt designs on the stress/strain field near the shell-to-skirt junction weld, as well as any other critical stress locations in the overall skirt design. The results confirm that the inclusion of the conventional slot design effectively reduces stress in the junction weld. However, it has also been found that the critical stress location migrates from

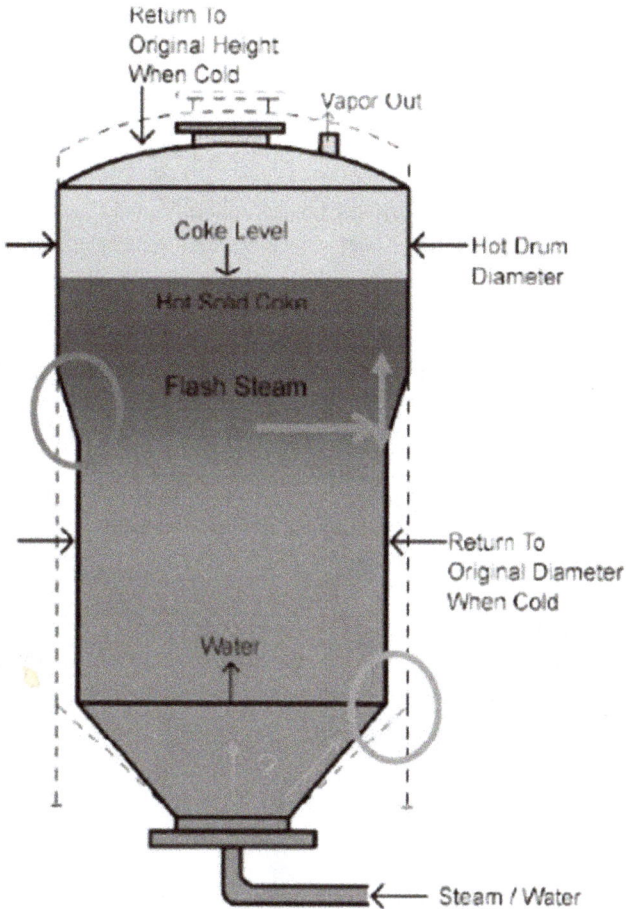

FIGURE 3.17   Typical drum deformations.

the shell-to-skirt junction weld to the slot ends. A method is used to estimate the fatigue life near the critical areas of each skirt slot design. Wwider skirt slots provide a significant improvement on fatigue life in the weld and slot area.

A study done by experts on the mechanical and metallurgical behavior of the skirt and the coke drum weld joint junction system that circumferentially cracked at the skirt junction found the following:

1. The immediate consequence of the skirt junction cracking was the uneven load distribution in the skirt due to the lateral and vertical displacements of the coke drum. This nonuniform load distribution at the weld skirt weld junction produced a high stress zone on the skirt weld joint, which was sensitized the material reducing the ductility. This leads to plastic deformation

that may lead to the local tilt in the coke drum and collapse, causing the interruption of the operation sometimes even a catastrophic failure.

2. The output of the finite element stress analysis of a coke drum with a cracked skirt and misaligned skirt and drum showed the creation of two local zones of high stresses in diametrically opposite positions in the skirt. However, the maximum effective stress was 36.7 ksi, which was lower than the design yield strength of the skirt material, but higher than the maximum allowable stress, prompting necessary modification in the existing design.

3. A wrapped skirt design was developed using a vertical plate along the circumference of the skirt to coke drum shell, which is known to reduce vertical displacement of the skirt joint and tilting. Using a longitudinal weld would reduce the problem of cracking. A *wrapped skirt design and longitudinal weld*, and use of a vertical plate in the drum with the proposed structure repotedly demonstrated more vertical stability to the skirt. A wrapped skirt arrangement is shown in Figure 3.18.

Other designs of drum to skirt joints include the following:

- Lap joint
- Weld bulidup
- Forged Y ring

The weld buildup is known as an improved version, and the forged Y ring is the latest version and known to demonstrate better performance with regard to life of the drum (shown in Figure 3.19 and Figure 3.20).

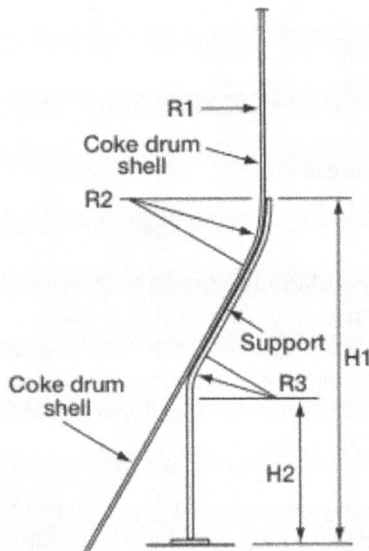

FIGURE 3.18 A wrapped skirt design. (From Cobby W. Steward and Aaron M. Sryke, Chicago Iron and Steel, CBI, 2006, coke drum design.)

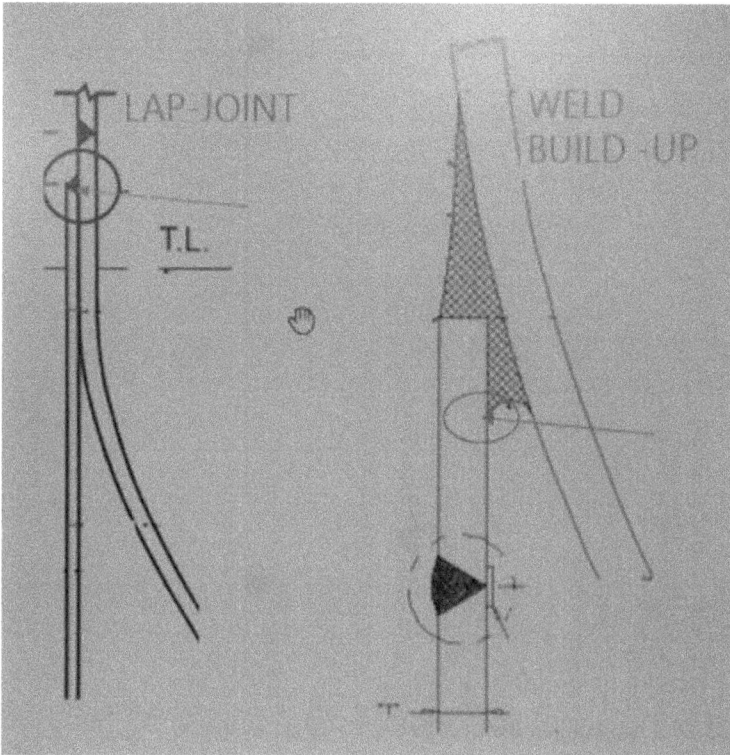

FIGURE 3.19 Two designs of drum to skirt joint. (From Amec Foster Wheeler catalog.)

4. Another way, from the metallurgical point of view, for the bulging effect, which is caused by stiffening and plastic instability, to be reduced is to use a small amount of phosphorus to the very low sulfur composition in the material. This increases the toughness and avoids the cause of temper embrittlement (due to low sulfur content). Vertical plate welding is also known to avoid drum bulging. Coke drum bulging is shown in Figure 3.21.

The author has witnessed similar bulging in coke drums that was strengthened/repaired by filling nickel alloys/alloy overlays in the bulged locations.

Shell cracks can be due to the following:

- Bulging-induced cracks
- Weld cracks
- Combination of the two

3.7.6.1 Summary by Experts

- Bulging is a common and recurring problem in coke drums that is linked to their design, fabrication, and operation.

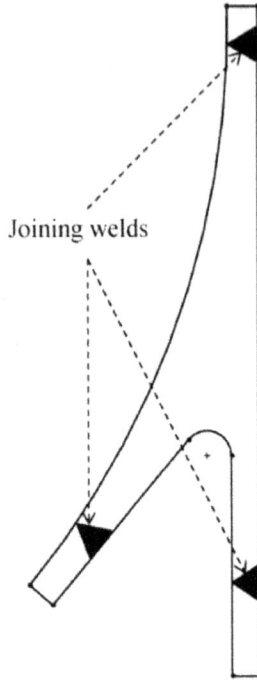

FIGURE 3.20   A simplified sketch of forged Y ring skirt support.

- Stress analysis of bulging is based on the assumption that stress concentration is an indication of severity. The method uses unit or design loads to calculate stress concentrations.
- The Plastic Strain Index (PSI) is a strain assessment method based on failure limits of API 579/ASME FFS and known to correlate well with buldging induced cracking.

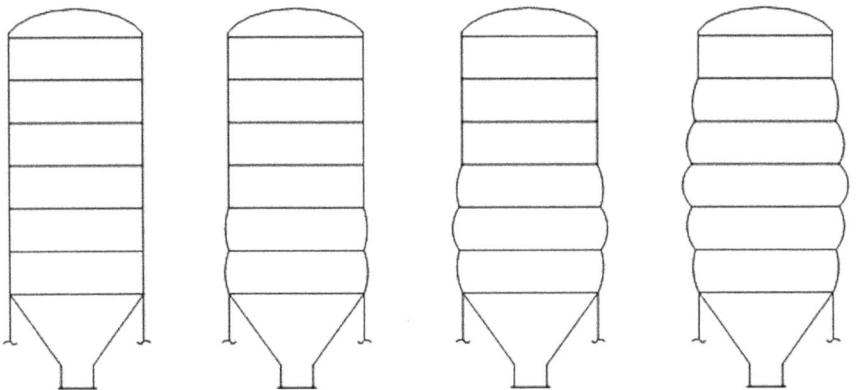

FIGURE 3.21   Coke drum bulging. (From Weil and Rapasky, 1958, the constrained balloon.)

3.7.7 THE COKE DRUM BANANA EFFECT SYNDROME

Coke drum susceptibility to bulging and cracking are a common concern among most refiners operating coker units. Different metallurgies react to coke drum thermal stresses in surprising ways. Depending on metallurgical quality and design, thermal stresses can lead to the "banana effect," typified by the harmful development of a cold spot on one side of the drum and a hot spot on the drum's other side, thereby contorting the drum into the shape of a banana.

In modern coker units, slides valves are normally used for heading and unheading of the drums fitted at the top of drums and bottom of drums. Thus, the feed enters from one side and not traditionally from the bottom (center feed). This may result in a nonuniform flow path for the feed inside the drum compared to a bottom feed system. Subsequently, during the water quenching step when water is introduced in the drum may follow a similar path. Thus, one side of the coke drum may cool faster than the other side producing stress and bending. The author has seen coke drums bend toward the opposite side of the feed entry. If feed enters from the left side to the drum, the drum bending is observed toward the right. The top end of the drum should be free to move and should not get restricted by piping/supports, etc., to avoid any damage. Banana bending is shown in Figure 3.22.

Increasing drum pressure during water cooling can make the cooling more uniform and can reduce the likelihood of deformities and the banana effect of the drum, and likely reduce hot spots in the coke beds and associated hazards. Accordingly, some units have a control valve at the inlet of the BDS (outlet of the coke drum) to increase the drum pressure and while cooling to effect more uniform cooling.

Another common problem experienced in a coking unit is passing and stuck ball valves deployed for controlling operations of drums at different junctures of the

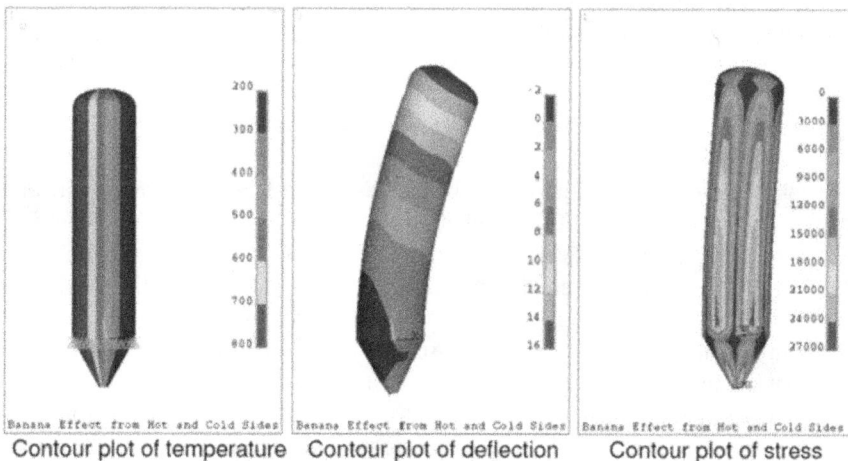

Contour plot of temperature   Contour plot of deflection   Contour plot of stress

FIGURE 3.22 Banana bending. (From Delayed Coker, "Delayed Coker Operations and Reliability" training session delivered by Gary Pitman during the RefComm, Galveston, 2016, refining conference.)

operating cycle. If the valves cannot be brought back to its desired position by external means, the shutdown of the unit cannot be avoided.

3.8   COKE MORPHOLOGY AND EFFECT ON OPERATION

3.8.1   FACTORS INFLUENCING COKE MORPHOLOGY

Delayed coking units deployed for production of fuel-grade petroleum coke are mostly operated to maximize the liquid products. When feedstocks for these units are high in asphaltene, the coke yield is heavily influenced by the asphaltene portion of the feedstock, frequently resulting in the shot coke structure. Shot coke is characterized by green coke forming as small round pellets (beebees or BBs) typically ranging 1.5 to 5 mm in diameter that are frequently loosely bound together. The name shot coke came into existence due to the resemblance of the type of coke with gunshot ammunitions. Frequently, the BBs bind into bigger pieces. While shot coke may look like it is entirely made up of shot, much shot coke is not 100% shot. Interestingly, even sponge coke may have some embedded shot coke. A very low shot coke proportion is often specified for anode grades of petroleum coke, typically less than 5%. Shot coke types generally fall into several categories:

- *Bonded (or agglomerated) shot coke*—This form is characterized by the shot BBs stuck to one another and, mostly, not loosened until broken apart by mechanical or hydraulic forces. These BBs can be formed into a cannonball structure, which when broken will be seen as a mass of bonded shot BBs.
- *Loose shot coke*—Free-flowing shot BBs that form loosely without bonding in the drum. Loose shot BBs are also frequently formed by attrition of bonded shot coke by flow dynamics, especially hydraulic cutting, and by shattering as the coke drops from the drum.
- *Embedded shot coke*—Shot BBs embedded in and often hidden in a matrix of sponge or amorphous coke. The latter mixture can be expected to have poor quench characteristics due to low porosity and permeability.

Shot coke formation can be suppressed by increasing the delayed coker pressure and/ or the recycle, or inducting FCCU high aromatic clarified oil along with feed.

The problem of shot coke is more pronounced when processing residues from heavy crudes. If operating conditions are not properly controlled, the production of shot coke can cause operational issues like high carryover from coke drums to the column, loading the column cleanup section heavily. A shot coke avalanche can cause hot steam/shot coke discharges, which can be a hazard.

While the operating conditions can control the of formation shot coke to some extent, mainly the characteristics of a feedstock is known to govern the formation and is considered as the major contributing factor. Parameters such as micro-carbon residuum (MCR), heptane insoluble asphaltenes (HI), metals and heteroatom content, especially nitrogen content, solids content, and the origin of crude are known to contribute to formation of shot coke. The rule of thumb is that shot coke is likely to be produced when the ratio of MCR:HI is less than 2. Shot coke can disrupt coke drum operations because it has a high bed density and is frequently loosely bound:

- Quenching is never an ideal operation. Water cools a part of hot coke, and at locations the coke may remain unquenched and hot spots can exist in the coke bed. This issue is often seen as more prevalent in cases of shot cokes. Specific operational issues that canoccur from include
- *Blowbacks and eruptions* of steam and coke results when cutting/drilling water comes in contact with unquenched hot coke. After the drum is filled with water (during quenching), a 30-minute soak is advised to reduce hot spots inside coke beds. An additional coke drum level detector near the top can be used to confirm that the coke bed is sufficiently submerged below the quench water level.
- *Plugging of the bottom nozzle* of the coke drum during drum drains resulting in frequent incomplete drum drains and may lead to higher draining time. This can be reduced by lowering the temperature soon after switching.
- *Coke beds can dump* during drilling and trap the drilling head. This can often result in an avalanche of shot coke balls flying with steam when the bottom cover is opened or during drilling/cutting operations. Appropriate installation of the bottom unheading slide valve system may be necessary to provide a fixed chute. The fixed chute has the benefit of containing and diverting shot coke fallouts from the drum and reducing the likelihood of a shot coke avalanche.

An enclosed operator shelter is provided with ventilation, personnel protection, and facilities for sighting the coke drum top head and the exit through coke chute. New units provide a station for remote operations for top and bottom un-heading devices as well as remote coke cutting are now being employed. These remote operations can be monitored through camera.

The appearance of the three types of cokes are presented in Figures 3.23 to 3.25.

FIGURE 3.23   Shot coke appearance.

FIGURE 3.24   Sponge coke appearance.

FIGURE 3.25   Needle coke appearance.

There are rules of thumb (observed by experts) and often followed (for fuel grade coke production). Raising the heater outlet temperature for the last 2 to 4 hours of the coking cycle produces lower VCM coke in the bed top, which is harder, easier to quench, and less prone to create hot spots. Conversely, using lower temperatures following switching and ramping the heater outlet temperature over the coking cycle is also advised by experts and sometimes practiced. Increasing the COT 2.8°C–3.9°C in the final hours of coking is known to reduce the VCM by around 1%. But change of the COT during operation needs to be done very carefully. Some people engaged in coker operation do not support changing the COT. This may be done only to address specific issues and resolve the problems.

3.8.2 Fouling in Coker

Major fouling problems experienced in a coker unit are summarized next.

Feedstock-related fouling of heater tubes is caused by high sodium in feed, high asphaltenes, and incompatible crudes or residue blends. Use of residues derived from cheaper heavy crudes normally results in high asphaltene in the feed. Coker heater fouling is also caused by low velocity in tubes, low velocity injection media rate and/ or wrong location, excessive temperatures (crossover temperature and/or coil outlet temperature (COT), or poor burner operation).

Coker fractionator fouling is caused by low overhead temperatures and consequent deposition of NH4Cl salts (salting) inside the column top section. In the coker fractionator, salting can be arrested by increasing the overhead product endpoint and providing a water wash system at the top of the tower to wash the deposits. The overhead coker naphtha produced may be split in a naphtha splitter to avoid lowering the fractionator top temperature and resultant salting inside column. The splitter top naphtha can be hydrotreated in the naphtha hydrotreater with a silicon guard catalyst ahead of the hydrotreater catalyst (to protect hydrotreater catalyst from deactivation by silicon), and bottoms can be routed along with LCGO to DHDT (normally provided with a silicon guard).

3.9 COKE STORAGE AND HANDLING SAFETY

A potentially catastrophic hazard in coke handling operations is the possibility of a *combustible dust explosion or deflagration*. Petroleum coke is readily able to form dust clouds of finely divided, combustible material. OSHA's National Emphasis Program for Combustible Dust (CPL 03-00-008) emphasizes the need for administrative and engineering controls when a combustible dust layer exceeds 1/32 of an inch (about the thickness of a typical paper clip) on more than 5% of the floor area of a facility or any given room.

In summary, petroleum coke is a material that needs to be stored and handled in a safe manner. Petcoke is stored in many locations, including at the point of production, at an intermediate terminal, and at the end-user facility. At each of these locations, maintaining the optimum moisture content of petcoke is a key practice required to control petcoke fugitive emissions. There are three primary techniques that should be considered to maintain optimal moisture content and control potential petcoke emissions (i.e., fugitive dust) during storage:

1) Pile maintenance
2) Wetting with water
3) Treating with an antidust chemical agent

All the aforementioned points are covered in the Guidance Document for the Storage and Handling of Petroleum Coke by API (API Guidance Document PC1 First Edition, December 2014).

In a nutshell, the coke storage system needs to be maintained to reduce fugitive coke dust and fines, and one effective way is to wet the coke with water.

Transfer of "moist" material can be considered as a measure for fugitive dust control.

As petcoke is hydrophobic, additional moisture can be applied without impacting product quality and helping transfer and handle the material without creating fugitive dust emissions. The moisture content required for fugitive dust control is dependent on the properties of the material and is best understood by the operator engaged in operation. The typical moisture maintained is 6%–9% wt. to control fugitive dust.

3.10   CASE STUDIES

3.10.1   Case Study 1

In a coker fractionator having shed decks in the wash zone, it was observed that the flow of wash oil and quench (at the drum overhead) was drying less than envisaged in the design. The design envisaged a drying ratio as half. To contain the recycle quantity, the wash oil (HCGO) flow to column wash zone was reduced. This may coke the shed decks and adversely affect the HCGO quality. The coke drum outlet temperature was 440°C; after quenching it was cooled to 425°C. A simulation showed the overhead quench liquid and wash oil HCGO should dry at least by half and this dried wash oil would have constituted the recycle (there was no condensation in drum overhead line).

Based on the finding of the simulation, the likelihood of potential heat loss was investigated. Later, the drum overhead outlet line insulation was checked and found to be not at all satisfactory. At places the line was found bare. So, a lot of cooling is inadvertently taking place and the vapor is entering the column at a much lower temperature than 420°C. It may be pertinent to mention that there was no temperature indication on the line at the inlet to the column. The temperature at the column inlet was estimated by measurement of the surface temperature (placing insulation on top of temperature-measuring element). It was understood that the wash oil was not drying as expected, as the vapor was admitted at very low temperature to the column. operators reduced the wash oil flow to contain the recycle. This may not be good practice for HCGO quality and can lead to heavy coking on shed decks in the wash zone. The insulation was repaired during a shutdown and drying of wash oil achieved. It may be pertinent to mention that some designers reduce thickness of insulation of the drum overhead line to allow some cooling and promote condensation to take care of loss of quench oil during an emergency. In the unit, all the slops generated in the refinery was processed through coker, in addition to slop oil generated from its own BDS tower.

3.10.2   Case Study 2

A higher pressure drop in the top ten trays were observed in a coking unit fractionator. There was no water wash facility in the column to wash the NH4Cl salt deposits. The unit had three side cuts with side strippers, viz., LCGO, HCGO, and HHCGO, with respective pumparound refluxes. The HCGO product and HHCGO pumparound reflux were deployed to preheat the feed. HCGO pumparound was cooled only by steam generation. So, HCGO pumparound duty was reduced to some extent to increase the reflux ratio in the column to reduce partial pressure of ammonia and HCl to reduce the salting temperature. The top temperature of the fractionator was also increased to the extent possible. Gradually, pressure drop normalized. It may be always advisable to maintain a high end point of coker naphtha and deploy a coker naphtha splitter to avoid the problem of higher pressure drop due to salting in the column top section.

3.10.3   CASE STUDY 3

In a four-drum coking unit inferior quality of HCGO was noticed along with higher pressure drops in drum overhead lines to fractionator. A sour water balance was performed to identify the root cause. The difference between sour water flow from main fractionator receiver and coil steam flow, wash water flow (upstream to overhead condenser) indicated use of large quantities of purge steam and steam for valve gland purging. The higher steam flow was not reflected in the indication of flow meter. The pressure drop through the overhead lines were also computed through simulation of a line (36 inch) with flow of the vapor products (with overhead quench), coil steam, and steam to valve glands (assuming an inch thick coke layer inside the line). The computed pressure drop was found lower than observed. This indicated higher coking inside the drum overhead line. Based on the observations, restricted orifices (ROs), total steam flow meter in the purge stream lines were checked and good numbers of ROs were found missing and steam flow indication found faulty. It was suspected that higher steam flow may be loading the wash zone of the fractionator. A simulation rating of the fractionator also corroborated higher loading of wash zone. Higher vapor velocity may be also contributing to higher pressure drop in the drum overhead vapor lines. However, higher pressure drop in vapor line can also result from higher thickness of coke in the line than considered for computation of pressure drop. The purge steam/gland steam to the valves were regulated, and the overhead quench restarted mostly with HCGO and heavy bottoms from the BDS column to increase the dew point of the vapor to induce condensation, if possible. The situation improved and the overhead line pressure drop remained under control until the planned shutdown of the unit.

3.10.4   CASE STUDY 4

A big coking unit (having six coke drums and three coking furnaces) was to be revamped for a capacity increase by around 25%. The revamp primarily envisaged a reduction of the decoking cycle time from 20 hours to 16 hours, the addition of a preheat furnace to contain the duty requirement of coking heaters, column internals change, the addition of a fractionator overhead condenser, a change of some pumps, and an increase in sizes of pipelines in some circuits.

The refinery changed the fractionator internals [fixed opening valve trays changed to trays with fixed valves tray having more open area (antifouling type)] during one shutdown. Later, the crude mix to the refinery has undergone a change producing residue with higher CCR due to nonavailability of earlier crude mix. The higher CCR residue would have resulted in higher coke make and reduced the clearance above the maximum coke height. Thus, the revamp was dropped. But the column internals remained same. It was found that when one block comprised of one furnace and two coke drums was taken out for furnace tube pigging, severe overlap results between adjacent products . This became more pronounced during vapor heating of a drum as a result of abstraction of more vapor from the fractionator feed resulting in severe mixing of all products due to dumping and resultant inefficient separation.

The total pressure drop across the column reduced, corroborating dumping of liquid from trays at low vapor velocities. In the unit the HCGO pumparound was not deployed for feed preheating (was cooled by steam generation) in design. The HCGO pump around duty was reduced, the stripping steam to strippers was increased limited by the overhead condenser duty and reflux temperature,and the column pressure was slightly reduced to increase vapor volume and velocity in the column to reduce dumping. The trick worked. This practice became a standard operating procedure when operation is reduced to four drums during outage of one block.

3.10.5 CASE STUDY 5

A very low recycle coker was in operation for 36 hours total cycle time. In between, the feedstock underwent a change with induction ofmuch higher proportion of cheaper heavy crudes in the refinery crude mix. The residue quantity increased and became much richer in asphaltene (MCR/HIS proportion reduced below 2). This resulted in difficulty in coke cutting and water draining. Both operations needed more time than usual, resulting in an increase in decoking cycle time. The fresh feed to the unit had to be reduced to handle the situation. Later, the furnace outlet temperature was reduced by 2°C–3°C which somewhat reduced the coke cutting time (did not affect the VCM). But water drainout continued to be a problem necessitating more time for draining. It was difficult to maintain the decoking cycle time of 18 hours. Later, the temperature was reduced by 2°C again, following switching to a new drum and maintained low for at least half an hour in an attempt to reduce the shot coke formation and accumulation at the bottom. The procedure followed worked. The situation improved and the throughput could be normalized and operation continued until the inventory of high asphaltene residue was consumed. Later, the operation team tried a normal residue as feed with lower asphaltene (having MCR/HIS ratio of 2) switched to normal residue as feed. At times when the feed becomes inferior the same operation philosophy can be followed and problem of drum draining can be managed.

BIBLIOGRAPHY

API. *API Guidance Document PCl*. 1st ed., Washington, DC: API Publishing Service, 2014.
Cobby W. Steward and Aaron M. Sryke. *Chicago Iron and Steel*. CBI, 2006. *Coke Drum Design*.

Darius Remsat. Reliability vs. recovery for delayed coking fractionators. *PTQ*, Q2, 2009. Koch-Glitsch, Canada, pp. 109–113.

Gary Pitman. These identified thermal stresses are among a variety of topics relevant to Coker process safety and reliability. In *Galveston 2016 Refining Conference, Delayed Coker Operations and Reliability Training Session*. Galveston, TX: Ref.Comm, 2016.

Mahmod Samman, Ediberto B. Tinoco, Fábio C. Marangone, and Hezio R. Silva. *Bulging Assessment of Coke Drums*. Galveston, TX: Coking.com Conference, May 2013.

Publications of Coking.com, Bellingham, WA: RefComm.

Richard Boswell. Fast quench problems and how they damage coke drums. Presented in *Coke Drum Reliability Workshop*, Galveston, TX. Stress Engineering Services, Inc., March 24, 2009.

S. S. Deshmukh and V. G. Arajpure. Analysis of the failure of a delayed coke drum and cracked skirt weld. *International Journal of Scientific Research and Review*, 7, no. 3, 2018, pp. 367–373.

4 Hydrotreating/ Hydrocracking

4.1 INTRODUCTION OF THE PROCESS AND BASIC FUNCTIONS

The process of hydroprocessing or hydrotreatment is required for removal of con-taminants from crude oil fractions. During hydrotreatment, heteroatoms (sulfur, nitrogen, and metals) are removed from crude oil fractions and unsaturated hydro-carbons get saturated. During hydrocracking, heavier hydrocarbon molecules are also cracked into lighter molecules. Both of these reactions occur in the presence of catalyst at higher hydrogen pressure and elevated temperatures. The general scheme of a hydrotreater is shown in Figure 4.1A and the typical flow scheme of a diesel hydrotreater in Figure 4.1B.

The process essentially comprises of the hydrotreating reactor, feed–effluent heat exchangers, reactor effluent air cooler (REAC), separator, recycle gas amine contac-tor, recycle gas compressor (RGC), makeup hydrogen compressor (MUG), and strip-per or fractionator. The feed is typically filtered and mixed with recycle hydrogen, preheated in feed effluent heat exchangers, further heated in the reactor furnace, then fed to the reactor. The reactor effluent exchanges heat with feed and is further cooled in the REAC and fed to the cold separator. The hydrogen rich gases from the cold separator are typically amine washed and recycled to feed and to provide quenches between catalyst beds through the RGC. The liquid from separators is fed to the stripper or fractionation section.

The chemistry of hydroprocessing is quite complex due to the varieties of con-stituents present in the feed. The main reactions taking place during hydroprocess-ing are

- Sulfur removal (hydrodesulfurization, HDS)
- Reduction of nitrogen (hydrodenitrification, HDN)
- Metal removal (hydrodemetallization, HDM)
- Saturation of aromatics (hydrodearomatization, HDA)
- Oxygen removal (hydrodeoxygenation, HDO)
- Hydrocracking and isomerization

4.1.1 HYDRODESULFURIZATION (HDS)

The reduction of sulfur in the final product is necessary, primarily from an envi-ronmental point of view. Another important reason for sulfur removal is that it can act as a poison for catalysts for the plants downstream and thus can adversely affect the performance of the plants. During hydrodesulfurization, sulfur compounds are

DOI: 10.1201/9781003268246-4

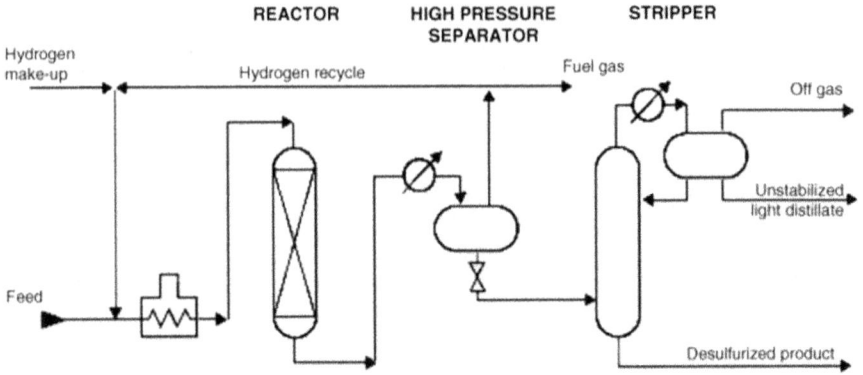

FIGURE 4.1A   General scheme of a hydrotreater.

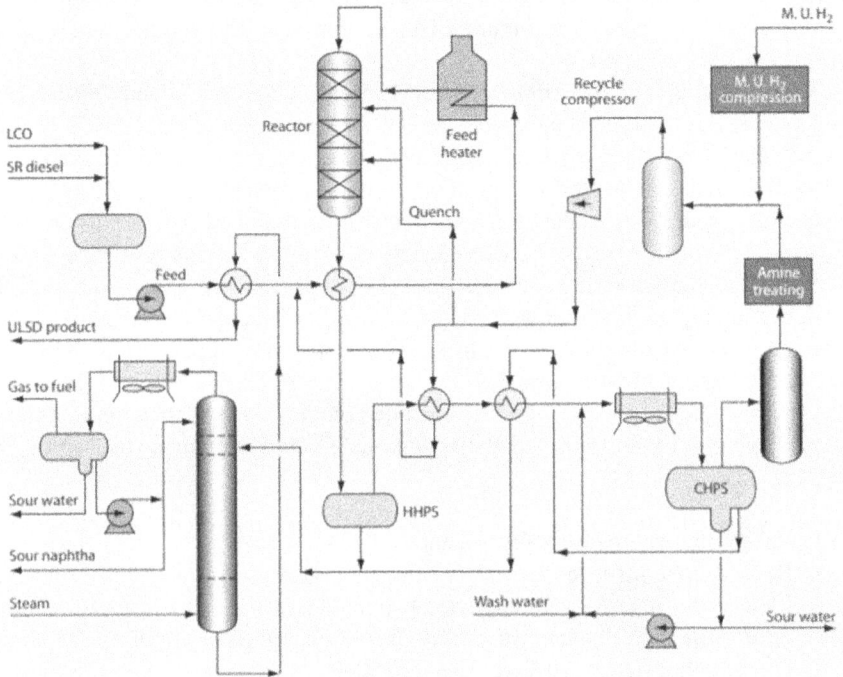

FIGURE 4.1B   Typical flow scheme of a diesel hydrotreater.

converted to hydrogen sulfide and removed from the feedstock. Carbon–sulfur bonds in alkyl sulfides and polysulfides are relatively weaker and can easily be removed during hydrodesulfurization. But the sulfur present in thiophenes and aromatic ring systems is difficult to remove. Typical sulfur-bearing compounds in petroleum stocks are mercaptans, sulfides, thiophene, benzothiophene, dibenzothiophene, and thioxanthene, as shown in Figure 4.2A.

Sulfur compounds

Thiols (mercaptans), sulfides and disulfides	R-SH	R-S-R'	R-S-S-R'

Thiophenes, benzothiophenes and dibenzothiophenes

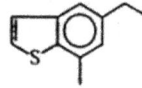

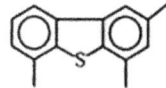

FIGURE 4.2A Typical sulfur compounds present in petroleum feedstock.

Hydrodesulfurization can proceed via two pathways, as shown in Figure 4.2B: hydrogenolysis (direct desulfurization, DDS) and partial hydrogenation of the aromatics prior to desulfurization (HYD). During hydrogenolysis both carbon and sulfur bonds in the ring compounds are substituted by carbon hydrogen bonds, which leads to ring openings.

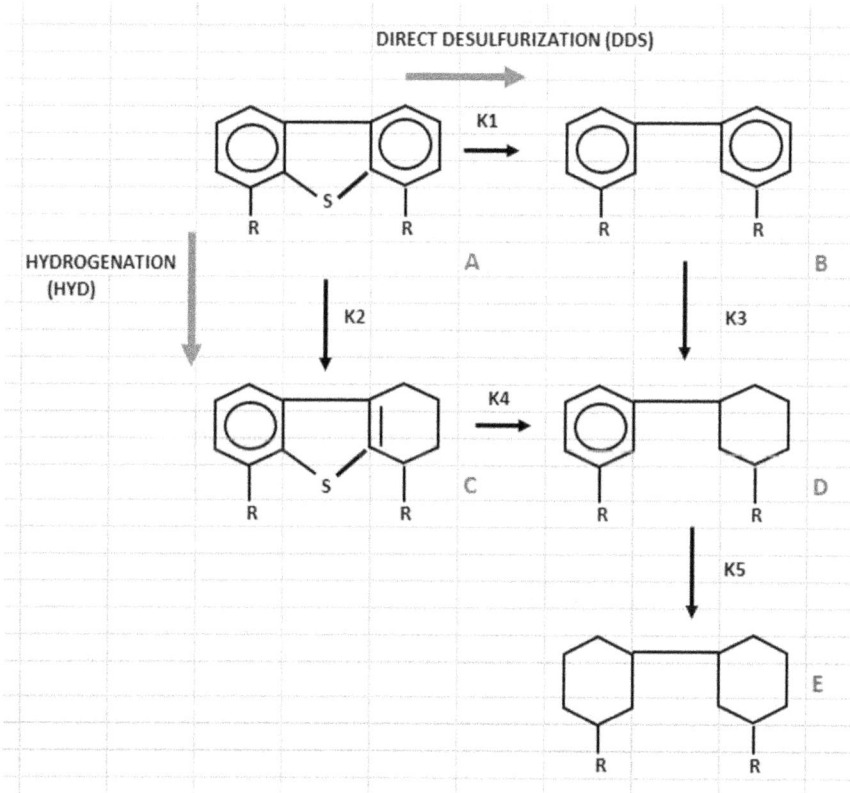

FIGURE 4.2B Reaction pathways for desulfurization (HDS).

The other pathway starts with the partial saturation of one of the aromatic rings, which may lead to tetra- or hexahydro intermediate product. The intermediates then undergo the desulfurization process.

Consumption of hydrogen is much lower in the hydrogenation pathway (DDS) compared to the pathway of partial hydrogenation (HYD). The reaction pathway is dependent on the catalyst deployed. Cobalt–molybdenum catalyst favor the hydrogenolysis route, while nickel–molybdenum catalyst favor the intermediate saturation of the aromatic rings followed by the desulfurization process. The reaction pathways are shown in Figure 4.2B.

Hydrogen sulfide formed during the HDS reactions needs to be removed from the recycled hydrogen stream since it has a large inhibiting effect on both HDS and HDA reactions. Thus, for deep desulfurization of petroleum fractions amine wash of the recycle gas stream is necessary.

4.1.2 HYDRODENITROGENATION/-DENITRIFICATION (HDN)

The most commonly found nitrogen compounds in the heavy petroleum stocks (distillates) are indoles, carbazoles, quinolines, pyridines, acridines, aniline, pyrrole, and pyridine (Figure 4.3). The nitrogen content of the heavy distillates usually varies

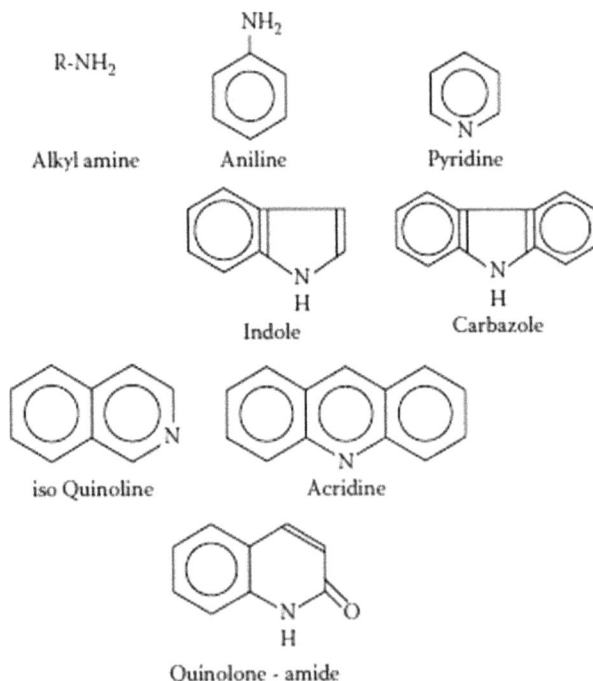

FIGURE 4.3 Most common nitrogen compounds in the feed to hydrotreaters.

based on the crude source but usually lies in the range between 300 and 3000 ppm. Nitrogen compounds are responsible for the development of color and lacquer formation during its use as lubricant.

Moreover, these compounds can contribute to a loss of catalyst activity due to the adsorption on the active sites on the catalysts. These nitrogen compounds may lead to a huge production loss if not properly taken care of. Some nitrogen compounds (nitrogen associated with six-membered rings) are known to be more basic in nature and primarily contribute to catalyst deactivation.

Usually, the process conditions are more severe for the HDN reactions than the HDS reactions. Usually HDN needs a higher temperature than HDS. During HDN reactions, the aromatics get saturated first prior to nitrogen removal. For the HDN reactions to proceed and to remove nitrogen, it is necessary to saturate the five-membered rings in the pyrrole derivatives and six-membered rings in the pyridine derivatives. Thus, the nitrogen removal reactions consume more hydrogen than the sulfur removal reactions. Saturation of the aromatic rings is a reversible process, and it is necessary to have a very high hydrogen partial pressure to drive the reactions forward and lower temperature favors saturation of aromatic rings. Thus, a very high reactor temperature can lower conversion of the HDN reactions. The reactions of HDN are shown in Figure 4.4.

Nitrogen compounds present in a typical petroleum stock can be classified into three categories, namely, neutral organo N-compounds, basic organo N-compounds, and weakly basic organo N-compounds. The most commonly occurring nitrogen compounds are the pyrrole and the pyridine derivatives shown in Figure 4.5.

(A) Aromatic hydrogenation

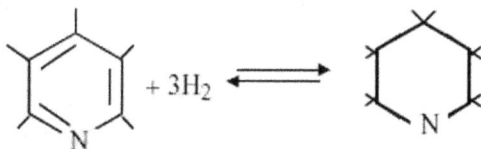

(B) Hydrogenolysis

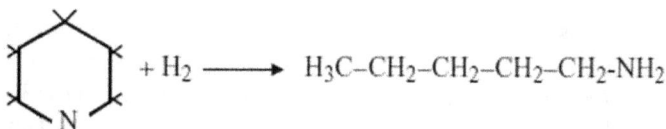

(C) Denitrogenation

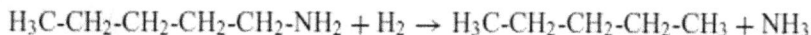

$$H_3C\text{-}CH_2\text{-}CH_2\text{-}CH_2\text{-}CH_2\text{-}NH_2 + H_2 \rightarrow H_3C\text{-}CH_2\text{-}CH_2\text{-}CH_2\text{-}CH_3 + NH_3$$

FIGURE 4.4 HDN reaction mechanism: aromatic hydrogenation followed by hydrogenolysis and denitrogenation.

Nitrogen compounds
Pyrrole, indoles and carbazoles

Pyridine, quinolines and acridines

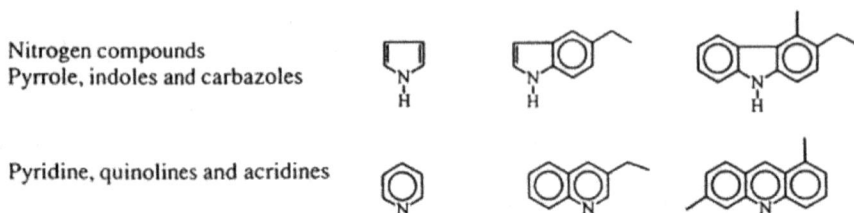

FIGURE 4.5   Nitrogen compounds pyrrole and pyridine derivatives.

4.1.3   HYDRODEMETALLIZATION (HDM)

The majority of the metallic impurities are present in the petroleum fractions as organometallic compounds and are found in ppm or ppb levels. These organometals can be converted into their respective metal sulfides in the hydrotreater. These metals can lead to catalyst deactivation once deposited on the catalyst pores and permanently deactivate the catalyst. The most commonly occurring metals are arsenic, mercury, nickel, vanadium, and silicon. In the guard bed designed and placed upstream of the active catalytic section, most of the metals are trapped within the guard catalyst pores. Thus, their harmful effect on active catalyst can be avoided. The catalyst used in guard beds has mild hydrogenation capabilities that help in hydrogenation to free the metal and trap them inside the pores.

4.1.4   AROMATIC SATURATION (HDA)

Aromatic saturation is necessary to improve the quality of products from a hydrotreater like smoke point of aircraft turbine fuel (ATF) or superior kerosene (SK), and to improve the cetane number of diesel. Aromatic saturation is also required to remove impurities like sulfur and nitrogen. Aromatic saturation is also accomplished to some extent to capture the metals from organometallic compounds. Aromatic saturation is carried out by the metal function of the catalyst, is a highly exothermic reaction, and consumes an appreciable amount of hydrogen.

4.1.5   HYDROCRACKING

Hydrocracking is a catalytic cracking process whereby heavy hydrocarbons are broken down into simpler and smaller ones. During hydrocracking both hydrogenation and cracking take place together. The cracking reaction is known to take place through formation of carbenium/carbonium intermediates. During catalytic cracking, in the presence of the acid function of catalyst, bonds are being broken asymmetrically. This heterolytic breakage of bonds yields carbocations and hydride anions. These ions are very unstable and undergo chain arrangements, C-C scissions, and intra- and intermolecular hydrogen transfer processes. These reactions follow self-propagating chain mechanism. Finally, these reactions are terminated by radical or ion recombination reactions. Hydrocracking requires a large quantity

of high purity hydrogen, high temperature, and high pressure and presence of catalyst. Typically, all the olefin produced get saturated with hydrogen. This addition of
hydrogen makes it an overall exothermic reaction. Recycle cold hydrogen streams
are provided to control the inlet temperature to the subsequent beds as quenches.

4.2 CATALYST FUNCTIONS AND MECHANISM OF DEACTIVATION

Hydroprocessing catalysts consist of an active phase (generally combination of Ni, Co,
Mo, and W) dispersed and supported on a silica–alumina carrier. Hydroprocessing
catalysts are active in their sulfided form. Typically, the catalysts are sulfided in situ
by injection of a sulfiding agent (dimethyl disulfide, DMDS). At times, presulfided
catalysts are used to reduce the time required for in situ sulfiding. The catalyst has
a high surface area and the average pore size is known to vary between 75 and 300
Å. The catalyst are normally bifunctional having metal sites for hydrogenation and
alumina and silica as acid sites that contribute to cracking and isomerization.

4.2.1 CATALYST SULFIDING

Catalyst sulfiding is done at a fixed pressure (hydrogen partial pressure) with the
recycle gas compressor in operation and the reactor furnace lighted to heat up the
catalyst. Gaseous phase sulfiding and liquid phase sulfiding are commonly deployed.
In liquid phase sulfiding, oil essentially, straight run gas oil low in aromatics, sulfur,
and nitrogen is fed by the feed pump and circulated through the reactors, and along
with recycle gas circulated by the recycle gas compressor (RGC). DMDS is typically
injected at the feed pump suction at a controlled rate. The reactions during sulfiding
are presented next.

The reaction of decomposition of DMDS to H_2S can be written as $CH_3 S SCH_3$
(DMDS) $+ H_2 \rightarrow 2H_2S + 2CH_4$. The reaction typically occurs at 200°C–230°C. This
is an exothermic reaction and is identified with an increase of temperature in the
upper part of the first catalyst bed indicating decomposition of DMDS. The reaction of sulfiding of metal is $6NiO + 3H_2S \rightarrow 3Ni_2 S_3 + 3H_2O$. This reaction is also an
exothermic reaction and can be marked with a rise in temperature of the beds from
the top bed to bottom bed sequentially. Initially, exothermicity can be seen in the
top bed and once the top bed sulfiding is complete it starts in subsequent beds and
so on. Water appears in the cold high-pressure separator (CHPS) and conforms the
sulfiding of catalyst. When all the beds are complete as indicated by no temperature
increase across catalyst beds, then the reactor inlet temperature (RIT) is increased
gradually to 380°C in steps for further absorption of H2S by the catalyst, known
as high temperature soaking and is continued until exothermicity continues and
water continues to appear in the separator drum. Sulfiding is considered complete
when no exothermicity is experienced across the catalyst beds and water formation
discontinues.

During sulfiding, recycle gas H2S content is monitored and never allowed
to drop too low (below 0.5% by volume). The H2S percentage in recycle gas is
maintained by adjusting the DMDS dosing. During sulfiding care is always taken

to ensure that metal oxides get directly converted to metal sulfides by reacting with H2S produced from decomposition of DMDS. The metal oxides should never get converted to metals by reacting with hydrogen to avoid an adverse effect on catalyst activity. The hydrogen partial pressure is controlled during sulfiding to arrest the reduction of metal oxides. There is a likelihood of temperature runaway during sulfiding that can cause activity loss of catalyst during sulfiding. In case of runaway, furnace firing needs to be stopped, DMDS injection needs to be discontinued and the unit may need to be depressurized to control excessive reaction and temperature excursion.

4.2.2 Deactivation of Hydroprocessing Catalyst

Lighter feeds contain less contaminants, consequently the deactivation of the catalyst is minimal and the process can be operated for longer periods of time before catalyst replacement becomes necessary. The lighter petroleum stocks like naphtha and kerosene contain less sulfur, nitrogen, aromatic compounds, and metals. The heavier petroleum distillates like diesel, light coker gas oil, vacuum gas oil (VGO), heavy coker gas oil (HCGO), and residues contain a large amount of sulfur and nitrogen-containing compounds and aromatics and metals. As a result, reactors processing heavier stocks are often operated at a higher pressure and temperature. Sometimes, a longer residence time is provided to bring down the impurity to an acceptable level. A very high operating temperature over a long period of time can result in catalyst deactivation. Catalyst deactivation can be short term, or long term and permanent. The reasons for catalyst deactivation can be coking or fouling, sintering, poisoning by metals, and mechanical failure.

4.2.2.1 Coking or Fouling

Lay down of coke on catalyst is the most common factor that contributes to catalyst deactivation. Coke is a carbonaceous substance that deposits on catalyst and causes loss of pore volume and decrease in the number of active sites. Coke deposition in the pores of fresh catalyst is very rapid during the start of run (SOR) due to high activity of fresh catalyst. Deposition of coke has a strong adverse effect on the hydrogenation and on the cracking activity. Coke can be formed during hydroprocessing of almost all feeds, and coke formation increases as the molecular weight and boiling range of the feed stock increases. Polymerization or polycondensation are the main reactions which contribute to coking. Coke precursors can be present in the feed or can be formed during hydroprocessing reaction. Formation of coke precursors occur generally toward the end of run of operation when the catalyst get deactivated. Asphaltenes and heavier species have a tendency to precipitate on the catalyst surface, which can lead to coke formation and catalyst deactivation. Resins are polar molecules that stabilize asphaltene molecules and prevent the precipitation of asphaltenes. During hydroprocessing the structure as well as the functionality of resin molecules can change which may lead to precipitation of asphaltenes. Coke can also be formed from various oxygen-containing species. Coking increases with increasing surface acidity of the catalyst and higher precursor basicity.

4.2.2.2 Sintering

Sintering is caused by exposure of the catalyst to high temperature (a temperature over 500°C). Sintering occurs due to agglomeration and growth of metal deposition on the catalyst surface or inside pores by heat and pressure. During sintering, atoms diffuse toward each other, get fused, and form a solid mass of material. Can happen in case of runway of temperatures.

4.2.2.3 Mechanical Deactivation

Catalyst can get deactivated due to mechanical failure due to the following:

1. Crushing of the catalyst pellets due to a load
2. Attrition (reduction of size and formation of fines)
3. Erosion due to high fluid velocity

Mechanical failure results in the loss of catalyst activity and can induce channeling and increase in pressure drop in the bed of the catalyst.

4.2.2.4 Poisoning

Common hydrotreater catalyst poisons are nitrogen compounds, oxygen compounds, and metals. Contaminants can be adsorbed reversibly and irreversibly. When the poison is adsorbed reversibly, catalyst activity can be regained by removing the poison. Sulfur and nitrogen compounds can be a temporary poison. Metals compounds can be considered as permanent poisons. Another type is quasi-permanent poison. It lowers catalytic activity by being adsorbed on active sites and appears to be permanent due to its very slow desorption rate. It normally takes a long time to get rid of the quasi-irreversible poisons.

4.2.2.4.1 Poisoning by Nitrogen Compounds

Organic nitrogen compounds present in feed can act as temporary poisons . The effect of deactivation by nitrogen is very strongly exhibited during operation. This reduces sulfur removal in hydrotreaters and conversion in hydrocracking. Certain nitrogen (and other polar) compounds are also known to cause distillate product color and product instability. Lube color and its stability is improved in lube oil hydrofinishing with removal of nitrogen compounds (typically operated around 60 Kg/cm2 pressure in the separator). The color of the unconverted stream of a hydrocracker was found almost milk white by the author after the first startup of a high-pressure hydrocracker unit using VGO as feedstock. The improvement in color was attributed to a very high degree of denitrification of the feed in the presence of fresh catalyst deployed in the unit and a high recycle ratio maintained.

4.2.2.4.1.1 Examples of Nitrogen Poisoning The author would like to share his unique experience of nitrogen poisoning on a couple of occasions. A two-stage hydrocracker unit was designed for 100% conversion. The first stage design conversion was around 40%. The first stage was designed to remove all the sulfur and nitrogen for the second stage to get a feed free from impurities. The unconverted from the first stage is fed to the second stage for further conversion and aromatic saturation to improve the product qualities.

In a two-stage hydrocracker unit no conversion was achieved in the second stage even by increasing the weighted average bed temperature (WABT) or catalyst average temperature (CAT) of the second stage higher than the design. Although the design conversion of around 40% was achieved in the first stage. The absence of conversion in the second stage was corroborated by no temperature rise in reactor beds and no hydrogen consumption in the second-stage reactor. The first-stage effluent nitrogen was analyzed and found to have less than 1 ppm ensuring that the first-stage catalyst is performing fine and lowering nitrogen to the desired level. The absence of conversion in the second stage continued to be a mystery. Then, the second-stage feed sample was analyzed from the suction of the second-stage high-pressure feed pump and that showed a result of nitrogen was found at around 3–4 ppm. The increase in nitrogen was detected across the low-pressure pump delivering unconverted to the second-stage high-pressure feed pump suction for feeding to the second-stage reactor after feed preheat. Raw feed was used as sealing fluid in the mechanical seal (having a single seal) of the low-pressure pump. The increase in nitrogen was diagnosed as due to the ingress of N2-bearing sealing fluid to the pump suction. The sealing arrangement was changed to self-sealing. The problem gradually disappeared. It took almost two to three days in removing the nitrogen compounds with nitrogen lean feed to attain conversion of around 60% in the second stage. This indicates how a small nitrogen ingress can cause a problem of this magnitude.

In another instance, in a VGO hydrotreater conversion dropped below the design of 30% for no apparent reason and was not picking up with an increase in WABT or CAT. The unit used to process part HCGO from coker and virgin VGO from two crude distillation units. One crude unit processes nitrogen-rich crude with low sulfur. The other processes high-sulfur crude with low nitrogen. The hydrotreater unit had two parallel reactor trains terminating at the hot high-pressure separator (HHPS). The feed to the unit was analyzed initially for flash point and then for nitrogen and found to have low flash point but the nitrogen content was found much higher than design. Then feed VGO stream coming from the crude unit processing high-nitrogen crude was analyzed and a low flash point and high nitrogen were found. So, a leak in the VGO/crude heat exchanger was suspected resulting in crude ingress to VGO, adding nitrogen and metal to VGO feed to the hydrotreater. The leaking heat exchanger was identified and isolated. The unit conversion slowly picked up and became normal. Nitrogen compounds can block the HDS sites as well and by removing these nitro compounds, the performance of HDS and conversion can be restored.

4.2.2.4.2   Other Inhibiting Agents

Oxygenated compounds may inhibit HDO reactions. This inhibiting effect can be reduced by maintaining a considerable amount of H2S partial pressure so that the catalyst can continue to remain in sulfided form. Otherwise, there is a risk for catalyst surface modification by oxygen, which can contribute to a loss in catalyst activity. The poisoning effect of water on HDS and HDN has also been found and can be significant with the increased concentration of oxygenated molecules in the feed. This inhibiting effect is quite significant during hydrotreatment of bio-oil and may not have any significant effect on hydrotreater catalyst deployed for hydrocarbons in refineries.

4.2.2.4.3 Poisoning by Metals

The hydrotreater feed can contain a lot of contaminants, particularly the heavier petroleum stocks, and can poison and deactivate the catalysts. Typically, guard catalysts are provided at the upstream of HDS, HDN, and cracking catalyst to protect them from deactivation by the contaminants. But once the guard beds get saturated/exhausted, the contaminants can get passed to the active catalyst located downstream and can cause irreversible damage/deactivation of the catalyst. The other common contaminants are described in the following sections.

4.2.3 CONTAMINANTS IN HYDROTREATER FEED

Common metals contaminant in hydrotreater feed are Silicon, Arsenic, Sodium, Calcium, and Phosphorus, Iron, Ni and Vanadium.

4.2.3.1 Silicon

Silicon (Si) may be the most common catalyst poison encountered in distillate hydrotreater feeds. The common source of silicon is from a delayed coker unit that uses an anti-foam agent based on polydimethylsiloxane to suppress foaming in the coke drums. The complex is known to break down in the coking process to lighter molecular weight components and end up primarily in the naphtha range. However, small quantities have also been reported in the kerosene and diesel fractions. As a result, silicon contamination is a major concern in units treating coker naphtha. As the silicon builds up on the catalyst, it begins to restrict the catalyst pores and eventually blocks access of feed to the active sites. This phenomenon is often referred to as pore mouth plugging.

4.2.3.2 Arsenic

Arsenic (As) is found in many crudes, including some crudes from West Africa and Russia as well as in many synthetic crudes, particularly from Canada. Arsenic is believed to bind with the metal sulfide sites, and in particular the active nickel on the catalyst forming nickel arsenide. This has a dramatic impact on lowering catalyst activity. Shale oil-based hydrocarbons (from Colorado) that can be processed are known to be relatively low in sulfur, but high in As and Ni.

4.2.3.3 Sodium and Calcium

Sodium (Na) is a severe catalyst poison that can cause significant activity loss even at low levels. It works by promoting the sintering of catalytic metals and neutralizing acid sites. Typical sources of sodium can originate from desalter malfunctioning, sea water contamination, or caustic contamination. The signs of poisoning include rapid activity loss and an increase in pressure drop.

4.2.3.4 Phosphorous

Phosphorous (P) contamination in oil has been traced to injection of corrosion inhibitors often in the fractionation section for processing of crudes from Western Canadian sources. The source is diphosphate esters used to reduce naphthenic acid corrosion, which are soluble in the crude oil. Chemical dosed to crude columns to minimize fouling have resulted in the depositing of phosphorous to the downstream hydrotreaters. Other sources of phosphorous can be imported feeds, and lube oil wastes. If phosphorous enters the hydrotreater, it will poison the active sites of the catalyst causing a loss in activity. A level of 1 wt.% of phosphorous on the catalyst is known to result in roughly 10°F loss in activity. Experts opine that a feed content of <0.5 wt. ppm be maintained without appreciable concern and recommend the use of feed filters to assist in trapping of phosphorous sediment.

4.2.3.5 Iron

Iron (Fe) enters in the hydrotreater feed in the form of rust or iron scales as a result of corrosion from upstream equipment and piping. The particulates can result in iron content in the feed due to improper filtration. Iron contamination can also result due to naphthenic acid corrosion and formation of iron naphthenate as the corrosion product (when processing crudes with high TAN). These iron particulates can fill the interstitial spaces in the catalyst bed, which may result in a higher pressure drop. Special grading materials having high void volume to trap and accumulate these particulates are used to mitigate the pressure drop associated with iron. The grading materials are helpful in delaying pressure drop buildup, but they do not prevent or eliminate it. Effective feed filtration helps to remove particulates, and use of high void grading material provides a way to help mitigate fast pressure drop buildup and prolong cycle length.

4.2.3.6 Nickel and Vanadium

Nickel (Ni) and vanadium (V) contamination have been found in heavier petroleum fractions like vacuum gas oils and heavy coker gas oil and in residues. They can pose significant problems in hydrotreaters installed for pretreatment of fluid catalytic cracking (FCC) feed . The problem of Ni and V is not encountered in diesel or light feeds. The deactivation mechanism of these poisons are known to be due to pore mouth plugging. Nickel and vanadium are usually present in large porphyrin molecules that are too big to penetrate into the pores of typical hydrotreating catalysts. Therefore, the nickel and vanadium end up depositing outside the catalyst surface and eventually blocking access to the active sites within the pores. Table 4.1 contains the different type of contaminants, their likely source, and potential remedies.

4.3 PROCESS VARIABLES

The process parameters that have significant impact on hydrotreatment reactions are

1. Total pressure and hydrogen partial pressure
2. Reaction temperature
3. H2/oil ratio and recycle gas rate (or H2 availability)
4. Space velocity

TABLE 4.1

Contaminants, and Their Limits, Possible Source, and Likely Remedies

Contaminant	Limits in Hydrotreater Feed	Likely Source	Remedial Measures
Si	<1.0 wt. ppm	Antifoam used in DCU	Guard catalyst
As	<200 ppm	Crudes from West Africa and Russia, and synthetic crudes	Guard catalyst with higher Ni
P,	<0.5 wt. ppm	Spent lube oil, Corrosion inhibitors added in CDU to handle high TAN crudes etc.	Use of next generation/ improved corrosion inhibitors with lower phosphorus and to carry out the programme by expert agencies
Ni, V	<1.0 wt. ppm	Entrainment due to poor performance of upstream column wash zones	Control of entrainment and enhancing efficiency of upstream column wash zone and strengthing guard beds
Fe	<1.0 wt. ppm	Corrosion products and poor filtration of feed	Fe traps, inert guard, top bed skimming at intervals
C7 insoluble	<100 wt. ppm	Feed with high end point and poor wash section performance	Better wash zone efficiency in upstream columns and maintaining feed with lower end point

4.3.1 PRESSURE

The quality of the feed and the quality of the desired product primarily decide the requirement of total pressure and hydrogen partial pressure. The pressure maintained at the high-pressure separator (HPS) usually control the total pressure at the reactor inlet. A hydrotreater unit operating at a high hydrogen pressure has the following benefits:

- Longer catalyst cycle life
- Capability of processing heavier feeds
- Higher throughput and conversion capability
- Reduction in production of purity purge gas (from HPS) for maintaining the recycle gas purity

It is important to maintain the highest allowable hydrogen partial pressure in the reactor since low hydrogen partial pressure will result in coke formation and eventual catalyst deactivation. The extent of the removal of contaminants can also be increased by maintaining a high hydrogen pressure.

Increasing the partial pressure of hydrogen also results in higher saturation of polyaromatic hydrocarbons. Lower H2 partial pressure can slow HDN reactions. The unreacted nitrogen compounds can block the active sites that are available for HDS and can reduce HDS activity considerably. The drawbacks of maintaining a high hydrogen pressure are increases in hydrogen consumption and it can increase reactor costs.

4.3.2 TEMPERATURE

Higher reactor temperature increases the activity of the catalyst. The unit is started with lower temperature at the SOR. As catalyst activity declines, the temperature is increased to compensate for lower activity. A higher temperature than required can result in faster reaction and consequently higher exothermicity and can lead to runaway. This can also lead to thermal cracking and the production of considerable amounts of undesired low molecular weight hydrocarbons and coke. Olefins can be produced due to thermal cracking and the hydrogenation of olefins can release a substantial amount of heat. Thus, catalyst gets deactivated much more quickly at higher temperatures, which can lead to lower working life of the catalysts.

Exothermic reactions take place during hydrotreatment. As a result, the reactor outlet temperature is normally higher than the reactor inlet temperature. The weighted average bed temperature (WABT) is used to define the average reactor bed temperature. WABT can be calculated by using the following correlation:

$$WABT = SUM\left(W_i \times T_i\right)$$

where T_i is the average temperature of a bed equal to $(T_{inlet} + T_{outlet})/2$, and W_i is the fraction of total weight of catalyst in the bed.

A longer catalyst life is observed for the hydrotreatment of lighter stocks like naphtha, and the increase in WABT over time is of less concern compared to the increase in WABT during the processing of heavier feeds. During the hydrotreatment of heavy oils, WABT has to be increased constantly to compensate for catalyst deactivation and to ensure constant product quality.

4.3.3 GAS RECYCLE AND H2/OIL RATIO

Higher recycle gas circulation is used to maintain high hydrogen partial pressure within the reactor. Operation at higher H2 partial pressure helps reduce the temperature of the reactor at SOR. This can increase the cycle life of the catalyst. The recycle ratio is normally expressed as NM3/hour of hydrogen as recycle per kiloliter of feed (NM3/KL of feed). Alternatively, it is sometimes expressed in terms of hydrogen availability. A minimum value for hydrogen availability is 3.0, recommended by catalyst manufacturers. H2 availability indicates whether there are risks for hydrocarbon fouling. H2 availability is defined as the following correlation:

$$H_2 \text{ availability} = H_2 T / (\text{oil feed rate} \times H_2 \text{ consumed})$$

TABLE 4.2

Typical Recycle Ratios and Hydrogen Consumption in Hydrotreaters

Unit	Recycle ratio	Typical Hydrogen Consumption	Typical Hydrogen Consumption
	NM3 of H2/KL feed	% wt. on feed (SOR)	% wt. on feed (EOR)
NHT	150	0.3	0.25
KHDS	180–200	0.3–0.5	0.3
DHDS	250–280	0.6	0.5
DHDT	350–380	1.3	1.1
VGO HT	600–800	1.65	1.4
Hydrocracker	600–1000	2.2 to 2.6	2.0 to 2.4
Residue hydrotreater	1200–1500	2–2.8	1.8–2.5
Lube oil hydrofinishing	450–600	0.4–0.5	0.4–0.5

$$H_2T = \text{total hydrogen} = \text{Recycle} + \text{Make up} + \text{Quench gas}$$

During the processing of high-sulfur feedstocks, the concentration of H2S in the recycle gas stream can go high. High H2S in the recycle gas stream reduces the hydrogen purity of the recycle gas stream and the partial pressure of hydrogen. The reduction of HDS activity can be around 3%–5% for each 1 volume % of H2S in the recycle gas stream. So, an amine wash of the recycle gas stream is necessary to reduce the concentration of H2S in the recycle gas stream and increase the effective recycle ratio (normally a bypass across the amine absorber is provided to maintain the minimum level of H2S in recycle gas). A certain amount of H2S in the gas stream is essential to ensure that the catalyst remain in their sulfided form. Typical recycle ratios and hydrogen consumptions in hydrotreaters/ hydrocrackers are presented in Table 4.2.

4.3.4 SPACE VELOCITY

Space velocity is related to the feed rate to the reactor and the amount of catalyst loaded inside the reactor. A high feed rate corresponds to low residence time inside the reactor, and the oil and hydrogen have less time to be in contact with the catalyst. Thus, with a higher oil flow rate (higher space velocity or lower residence time), the temperature needs to be increased to achieve the removal of impurities to the same extent or to maintain a similar product yield. Conversely, lowering of the feed flow rate will need lowering of the reactor inlet temperature or WABT. Typical space velocities in a hydrotreaters are presented in Table 4.3.

Operating concerns in different hydrotreating units are described next

4.4 OPERATING CONCERNS IN NAPHTHA HYDROTREATER UNITS

The flow scheme of a naphtha hydrotreater (NHT) is shown in Figure 4.6.

In several naphtha hydrotreaters preceding a catalytic reforming unit (CRU/CCRU) and isomerization unit, a frequent increase in pressure drop across reactors is observed,

TABLE 4.3

Typical Space Velocities in Hydrotreaters

Unit	Typical Space Velocity (HR)$^{-1}$
NHT	6–7
DHDS	0.83 (based on quantity of active catalyst)
DHDT higher pressures	1.1 (based on quantity of active catalyst)
VGO HT	0.57–0.65 (based on quantity of active catalyst)
Hydrocracker high pressure	0.55–0.62 (based on quantity of active catalyst)

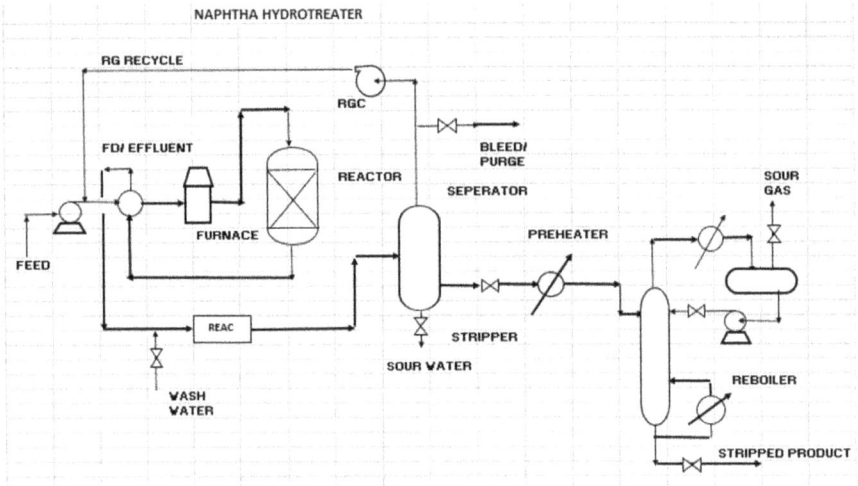

FIGURE 4.6   Flow scheme of naphtha hydrotreater (NHT).

particularly after commissioning. An increase in pressure drop across the feed side of feed/effluent exchanger is also very common. Both issues result in outage of the unit for attending the problems. The actions to resolve the problem of higher pressure drop across the reactor is to carry out scheming of the catalyst layer from top of the catalyst bed, cleaning trash baskets, loading fresh improved grading material/catalyst at the top, etc. It may be necessary to analyze the factors contributing the problem of higher pressure drop across the reactor and feed side of the feed/effluent exchanger. A very common reason for the higher pressure drop across an NHT reactor is deposition corrosion products on the top bed of catalyst or coke buildup on catalyst.

4.4.1   INCREASE IN PRESSURE DROP ACROSS REACTORS

There are two possible reasons for increase pressure drop across a NHT reactor.

The one that is more common is mechanical plugging of catalyst by particulates in the feed to the reactor. External particulates in the feed to the unit may cause plugging of the catalyst beds resulting in additional pressure drop across the catalyst bed(s). The particles may be iron oxidation (rust) particles from feed storage tanks or from corrosion products from piping or from anywhere in the circuit. Fine feed filters may be the best solution to protect from catalyst plugging.

The second reason is formation of particulates in situ. If there are organic acids present in the crude oil or naphtha feedstock, iron can get chemically dissolved in the naphtha feedstock. The hydrotreating reactions can reduce the organic acids causing iron (or other minerals) to precipitate out of solution and in turn can lead to catalyst plugging. Potential reasons for higher reactor pressure drop are presented along with case study examples.

1. Absence of filter or coalescer in feed circuit at the inlet to the unit: This is necessary to arrest external particulates entering the unit and depositing in the feed side of the feed/effluent exchanger and increasing pressure drop across the feed side of the exchanger. This can also deposit at the top of reactor and increase pressure drop. In one NHT unit there was no feed filter and feed coalescer. Soon after startup the higher pressure drop in the feed side (shell side) was observed in the feed/effluent exchanger and also across the reactor. The exchanger was cleaned, catalyst scheming done in the reactor, and a filter installed (5 micron after analysis of distribution of particulates in feed). The problem got resolved.

2. In many NHTs, the root cause of delta P increase could be oxygen ingress in the feed tank, if the feed tank does not have nitrogen blanketing and feed had some olefin content. It may be possible that some diolefins present in feed cause polymerization and thereby increases reactor pressure drop (DP). Hence, as a typical guideline a diene number of <1–2 is maintained in the NHT feed.

   Severe coking is still observed across the feed side (shell side) of the feed/effluent exchangers and reactors of quite a few NHT units even though the feed diene content (iodine number) was within limits. This can be attributable to dissolution of oxygen in feed and formation gums in the process. In a NHT unit, having floating roof feed tanks, the problem was partially resolved by operating the feed tanks at a higher level, so that the level never drops below the minimum leg height of the floating roof to avoid contact of air with oil. Fixed cum floating tanks are used in the majority of units and N2 blanketing is provided below the fixed roof to totally eliminate oxygen contact with feed.

3. It may be worth noting that reactors in NHTs typically operate in vapor phase and hence no inlet tray distributor is required and normally not provided. Feed directly enters the catalyst beds through the inlet diffuser. Now, if the reactor inlet temperature is maintained close to the dew point temperature, there is a possibility of liquid formation at the reactor inlet that can create maldistribution/channeling in the reactor catalyst beds and may result in coking and pressure drop in long run.

   Particularly, during startup, reactor inlet temperature should be maintained high enough to have feed in total vapor phase at the inlet to the reactors and to avoid two-phase generation. In one unit the operators reduced inlet temperature as feed contained less sulfur than designed and that resulted in heavy coking in the reactor after two to three months of operation and increased the pressure drop across the reactor. Later the dew point of the feed+hydrogen mixture was computed and an inlet temperature maintained about 10°C higher than the dew point and that resolved the problem.

4. It may be possible to have leakage in the feed filter and that would allow scales to the reactor and cause a reactor pressure drop. In a NHT, typically there is no grading material provided to arrest scales which may lead to an increase in pressure drop unless arrested by feed filter. So, health of the feed filter is very important and needs to be ensured.

5. When the pressure drop (delta P) goes up at constant feed intake, it indicates that the catalyst bed is fouling up. Most often this is due to corrosion products (FeS) from upstream piping. Sometimes an upset causes these corrosion products to be dislodged and dumped on the bed. Crushing of the catalyst itself is seldom a cause. A simple top tray and/or top bed scheming/grading can take out these particulates before they clog up the entire bed. Sampling the bed before the next reload can confirm the source of fouling and can act as a guide to arrest the recurrence of the problem.

6. Naphtha hydrotreaters normally utilize makeup hydrogen from the catalytic reformer. If there is no chloride guard in the makeup hydrogen to the NHT, then the likelihood of delta P increase may be due to the deposition of corrosion products (due to HCl corrosion) carried over from the pipes and depositing on the catalyst beds.

If the NHT catalyst is not designed to handle high olefins in feed and feed with high olefin is processed it may lead to deposition of polymeric material and increase pressure drop. One NHT had issues of high delta P due to carryover of sludge from the feed tank. Typically, a significant amount of iron rust is always found whenever a reactor is opened and scheming always helps to reduce the delta P. Chemical dosing is sometimes practiced online to dissolve this iron rust over the catalyst bed thus reducing the delta P.

4.4.2   Increase in Pressure Drop in the Reactor Effluent Circuit

A pressure drop increase in the high pressure circuit often is caused by NH4CL deposit ahead of the REAC. It is advisable to inject the water for some time at the inlet of the heat exchangers ahead of the REAC at periodic intervals. Provision for additional water injection may be made based on computation of NH4CL deposition temperature and identifying the location of likely deposition.

4.5   DIESEL HYDROTREATERS AND COMMON OPERATING PROBLEMS

The process flow diagram of a diesel hydrotreater is almost similar to that of a naphtha hydrotreater. The reactor is provided with multiple beds with the provision of quench gas flow between the beds to limit the exotherm in each bed. Further, a

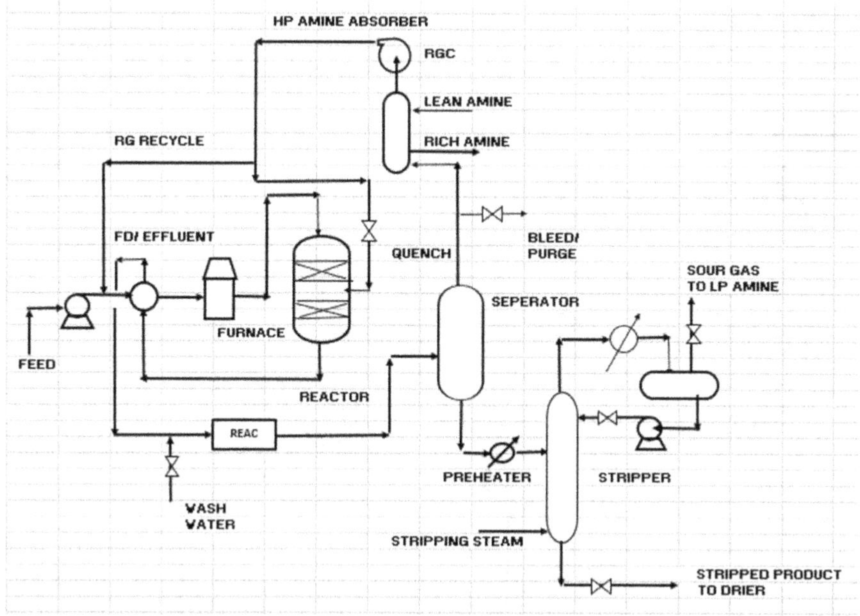

FIGURE 4.7 Scheme of a diesel hydrotreater.

recycle gas amine wash is typically provided to lower H2S in the recycle gas to help deep desulfurization of diesel. In units up to 40 kg/cm2 separator pressures recycle gas amine wash is followed by a water wash to arrest carryover of amines along with the recycle gas. Typically, the water wash chamber is placed above the amine wash column. In higher pressure hydrotreaters water wash downstream to amine wash is normally not provided as the water circulation pump suction pressure will exceed the limit of suction pressure indicated by API 610. The stripped diesel is typically routed to the vacuum drier to lower its moisture content less than 100 ppm wt. Figure 4.7 shows a diesel hydrotreater.

A diesel hydrotreater with four drum system in the effluent circuit and high pressure recycle gas amine wash is shown in Figure 4.8.

4.5.1 Common Problems in Diesel Hydrotreaters

Common problems experienced in diesel hydrotreaters include desulfurized product sulfur not achieved. This may be due to higher sulfur and nitrogen in feedstock. A higher sulfur in feed will increase the load and higher nitrogen in feed may inhibit desulfurization and may need higher temperature (WABT) for operation. The presence of higher content of more difficult sulfur components, e.g., benzothiophenes (BT) and dibenzothiophenes (DBT), in feedstock can also result in higher sulfur content in the product. Lower hydrogen partial pressure at the reactor inlet due to lower recycle gas purity, lower recycle ratio, lower system pressure, and lower activity

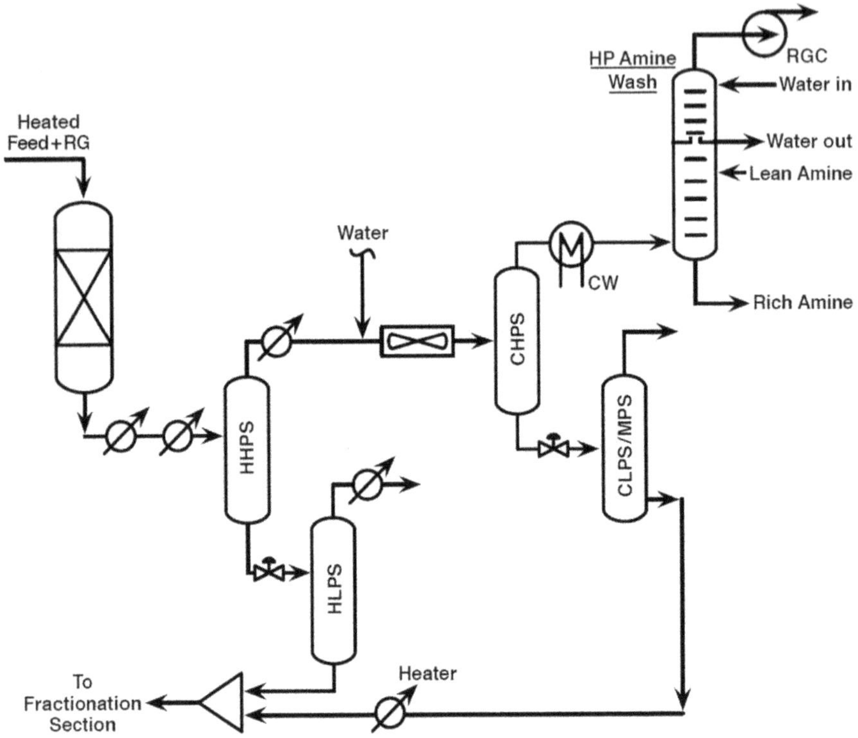

FIGURE 4.8   A simplified scheme of diesel hydrotreater showing four drum systems and recycle gas amine wash and water wash above amine wash.

of catalyst can also lead to higher sulfur content in the treated diesel. Lower activity of catalyst can be caused by poisoning by contaminants like nitrogen, and coking. Coking can result from emergencies like recycle compressor failure, temperature excursions during emergencies, operation with lower H2 partial pressure, processing of heavy feed/contaminants in feed, and olefinic feedstock.

4.5.2   COLOR IN TREATED DIESEL

Color in treated diesel is usually caused by presence of color contributing aromatic compounds like anthracene and fluoranthene etc. A case study of coloration of ULSD is presented later in 4.10.3

4.6   HEAVY DISTILLATE HYDROTREATERS AND COMMON OPERATING PROBLEMS

Heavy oil hydrotreaters include hydrotreaters for vacuum gas oil (VGO), heavy coker gas oil (HCGO), dewaxed oil (DAO), and VGO produced from residue hydrotreaters or a blend of the heavy distillate components. The hydrotreaters are essentially similar to diesel hydrotreaters with additional loads of metal guard catalysts upstream to protect the active catalysts and operate with higher recycle (nM3/KL feed) and consume more hydrogen in processing. They typically operate at higher pressures (80–160 kg/cm2) than diesel hydrotreaters. The common problems experienced are described next.

4.6.1 Common Problems in Heavy Oil Hydrotreaters

4.6.1.1 Higher Filter Backwash Frequency

Frequency of backwash increases in VGO/HCGO hydrotreaters and hydrocrackers with higher viscosity of the feedstock, presence of higher proportions of asphaltenes, and coke fines in HCGO, if included in feedstock. A higher feed viscosity or density can be resolved with little increase in filtration temperature, but contaminants (particles) in the feedstock create bigger problems. Typically, a higher VGO cut point or lower wash zone efficiency of the vacuum column does not contribute to the problem as much, but lower efficiency in the coker fractionator wash zone immediately results in an increase in the filter backwash frequency as a result of carryover of coke fines along with heavy coker gas oil (HCGO). Use of a higher wash oil rate in the coker fractionator may resolve the problem. Some units have provision to route the lighter coker gas oil as a wash as an alternative to HCGO. In case of chocking of the spray nozzles of the wash oil system, the only solution may be to clean the nozzles after bringing down the coker unit.

4.6.1.2 Higher Differential Pressure Drops across Reactors

The root cause of higher differential pressure drop across a reactor can be due to coking in reactors triggered by inferior feed quality (higher end point, higher proportions of olefinic feed, presence of contaminants in feed particularly in coker gas oils), temperature excursions, and caused by emergencies like recycle gas compressor failure. After a recycle gas failure, the reactor content should be thoroughly purged free of oil. Some technology suppliers suggest *hot stripping*, which is the stripping and removal of the balance oil in the reactors using higher reactor temperature than normal and full recycle gas flow (in absence of feed). Hot stripping is practiced to avoid an increase in pressure drop across reactors and is expected to reduce the likelihood of increase of pressure drop across the reactors post normal shutdown and emergencies. For heavy oil hydrotreaters/hydrocrackers, higher pressure drops result mostly after emergencies like RGC failure and unit trips. The higher pressure drop can be also due to particulate carryover due to malfunctioning of feed filters and coke fines along with coker gas oils. Contaminants in feed like phosphorus and silica depositing on top of bed/grading catalysts may also result in higher pressure drop.

4.6.1.3 Higher Differential Pressure in Reactor Effluent Circuit

Higher differential pressure in the reactor effluent circuit is experienced in many units with deposition of salts in the heat exchangers at colder temperatures zones (typically below 210°C). Chloride present in feed combines with ammonia to form ammonium chloride (NH4Cl) and deposit based on partial pressures of ammonia

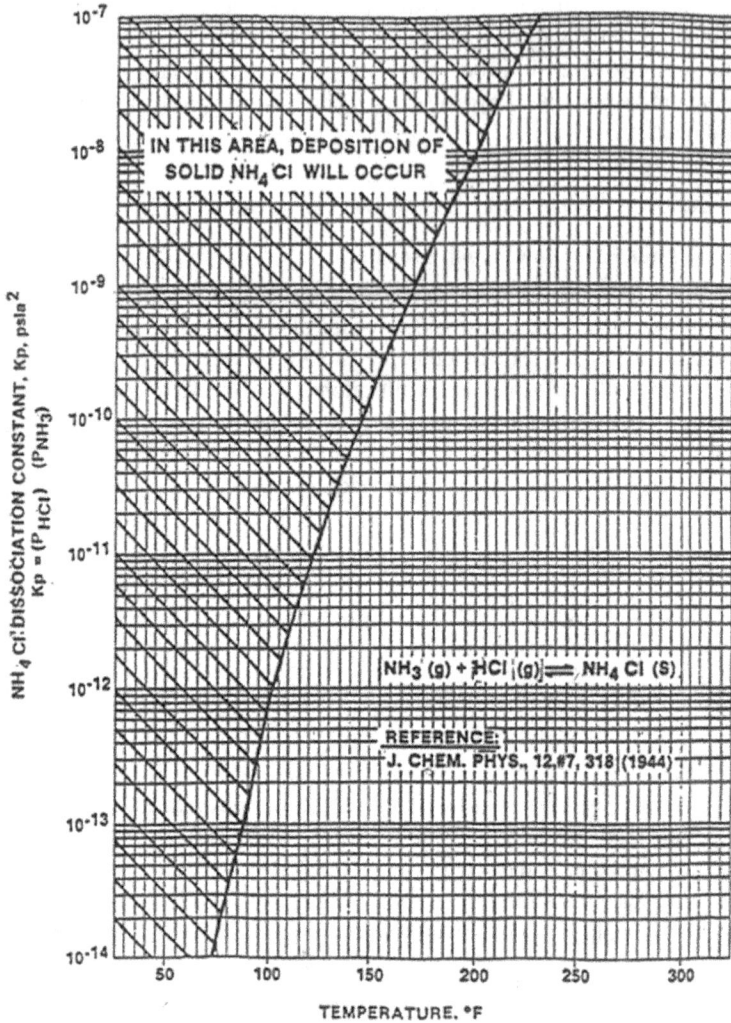

FIGURE 4.9A   NH4Cl deposition chart (lower temperature range).

and HCl in vapor phase far ahead of the REAC. Deposition charts for NH4Cl and NH4HS can be referred (Figure 4.9A and Figure 4.9B) for analysis of the cause of higher pressure drop. Usually, it is contributed by deposition of NH4Cl ahead of the REAC. The author has seen a refinery practicing hot stripping (after withdrawal of feed) at regular intervals to vaporize and remove the salts from the effluent circuit and get rid of the pressure drop in the effluent circuit. But this may lead to gasket failures due to temperature shocks in the process of hot stripping in the effluent circuit. Hot stripping may be only practiced to remove oil from reactors after normal shutdown/emergencies. The deposition salts should be washed with water after relocation of additional water injection point.

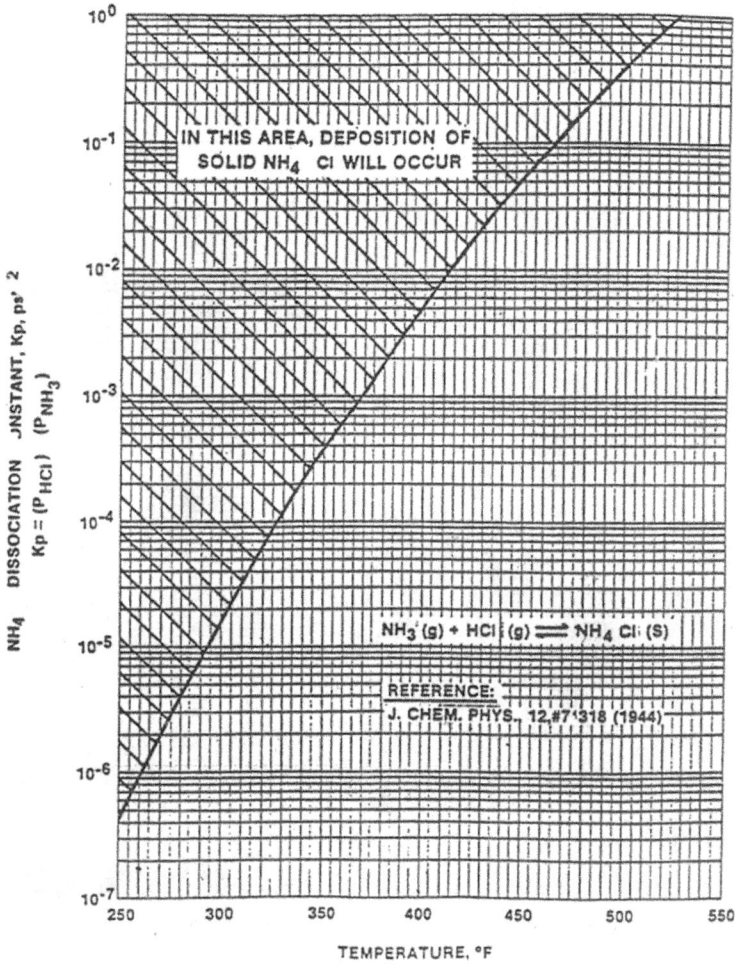

FIGURE 4.9B NH4Cl deposition chart (higher temperature range).

A NH4HS deposition chart is presented in Figure 4.10.

NH4HS deposition can also cause the pressure drop problem, but the deposition of NH4HS occurs at a lower temperature than NH4Cl and usually gets washed by continuous water injection provided at the REAC inlet. If the unit is operated for longer duration without water wash, the NH4HS deposition can lead to high pressure drop in the effluent air coolers. Normally, operation of the unit is not continued for a higher duration in absence water injection. The quantity of NH4HS can be computed based on feed sulfur and nitrogen. Operation of the unit beyond five to six hours without water injection may not be at all advisable. In the case of a problem in the water injection pump, the author had to operate one hydrocracker unit [having virgin VGO as feed with very low feed sulfur (0.23% wt.) but high nitrogen (around

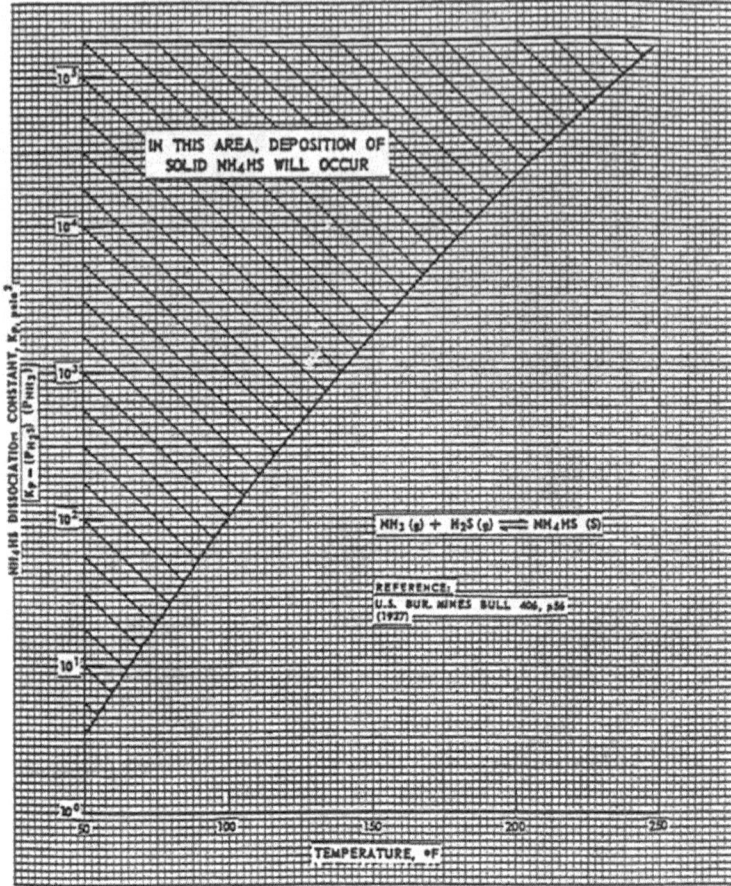

FIGURE 4.10   NH4HS deposition chart.

800 ppm wt.)] for about six hours without water injection. The deposition of salt may lead to RGC surge or opening of the anti-surge valve leading to abstraction of recycle gas flow in the furnace/reactors, and in turn can lead to catastrophes like temperature runaway, coking, and equipment damage.

4.7   OVERVIEW OF RESIDUE HYDROTREATERS

A vacuum residue (VR) hydrotreater is typically comprised of two sections: one for hydrotreatment of VR and the other for further hydrotreatment and conversion of the VGO produced from residue hydrotreatment. A feed filter is not usually provided in the residue hydrotreatment section. The preheated feed is directly routed to the furnace and joins the preheated recycle gas (usually heated in a separate furnace), and then is routed to reactors (ebullated beds are common with circulation of catalysts).

Reactors have provision to introduce fresh catalyst (in the form of slurry) and with-draw spent catalyst.

A hot separator is used immediately at the outlet of reactors to separate vapor and liquid streams from the reactor effluent. The oil vapors and hydrogen recovered from the top of the hot separator and are partly cooled in heat exchangers and routed to another separator (warm separator) at lower temperature. The hot high-pressure separator liquid is flashed at lower pressures, partly cooled with quench oil, and fed to the heavy oil stripper along with the liquids from other low-pressure separa-tors for recovery of converted fraction. The stripped residue is further preheated in the vacuum furnace and fed to the vacuum column for recovery of gas oil compo-nents and VGO. The residues from the vacuum column bottom are typically fed to the coker or RFCCU. The VGO components recovered from the vacuum column are again treated in a typical high-pressure hydrotreater along with virgin VGO, if desired. Typical yield of diesel, lighter VGO, and residue are around 30%, 35%, and 27% by weight, respectively, in the VR hydrotreatment section. An overall conver-sion of around 72% by weight can be achieved together with the VGO hydrotreating/hydrocracking facility. Issues in the hydrotreating section of the VGO, derived from residue hydrotreaters, are similar to those of heavy oil hydrotreaters (described in Section 4.6.1).

4.8   HYDROCRACKERS AND VARIOUS CONFIGURATIONS

Some simplified schemes of hydrocrackers are presented showing different configu-rations. A simplified scheme of a two-stage hydrocracker deploying a single recycle compressor is shown in Figure 4.11 and Figure 4.11A.

The unit with the configuration in Figure 4.12 was taken for capacity revamp, and a configuration having heat exchangers in between the reactors as in Figure 4.13 was considered as solution to reduce the requirement of quench gas flow. The new configuration suggested is presented below where the cold low-pressure separator (CLPS) liquid picks up part of the heat of reaction of reactor 1 (in heat exchanger E1) and reactor 2 (in heat exchanger, E2), thereby conserving the quench gas flow (quantity) requirement at the reactor 1 and reactor 2 outlets and reducing the require-ment of recycle gas flow.

The effluent circuit can have two-drum systems with a CHPS and CLPS, or four-drum systems with a HHPS, HLPS, CHPS, and CLPS. The fractionation section usually comprises of a stripper followed by a fractionator and gas concentration sec-tion for recovery of liquified petroleum gas (LPG).

4.9   METALLURGY OF HYDROTREATERS AND HYDROCRACKERS

The metallurgy of hydrotreaters and hydrocrackers is based on high temperature hydrogen attack (HTHA). Nelson charts or curves are normally used as a guide to specify the operating limits for steels in hydrogen service (these limits are also presented in API RP 941). Modern reactors are made with 2¼% Cr–1% Mo steel with 0.25%–0.35% V in order to maintain better mechanical properties at higher

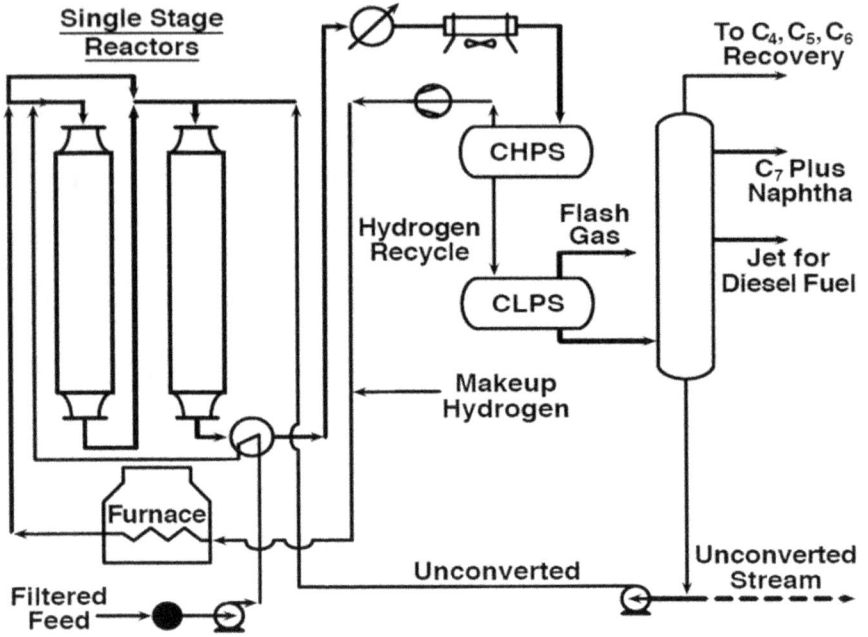

FIGURE 4.11   Simplified scheme of unconverted recycle to a hydrocracker reactor.

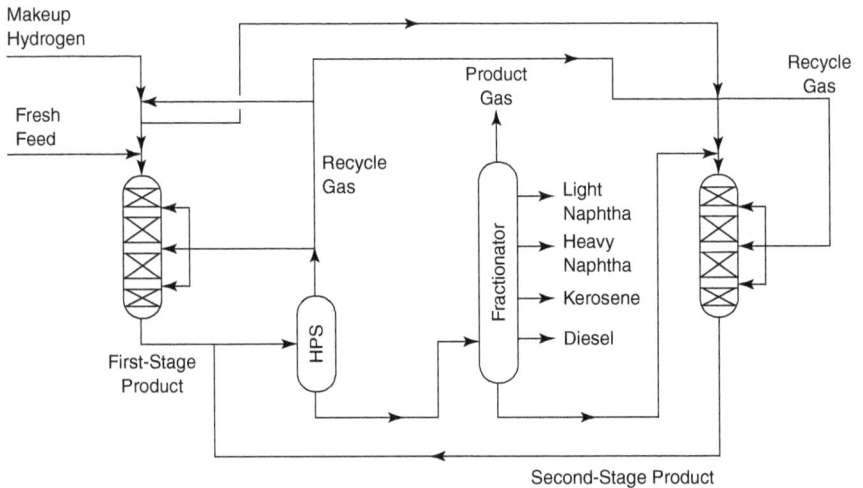

FIGURE 4.11A   A simplified scheme of a two-stage hydrocracker deploying a single recycle compressor.

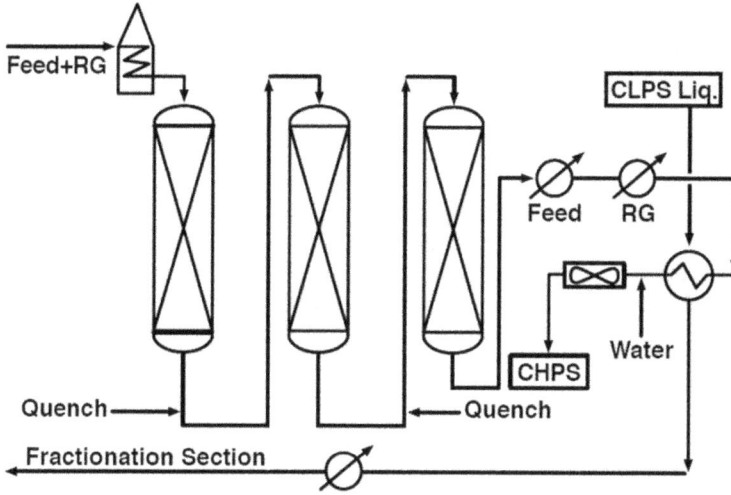

FIGURE 4.12   Simplified sketch of an actual hydrocracker deploying three reactors in series.

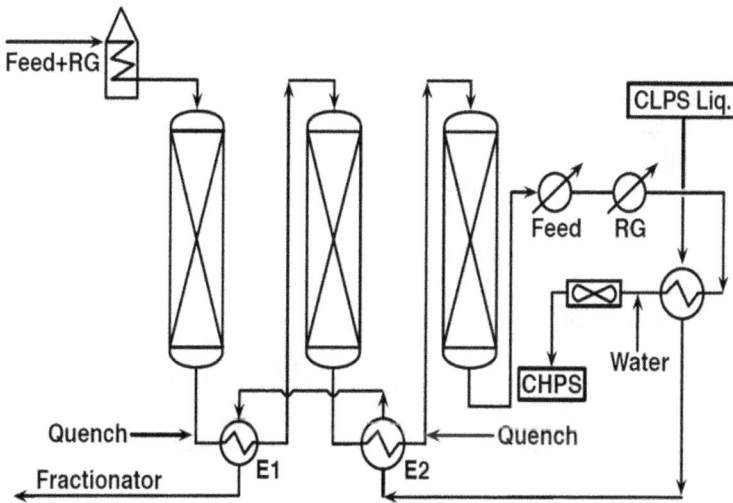

FIGURE 4.13   Simplified sketch with three reactors in series with heat exchangers in between to reduce quench gas flow.

temperatures and have 3 mm SS 347 lining inside. The addition of vanadium also increases the tensile strength of the reactor material. For higher operating temperatures (temperatures from 440°C to 500°C), API suggests the use of 3% Cr–1% Mo steel or 2¼% Cr–1% Mo steel with 0.25% to 0.35% vanadium (V).

Further, hydrogen also has permeability in steel, therefore reactor metal is also susceptible to hydrogen embrittlement. The inside lining of SS 347 provided in reactors is intended to reduce permeability of hydrogen in steel and likelihood of hydrogen embrittlement. During operation, the heating and cooling rates are carefully monitored to decrease the likelihood of embrittlement and fissuring. During the cooling of reactors, the temperature is not decreased below a defined limit until the pressure is also decreased to the prescribed level. This is also to avoid hydrogen embrittlement.

The feed effluent heat exchangers are made of 2.25 Cr and 1 Mo steel, and 1.25 Cr and 0.5 Mo steel, and carbon steel based on the prevailing temperatures and hydrogen partial pressure based on HTHA guidlines. The REAC can have metallurgy of carbon steel, alloy 825 (UNS N08825), or alloy 625 (UNS designation N0 6625). If the feed contains chlorides, NH4Cl can form and reduce the pH of the sour water and cause heavy corrosion in the REAC. The use of alloy 625 may be advisable to resist the chloride attack. The author has observed heavy corrosion in the REAC made of alloy 825 in a VGO/HCGO hydrotreater where feed had chloride content of 1–2 ppm wt. In hydrotreaters and hydrocrackers, while processing feed with chlorides the use of alloy 625 metallurgy for the REAC is becoming increasingly popular.

4.10   CASE STUDIES

4.10.1   Effluent Circuit High-Pressure Drop

In a VGO hydrotreater (having 80 Kg/cm2 pressure in the separator) the phenomenon of gradual increase of pressure drop was observed. The pressure drop in the effluent circuit was increased from 6–7 kg/cm2 to 14 kg/cm2 over a period of three to four months. The unit has a HHPS in the effluent circuit with an operating temperature of around 260°C. It has four exchangers upstream of the HHPS for heating recycle gas and feed separately and four heat exchangers in the downstream vapor circuit, namely, HE-01 (effluent/feed), HE-02 (effluent/RG), HE-03 (effluent/CLPS liquid), HE-04 (effluent/RG), and the REAC (Figures 4.14).

The increase in pressure drop resulted in a reduction of the recycle flow rate as the higher pressure is required by the RGC to push the recycle gas along with feed. Beyond 14 kg/cm2 pressure drop in the effluent circuit, the feed flow used to get lower. The refinery used to withdraw the feed and carry out hot stripping for almost 12–20 hours to volatilize the NH4Cl deposits in the circuit and introduce the feed again. Thus, the feed is withdrawn in the unit after a period of three to four months for almost a day. The existing reactor effluent circuit is shown in Figure 4.14.

This was practiced for several years since commissioning. The refinery used to advance the water injection at the inlet of the upstream heat exchanger (HE-04) once a week without any avail, as the likely deposition temperature of NH4CL was found around 205°C and deposition is expected much ahead in the upstream exchanger (in HE-02 tube side).

The metallurgy of heat exchangers HE-02, HE-03, and HE-04 is 2.25–1 Mo, 1.25 Cr–0.5 Mo, and CS, respectively, provided based on HTHA consideration. The licensor have advised to install an additional water injection point at upstream of HE-02 to wash the NH4CL deposit. The licensor was also advised to upgrade the metallurgies

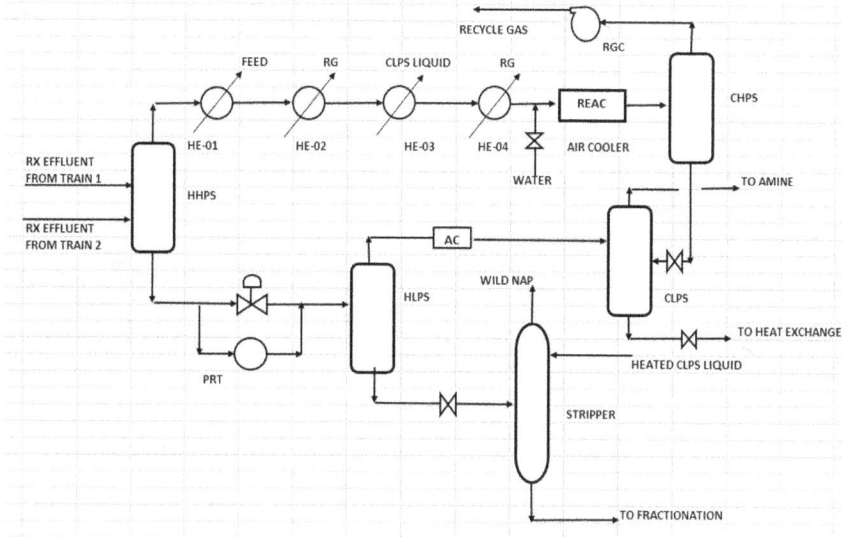

FIGURE 4.14   Existing scheme of the VGO hydrotreater.

of the HE-02, HE-03, and HE-04 tubes, channels, tube sheet, etc. to alloy 625 and also upgrade the metallurgy of the REAC to alloy 625 (from existing alloy 825), as the acidic water will flow along the circuit up to the CHPS. They also advised to provide Monel lining in the CHPS and CLPS. Piping metallurgy of the entire circuit from the E-02 outlet was advised to be upgraded to alloy 625. Thus, the piping metallurgy change was suggested from E-02 downstream to the REAC inlet and REAC outlet to the CHPS, and the CHPS to CLPS to nickel alloys (alloy 625). It may be noted that NH4Cl is a salt of strong acid and weak base. Thus, NH4Cl solution in water will be acidic. The pH value computed with typical NH4Cl considering 1.5–2 ppm chloride in feed is 5.0–5.4. But if accumulated NH4Cl dislodged from somewhere joins the continuous generation of NH4Cl, then the pH can drop even further (to around 4.5). These are very low pH values and the existing metallurgy of chromium–molybde-num steel cannot withstand this low pH. So, alloy 625 is proposed by the licensor for heat exchanger tubes, the REAC, and pipe lines, and Monel lining inside for vessels (CHPS and CLPS). The alloy 625 (UNS designation N06625) is suitable for handling chloride solution and is essentially a nickel base alloy.

But the modification proposed will be very expensive and will need a long outage of the unit for implementation. An inhouse alternative was worked out using the unique configuration in Figure 4.15. An additional water injection point upstream of HE-02 will continue to wash out the likely deposits of NH4Cl (depositing at the operating temperature inside EE-02 tube side) over and above the injection point upstream of REAC. The resultant effluent from HE-02 having hydrocarbons, H2, and acidic water will be routed to a new warm separator (a three-phase separator, inside Monel lined) at a temperature of 180°C–190°C. The vapors will be routed back to the existing original circuit of HE-03 → HE-04 → REAC (and after cooling/condensation) → CHPS →

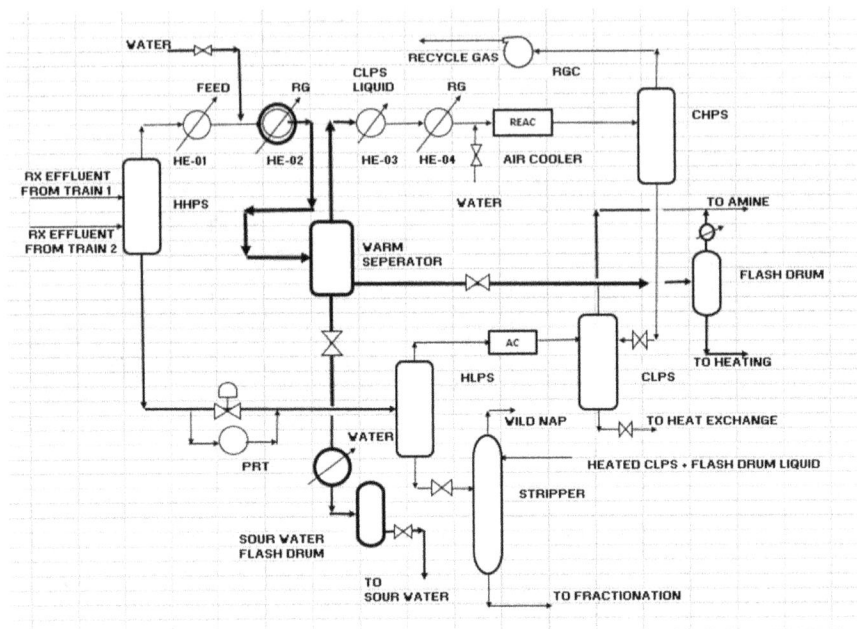

FIGURE 4.15 Modified scheme of the VGO hydrotreater modifications shown in bold.

CLPS as usual. The hydrocarbon phase from the warm separator will be routed to a new flash drum. The outlet of the new flash drum will be routed to a stripper along with CLPS liquid after further heating flowing through the original circuit. The acidic water from the warm separator will be cooled further and will be routed to the sour water flash drum at lower pressure and finally to the sour water system. The only metallurgy upgradation will be required for the heat exchanger HE-02 (tubes and channel that will come in contact with acidic water) and the entire balance system can be retained with the same metallurgy as acid water will leave the system from HE-02 to the warm separator without affecting the existing equipment and circuits downstream of HE-02. However, two small new medium pressure flash drums will add over and above the high pressure warm separator. This unique proposition was accepted by the licensor of the unit and is expected to need only 60%–65% of capital expenditure compared to the one envisaged in the original proposition. The likely drop in recycle gas hydrogen purity (as a result of some liquid getting abstracted from CHPS) and will be managed by a marginal increase of purity purge. The new equipment can be installed during operation of the unit and will need much lower outage of the unit to implement the scheme. The same is being implemented.

4.10.2 CORROSION IN DIESEL HYDROTREATERS

A diesel hydrotreater operating at a pressure of 110 Kg/cm2 experienced acute corrosion. The diesel hydrotreater was designed to process a mixture of straight-run gas oil (SRGO) and coker unit LCGO. It had two parallel trains of reactors and effluent heat exchanger trains terminating in a common HHPS. The HHPS outlet vapor was routed directly to the reactor effluent air cooler (REAC) and then to the CHPS, and the HHPS bottom liquid was routed to the HLPS and was fed to the stripper after preheating to about 250°C. The two parallel effluent trains had two Packinox exchangers heating the feed and recycle gas mixture before it was heated in the furnace prior to its entry to the reactors. The common HHPS is located downstream of the effluent vs. feed Packinox heat exchangers. The effluent temperature at the outlet of the Packinox was around 180°C; accordingly, the HHPS temperature was also 180°C. There was a gradual increase in the pressure drop across the effluent side of the Packinox. To resolve the problem, the refinery introduced water injection at the inlet of the Packinox in both trains. The unit already had the normal provision of a water injection at the inlet of the HPAC (REAC).

Severe corrosion was noticed in the stripper upper section. The corrosion product was analyzed and it was found to have been caused primarily by NH4Cl. The suction strainer of the pressure recovery turbine (PRT) located downstream of the HHPS going to the HLPS had frequent deposits of solids, making the operation of the PRT almost impossible. It was found that NH4Cl was likely to deposit in the effluent side of the Packinox, might increase the pressure drop across the Packinox, eventually get into the HHPS liquid, and choke the strainers of the PRT downstream of the HHPS. It was found from the NH4Cl deposition (Figures 4.9A and 4.9B) that a small amount of chlorides (1 ppm) in the hydrotreater feed can deposit as NH4Cl in the Packinox at the outlet temperature of the Packinox (180°C) and get into the HHPS liquid. From there it can easily get into the HLPS, which, when introduced in the stripper, can cause corrosion. The only solution to the problem would be to increase the Packinox outlet temperature and consequently the temperature of HHPS. A minimum temperature of around 235°C–240°C would be necessary to avoid the problem. A higher HHPS temperature could be achieved by sending part of the effluent through the bypass across the feed and effluent Packinox (facility already provided in the existing scheme), but this would reject additional heat to the REAC and reduce the feed preheat at the inlet of the reactor furnaces. The furnace heat duty requirement would increase and quickly reach its limit. Thus, as a solution, it was suggested that new heat exchangers be installed at the following locations (Figure 4.16):

1. At the hot vapor line from the HHPS. The vapor from the HHPS would heat the feed and the recycle gas mixture and cool the effluent to around 180°C (original level) before it is routed to the REAC provided with a water injection at upstream. The preheated feed would be routed to the feed and effluent Packinox in the one train for further heating.

2. At the downstream (in the HLPS liquid line) to heat the feed and recycle the gas mixture before it is routed to the feed and effluent Packinox in the other train for further heating. The REAC duty requirement increases as the HHPS vapor quantity increases with the increase in HHPS temperature. This increases the CHPS temperature by 5°C–10°C, so a water cooler was installed in the CHPS vapor line to cool the recycle gas back to the

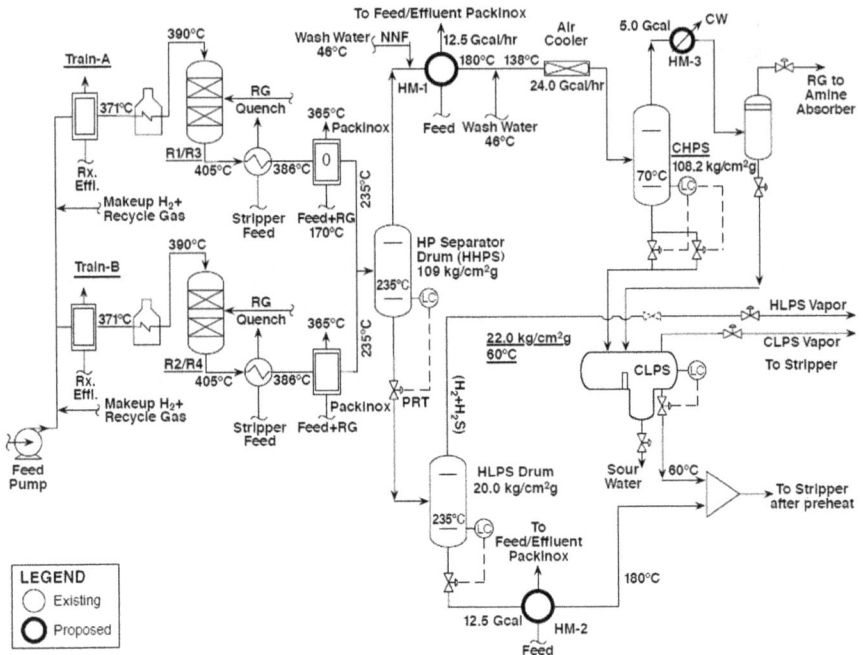

FIGURE 4.16 A diesel hydrotreater with a plate type exchanger (Packinox) in effluent circuit.

original temperature level of 55°C. Thus, two sets of new feed and recycle gas mixture vs. effluent heat exchangers were proposed: one at the HHPS vapor and the other at the HLPS liquid to the stripper. The installation of the new heat exchangers would also make it feasible to impart almost the same amount of heat to the feed and recycle gas mixture to maintain the reactor furnace inlet temperature. The stripper feed inlet temperature would likely reduce to some extent, as the HLPS liquid quantity would reduce as a result of the HHPS operating at a higher temperature. The stripper feed vs. stripper bottom heat exchanger have a margin in the surface area to maintain the stripper feed temperature at the same level by reducing the steam generated by the stripper bottom in the downstream circuit (Figure 4.16). (This case is also described in *Distillation and Hydrocarbon Processing* written by the author.)

4.10.3   COLOR IN ULTRA-LOW HIGH SPEED DIESEL (ULHSD)

Color in rundown diesel was observed on several occasions in diesel produced from low-pressure hydrodesulfurization units (DHDS). Introduction of ultra-low sulfur specification of diesel (ULSD) needs deeper desulfurization and often results in diesel product color, particularly toward the end of run (EOR). The typical ULSD unit cycle length is often limited by color degradation of the product. The typical ULSD unit has a deactivation rate in the range of 2°F–3°F/month, so an increase in temperature by 10°F–25°F may cause a problem of color before actual EOR is attained. It is well known that the color of distillate products is affected by the reaction conditions in the hydrotreater, especially temperature and hydrogen partial pressure. Species responsible for color degradation in distillates are identified as polynuclear aromatic (PNA) molecules in the finished diesel streams. Some of these PNAs are yellow/green/blue and fluorescent in color, which is apparent even at very low concentrations of these species. The author has observed fluorescent yellow to dark yellow color in ULSD on several occasions. As the (outlet) temperature increases and/or hydrogen partial pressure decreases, the product color degrades. The reactor inlet temperatures are increased as the catalyst approaches EOR to compensate for lower activity to achieve the specified product sulfur. This results in lower aromatic saturation as equilibrium conversion of aromatics reduces.

Reduction in the WABT can improve the color, as scope to increase the recycle rate was not available. But reduction of the WABT can affect the product sulfur. It was concluded that the specific species responsible for color degradation of diesel are anthracene, fluoranthene, and their alkylated derivatives (shown in Figure 4.17). These PNAs are readily saturated to one and two ringed aromatics under typical diesel hydrotreating conditions at start of run (SOR), but as the temperature of the reactor increases toward EOR, an equilibrium limit may be reached whereby the reverse reaction becomes more pronounced and the PNAs may remain in the diesel stream. The source of PNAs was primarily LCO stream in DHDS feed. In the unit the light cycle oil (LCO) proportion from FCCU in feed to DHDS was reduced to improve color. The LCO was dropped to clarified oil (CLO) in the FCCU main fractionator to reduce production of LCO. The CLO in turn is added to coker feed and eventually the same is recovered partly in HCGO that goes to the higher pressure VGO hydrotreatment unit and gets hydrotreated saturating the PNAs at higher pressure.

Some catalyst suppliers are claiming that incorporation of higher nickel proportions in their catalyst will improve hydrogenation of the PNAs under the same pressure and recycle ratio and will reduce the likelihood of product color in lower pressure units, however, with marginally higher hydrogen consumption. Higher nickel in catalyst will also reduce product sulfur.

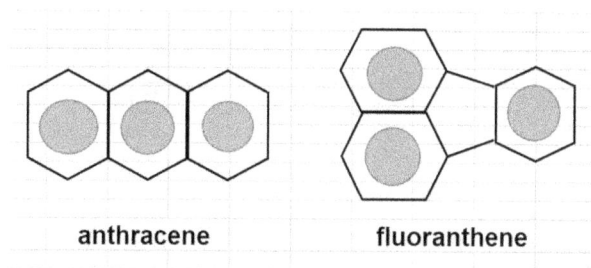

anthracene   fluoranthene

FIGURE 4.17 Primary fluorescence species in hydrotreated diesel.

4.10.4 CAPACITY AUGMENTATION OF A NHT

A NHT unit capacity was planned to be increased by around 30% by a refinery to process a high quantity of hydrotreated of coker naphtha (around 10%–12% on coker throughput). The unit had a configuration as shown in Figure 4.18. It had a feed/effluent heat exchanger (E1) with six bundles in series to heat the combined feed (feed+recycle gas) to furnace the inlet temperature (and to cool the effluent). The pressure drop in the recycle loop had reached a limit of 8 Kg/cm2. The refinery contacted heat exchanger specialists and planned to replace the six shell and tube heat exchangers with non-conventional ones (welded-plate type) to reduce the pressure drop to carry out the capacity increase. One specialist proposed replacing four bundles with a proprietary welded-plate type, while retaining the two existing heat exchangers in series with the proposed new one to deliver the required heat duty at higher capacity. But the refinery was not very sure about the performance of the proposed plate and frame heat exchanger and was hesitant to go ahead .

The problem was revisited by the author and team. The reactor effluent of a naphtha hydrotreater is expected to be in full vapor state. The simulation of the unit confirmed that it will be full vapor even at SOR and will be above the dew point by around 20°C (superheated vapor). Thus, the same can be divided in two streams. Accordingly, a scheme was developed considering splitting of reactor effluent in two unequal flows and routing them in two heat exchangers trains. All six sets of heat exchangers (shell and tube) in the existing series (E1) were retained, and in parallel

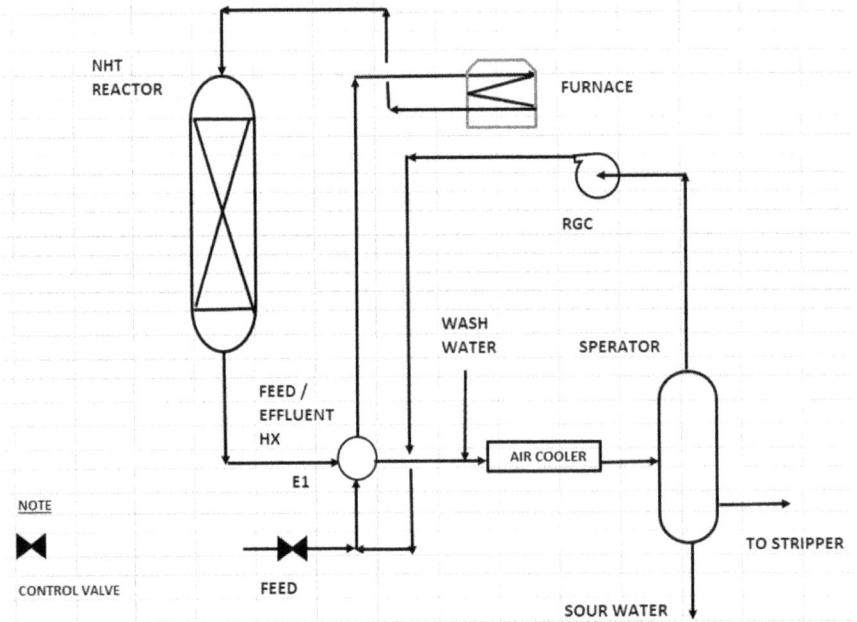

FIGURE 4.18 Existing scheme of the NHT.

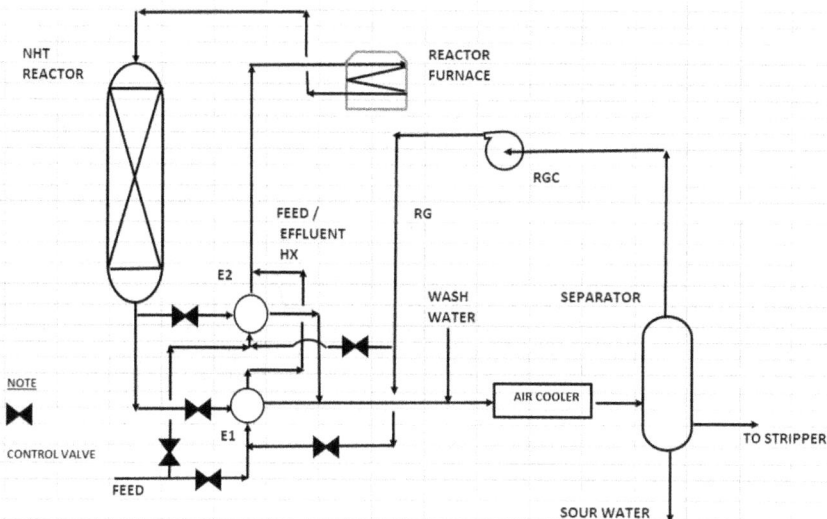

FIGURE 4.19 Modified circuit of the NHT.

six more small heat exchangers (shell and tube) were added in series (E2). The feed, effluent, and recycle gas flow was split by use of control valves, and the major problem in capacity enhancement was resolved (shown in Figure 4.19). The reactor was loaded with new generation catalyst having higher space velocity than existing to handle the increased feed flow. The recycle gas flow (recycle ratio) requirement was found adequate for the new catalyst (at higher feed flow). Some small modifications to the reactor furnace were carried out in the convection section (converting studded tubes to finned tubes as the furnace uses fuel gas firing only) to deliver higher heat duty. The unit was revamped and the desired capacity achieved without any problem. The recycle loop pressure drop rather reduced by around 10%–15% as effluent and feed flows through the existing heat exchanger train (E1) got reduced and transferred to the new parallel heat exchanger train (E2).

The example of the naphtha hydrotreater revamp was presented to emphasize the fundamentals that determine the modifications/configurations necessary in hydrotreaters for capacity increase. In a high-pressure diesel hydrotreater, VGO hydrotreater and hydrocracker the the reactor effluent will be in two phase and parallel splitting of the stream having the same vapor/liquid (V/L) ratios cannot be achieved. A low pressure DHDS can have reactor effluent in full vapor state and depending on the condition of the reactor effluent splitting of flow may be possible.

4.10.5 CAPACITY AUGMENTATION OF A HYDROCRACKER

A hydrocracker unit with two drums, namely a CHPS and CLPS, and originally having all the effluent circuit heat exchangers in series was considered for capacity expansion by around 30%. The hydrocracker reactor effluent is in mixed phase

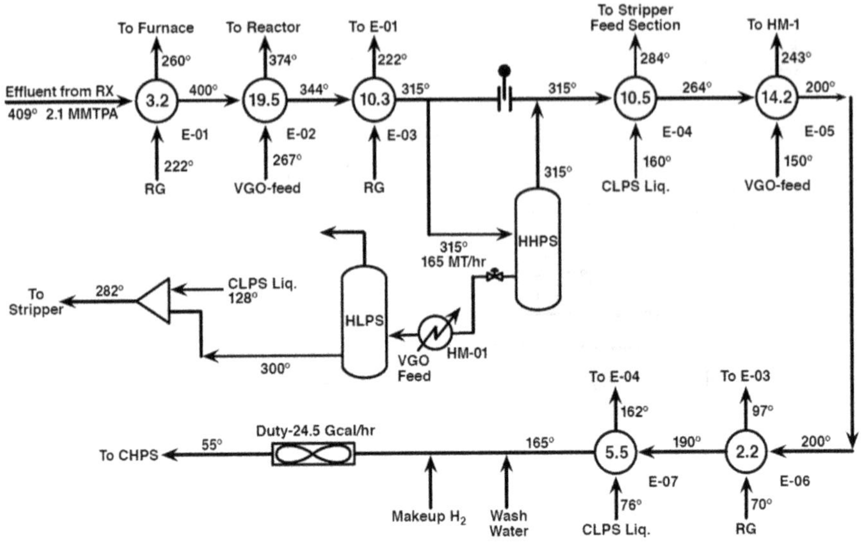

FIGURE 4.20 A simplified scheme showing modification to install a HHPS/HLPS.

(mixture of vapor+liquid. So, the effluent stream cannot be divided in two parallel passes having the same V/L ratios. A simplified scheme shows modification to install a HHPS/HLPS for capacity debottlenecking (shown in Figure. 4.20) with a two-drum configuration and converted to a four-drum configuration for capacity debottlenecking.

A hydrocracker fractionator can be preceded by a stripper or debutanizer to remove all the light ends (gas and LPG) as the top product or wild naphtha, as shown in Figure 4.21. The wild naphtha produced from the debutanizer can be stabilized to recover LPG and to produce debutanized naphtha. The installation of the stripper upstream can debottleneck the fractionator and fractionator charge furnace.

4.10.6 HYDROTREATED WILD NAPHTHA PROCESSING

A stabilizer system to process hydrotreated wild naphtha needs to be specially designed (Figure 4.22; also presented in Chapter 1). It should necessarily have a small absorber with two to three stages where gases produced from the stabilizer and feed can recontact. The wild naphtha will be fed to the top of the absorber and then will be picked up by the pump and would be fed to the stabilizer after preheating by the hot stabilized naphtha stream from the column bottom. The stabilizer configuration will be conventional with a reboiler at the bottom and condenser at the top. The non-condensable vapors from the reflux drum will be fed to the absorber bottom and finally non-condensables will be let out from the absorber top under pressure control of the absorber and stabilizer. The simplest version of the absorber can be a recontacting drum (i.e., having single stage for contact). But normally a higher number of stages helps in increasing LPG yield at lower reboiler duty.

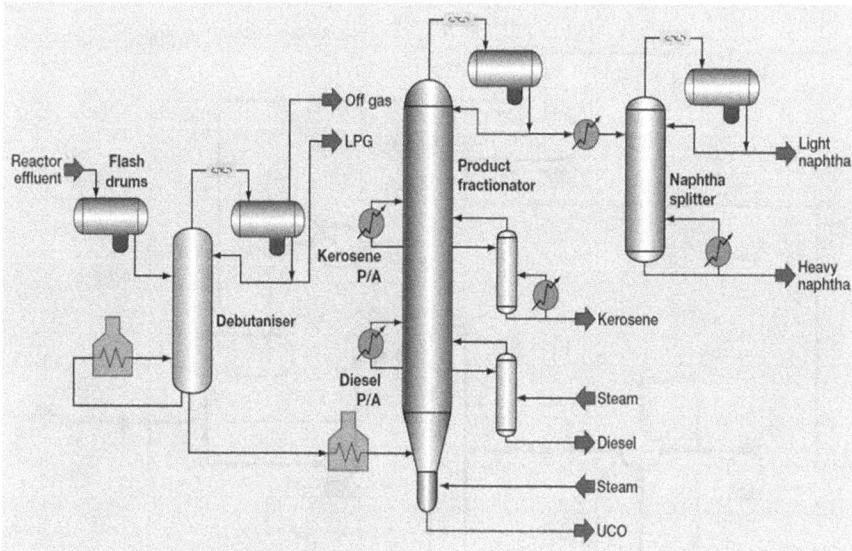

FIGURE 4.21 Simplified sketch of a hydrotreater fractionation section with stripper/debutanizer upstream of fractionator.

4.11 IMPROVEMENT OF RECYCLE GAS (RG) PURITY

Recycle gas (RG) purity is often a problem in heavy oil hydrotreaters and hydrocrackers. If the recycle gas purity goes down, the hydrogen partial pressure reduces at the reactor inlet and consequently results in adverse impact on product quality, catalyst life, and recycle gas compressor (RGC) power consumption, and can also result in reduction of the unit feed rate. The recycle compressor may not be in position to deliver adequate capacity to meet the prescribed recycle ratio and requirement of quench gas flow. Normally with a four-drum system, the recycle gas purity declines and is lower than a two-drum system owing to availability of less liquid in the CHPS to dissolve the light hydrocarbons that can lower the purity of RG.

Several approaches are followed to increase recycle gas purity:

- Use of higher purity of hydrogen in the makeup gas.
- Start purity purge (PP) from recycle loop downstream of the amine wash column. The purge gas stream can be cascaded with a lower pressure hydrotreater unit directly or can be routed through membrane for enhancing hydrogen purity of the permeate stream. The hydrogen-rich permeate can be routed to the makeup gas compressor (MUG) suction at the appropriate stage and reintroduced into the recycle gas loop of the same hydrotreater.

Alternatively, the permeate (hydrogen-rich stream) can be coupled with another hydrotreater at lower pressure in the recycle gas loop directly or can be compressed

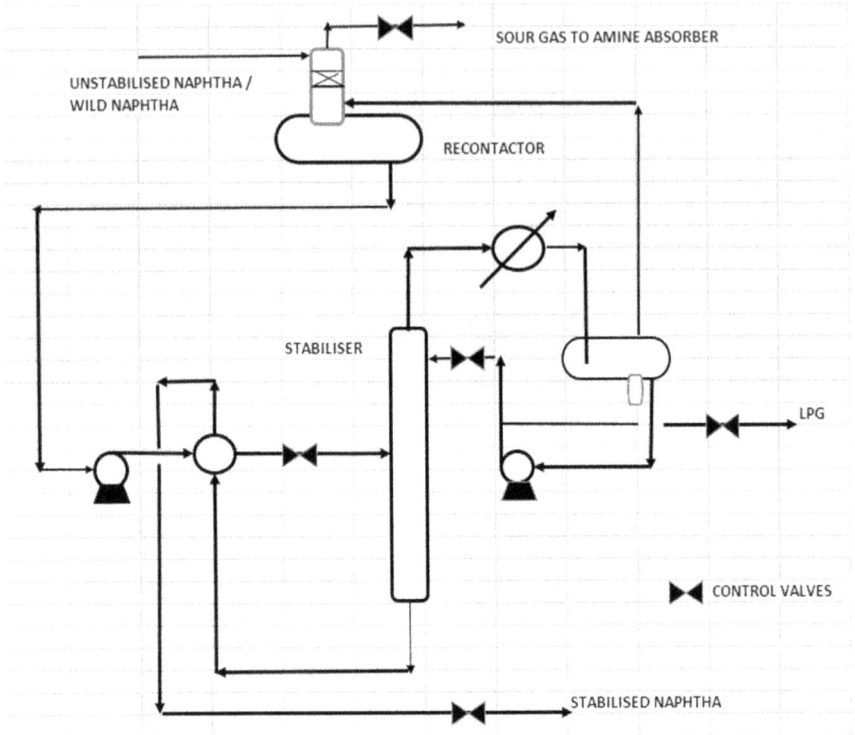

FIGURE 4.22   Scheme to process hydrotreater wild naphtha in stabilizer.

and introduced in the recycle gas loop of a hydrotreater at similar pressure with the help of MUG of the unit. The residue or retentate is typically injected to fuel gas. Alternatively, it can be fed to another pressure swing adsorption (PSA) unit, and pure hydrogen can be recovered if the retentate stream quantity is high and contain an appreciable proportion of hydrogen. The scheme is shown in Figure 4.23.

Other options for increasing recycle gas purity are described as follows:

- Lowering the CHPS temperature, if found higher than design. Much lowering of CHPS temperature may result in difficulty in separation of sour water from effluent hydrocarbon stream.
- Increase in CHPS pressure marginally, if cushion is available
- Introducing a CLPS liquid recirculation at the inlet to the REAC as in Figure 4.24. But it needs a high-pressure pump and is sometimes found in moderate-pressure hydrotreaters only.

A diesel hydrotreater with four drums is shown with the facility (in dark line) to recirculate CLPS liquid ahead of the REAC and this has shown to improve recycle gas purity (Figure 4.24).

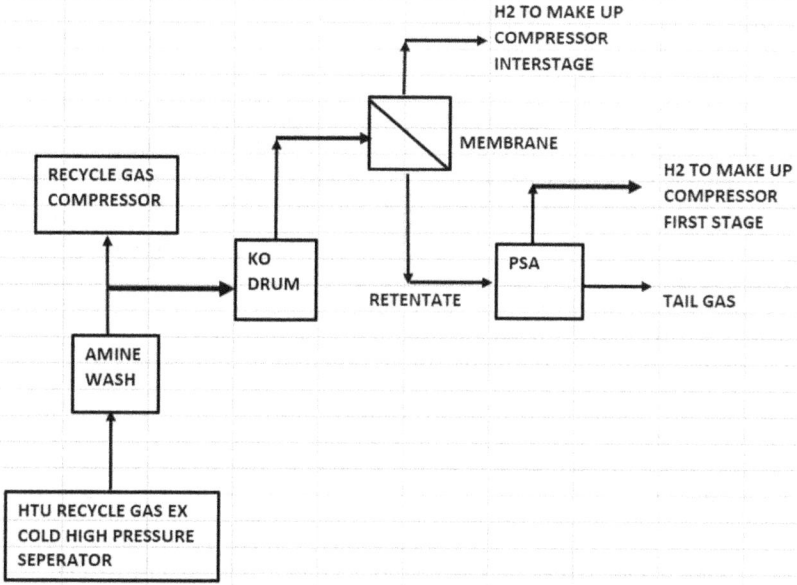

FIGURE 4.23 Recovery of hydrogen from purity purge gas.

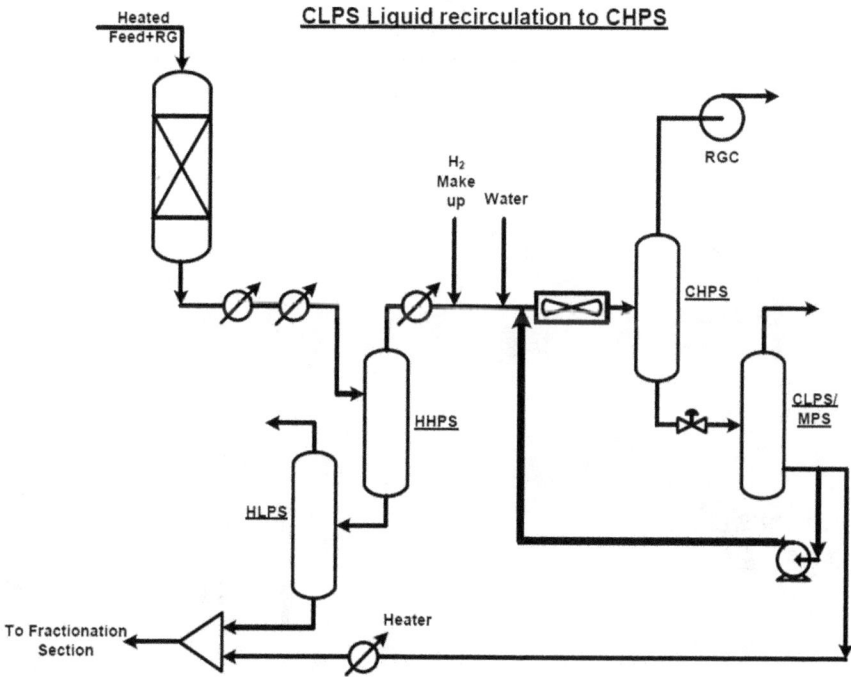

FIGURE 4.24 A scheme in a diesel hydrotreater with CLPS liquid recirculation.

4.12   ESSENTIALS OF SIMULATION

The simulation of the open art section of a hydrotreater starts with constitution of reactor effluent. The effluent can be generated be mixing the products using distillation and density of each or can be entered as a combination of pure components and NBP cuts, usually supplied by pilot plants or from mathematical models by experts. Required amount recycle hydrogen can be added to generate the combined effluent (effluent oil and hydrogen mixture) at reactor outlet. The same combined effluent can run through effluent cooling circuit with separators as applicable and to CHPS. In CHPS recycle gas (RG) and hydrocarbon liquid and water separates. The hydrocarbon liquid is fed to CLPS. The CLPS liquid is heated mixed with the liquid from HLPS and further heated and fed to the fractionator charge heater and to fractionator for recovery of products having side strippers for each. The hydrogen-rich gases recovered from CHPS are amines washed at high pressure and compressed in RGC and recycled with feed. A simulation flow diagram of an operating partial conversion (40%) hydrocracker is presented in Figure 4.25.

The feed to the unit is usually obtained with distillation and density and mixed with recycle hydrogen in simulation and heated in heat exchangers and reactor feed furnace for feeding to reactor (not shown).

The fractionator has a partial condenser and wet gases are picked up by the wet gas compressor (WGC) in the configuration and fed to the GASCON section. The column is provided and three side draws with strippers (namely HYN, KERO, GAS OIL) in design and with one pump around reflux (at gas oil draw location). Vapor/liquid profile of the original column is presented from simulation in Table 4.4.

Later, a kerosene pump was added in the column when conversion was increased by around 10–15% by change of reactor catalyst. This was done to contain the load

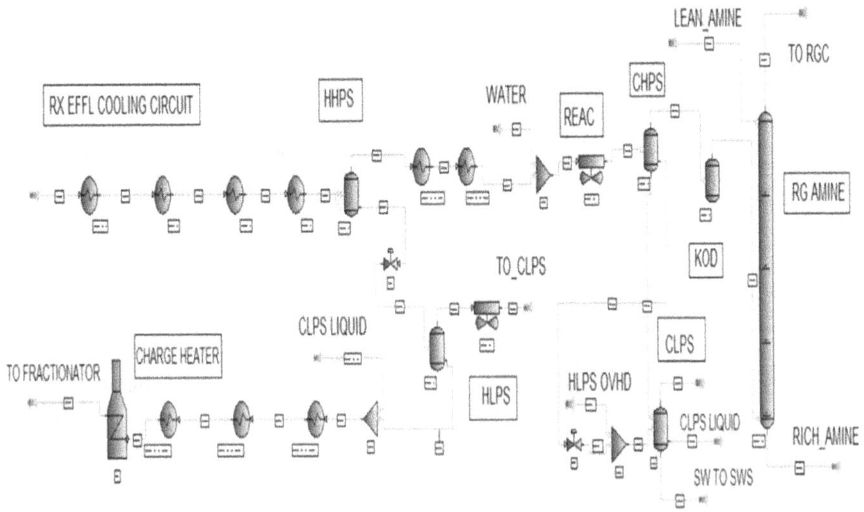

FIGURE 4.25   Simulation flow diagram of a hydrocracker.

TABLE 4.4
Vapor/Liquid Profile of Fractionator of an Operating Hydrocracker

| Tray | Temp | Pressure | Net Flow Rates | | | | Heater Duties |
| | | | Liquid | Vapor | Feed | Product | |
	DEG C	KG/CM2			KG/Hr		M*KCAL/HR
1C*	57.6	2.10	94106.2			8783.6V	−12.0731
						4500.0L	
						3918.7W	
2	98.0	2.10	114437.4	111308.5			
3	108.4	2.11	119063.3	131639.7			
4	114.1	2.12	120563.4	136265.6			
5	118.8	2.13	114334.1	137765.7	3463.2V	7000.0L	
6	122.9	2.14	114535.1	135073.2			
7	127.1	2.15	114524.2	135274.2			
8	131.5	2.16	113843.3	135263.3			
9	137.0	2.18	111321.8	134582.4			
10	145.8	2.19	103567.1	132060.9			
11	164.7	2.20	33808.0	124306.2	23944.2V	59489.2L	
12	202.4	2.21	32741.5	90092.0			
13	220.9	2.22	32589.2	89025.6			
14	229.3	2.23	31359.4	88873.3			
15	234.7	2.24	29362.8	87643.5			
16	239.6	2.25	25609.7	85646.9			
17	246.6	2.26	265932.6	81893.8	195887.7P		−7.0000
18	270.1	2.27	45425.8	126329.0	9059.2V	195887.7P	
						44404.2L	
19	309.3	2.28	42890.1	137054.8			
20	325.6	2.29	40321.4	134519.2			
21	333.1	2.30	37556.4	131950.4			
22	337.3	2.31	33524.9	129185.4			
23	341.0	2.33	26042.4	125154.0			
24	346.6	2.34	75489.2	117671.5	148858.6M		
25	343.9	2.35	71406.5	18259.7			
26	342.5	2.36	69799.2	14177.0			
27	341.3	2.37	68633.0	12569.7			
28	339.8	2.38	67347.7	11403.5			
29	337.1	2.39	65264.7	10118.2			
30	329.2	2.40		8035.2	3046.0V	60275.5L	

* Trays with free water.

of overhead condensers and overall loading of the fractionator. Alternatively, stripper upstream of fractionator can be installed. This will reduce the flow through the charge heater and will also reduce the load of the fractionator considerably and can also eliminate the requirement of WGC. A preflash drum (like the one in crude units) can be installed upstream of the fractionator charge heater to reduce the flow of vapors though the heater and will help to reduce the pressure drop across the heater.

BIBLIOGRAPHY

AIP Publishing LLC, NH4Cl deposition charts. *The Journal of Chemical Physics*, 12, no. 7, 1944, p. 318.

Ashis Nag. *Distillation and Hydrocarbon Processing Practices*. Tulsa, OK: PennWell, 2015.

Dave Krenzke. *Controlling Feedstock Contaminants in Diesel Hydrotreating Operations*. Richmond, CA: Advanced Refining Technologies. Special edition, Issue no. 110, 2011.

Greg Rosinski, Brian Watkins, and Charles Olsen. *Factors Influencing ULSD Product Color*. Chicago, IL: Advanced Refining Technologies, pp. 34–37. www.e-catalysts.com.

Josh Siegel and Charles Olsen. Feed contaminants in hydro-processing units. *Catalagram*, 104, Special edition, Fall 2008. Chicago, IL: Advanced Refining Technologies, pp. 2–6.

Khalid Hannan. Effect of organic nitrogen compounds on hydrotreater performance. Master of Science Thesis in Chemical Engineering (Civilingenjör), School of Chemical Science and Engineering, KTH Royal Institute of Technology, Stockholm, Sweden, November, 2014.

NH4HS deposition chart. U.S Burro of Mines, *Bull*, 406, 1937, p. 54.

ABBREVIATIONS

CAT	catalyst average temperature
CHPS	cold high-pressure separator
CLPS	cold low-pressure separator
DHDS	diesel hydrodesulfurization unit
DHDT	diesel hydrotreatment unit
HHPS	hot high-pressure separator
HLPS	hot low-pressure separator
HTHA	high-temperature hydrogen attack
KHDS	kerosene hydrodesulfurization unit
NHT	naphtha hydrotreater unit
WABT	weighted average bed temperature

5 Acid Gas Treatment

5.1 INTRODUCTION

Gas treating involves removing the contaminants from a gas stream to sufficiently low levels to meet desired specifications, with a primary focus on removal of the following acid gas components:

- Carbon dioxide (CO2)
- Hydrogen sulfide (H2S)
- Other sulfur species (CS2, COS, etc.)

Other contaminants can be mercury, ammonia, elemental sulfur, and arsenic. Problems with the acid gas component H2S is that it is toxic, produces SO2 on combustion, and can lead to acid rain. CO2, having no calorific value, is a diluent for overall calorific value of natural gas and corrosive in presence of water. Thus, removal of both H2S and CO2 are essential from natural gas to reduce corrosion and improve the calorific value of the gas. Acid gas removal from natural gas has been practiced for decades to meet the specification of the gas supplied. Further, acid gas removal from producer gas and synthesis gas has been in use for production of fertilizers and chemicals/petrochemicals.

Acid gas removal from sour gas streams in refineries and petrochemicals are rigorously followed to meet downstream process specifications and also to limit SO2 emission.

Purification levels. For pipeline-grade natural gas, H2S needs to be lowered to typically 4 ppm (volume) and CO2 to typically 4% volume. Cryogenic processes downstream need 50 ppm by volume of CO2, as CO2 freezes at –70°C. Of late, CO2 capture and sequestration has gained paramount importance for enhanced oil recovery (EOR) from underground reservoirs and enhanced recovery of methane from coal beds by injection of CO2 (CO2 is removed from gases and injected in oil reservoirs and coal beds for enhanced recovery of oil and methane) and also for carbon sequestration or carbon dioxide removal (CDR). CDR is the long-term removal, capture, or sequestration of carbon dioxide from the atmosphere to slow or reverse atmospheric CO2 pollution and to mitigate global warming.

Further, in the recent past, high emphasis has been given for gasification of heavy oils and petroleum coke for the purpose of their disposal and to produce synthesis gas, and in turn to synthesize a large number of chemicals, produce hydrogen, and also sometimes to produce low BTU gas (LBG) for firing in furnaces. The raw synthesis gas produced in gasification needs extensive treatment to remove the undesired components, primarily the acid gases. Thus, acid gas treatment alternatives and their benefits may need a relook and accordingly revisited with different alternative processing schemes with merits and demerits of each.

DOI: 10.1201/9781003268246-5

The different gas sweetening processes are applied depending on the quality and quantity of acid gas contaminants present in the feed gas and the extent of removal required. The following scenarios are likely to influence selection of the process:

- CO_2 when this is the sole contaminant of the gas
- H_2S when this is the sole contaminant of the gas
- CO_2 and H_2S removal simultaneously
- Selective removal of H_2S when both CO_2 and H_2S are present in the gas

5.2 OPTIONS FOR ACID GAS REMOVAL

5.2.1 PROCESSES

The existing processes are as follows:

- Chemical absorption
- Physical absorption
- Physiochemical absorption
- Permeation (membrane)
- Cryogenic fractionation
- Direct conversion to sulfur
- Physical adsorption

A number of methods are available for removal of acid gases from product gas streams. Some of the commonly used methods are *chemical solvents*, *physical solvents*, *membranes*, and *cryogenic fractionation*.

Conventional processes for removing acid gases with solvents typically involve countercurrent absorption of acid gas components from the syngas using a regenerative solvent in an absorber column. This process of gas–liquid contacting to remove acid gases is very commonly used in a wide range of process industries, including refining, chemicals, and natural gas production/processing. However, due to the significantly different process conditions and requirement of different degrees of acid gas removal, the choice of solvents and the process varies significantly.

5.2.2 CHOICE OF PROCESS

Physical solvents such as DEPG (Selexol™ or Coastal AGR®), NMP or N-methyl-2-pyrrolidone (Purisol®), Methanol (Rectisol®), and propylene carbonate (Fluor Solvent™) are popular as gas treating solvents especially for coal gasification applications. Physical solvents tend to be favored over chemical solvents when the concentration of acid gases or other impurities is very high. In addition, physical solvents can usually be stripped of impurities by reducing the pressure without the addition of heat in the solvent regenerators. For chemical synthesis applications that require syngas with less than 1 ppm by volume of sulfur, physical solvent processes using Rectisol and Selexol are normally employed, which allow essentially total removal of sulfur. These processes may operate at depressed temperatures (Rectisol operates at −40°C). This is discussed further in Section 5.3.

Mixtures of chemical and physical solvents, also known as hybrid or composite solvents, are also used. Sulfinol is a well-known example and discussed in Section 5.4.

For power-generation applications, which allow higher sulfur levels (approximately 10 to 50 ppm v sulfur), chemical solvent processes are normally practiced use ethanol amines as solvents. Processes involving ethanol amines are discussed in detail in Section 5.5.

Non-amine solvents (hot potassium carbonate) are described in Section 5.5.3. Recent development in chemical solvents involving solvent blends is described in Section 5.5.4. Introduction of ionic liquid (IL) solvents is described in brief in Section 5.6.3.

The *membrane process* is applicable for high-pressure gas containing high-acid gas concentrations. CO_2 recovery is accomplished by pressure-driven mass transfer through a permeable membrane where separation is due to the differences in the permeation rate of different compounds. *The acid gas components are recovered at low pressure as permeate.* A high purity product containing approximately 95% CO_2 can be achieved with one or two stages. This is discussed later in Section 5.7.

Cryogenic fractionation (Ryan/Holmes) has the advantage that the CO_2 can be obtained at relatively high pressure as opposed to the other methods of recovering CO_2. This advantage may, however, be offset by high capital investment and operating costs due to the large refrigeration requirement. Special materials are also required for cryogenic service adding to the capital cost.

Cryogenic fractionation is a process for the removal of CO_2 from natural gas. If H_2S is present and requires removal, a dedicated process for H_2S selective removal would be needed upstream of the cryogenic fractionation. Cryogenic fractionation does not remove H_2S from natural gas. This process, originally developed by KOCH Process Systems Inc., was titled the "Ryan/Holmes Process" after the name of two employees, inventors of the process (for lowering the CO_2 freezing point).

Direct conversion is an oxidative process for converting H_2S into sulfur. *Oxidative solvents* react with the H_2S and oxidize it to elemental sulfur in the solution. These processes are generally used for the removal of small quantities of H_2S from gas streams . They are based on the direct conversion of H_2S into sulfur by an iron-based catalyst. The ferric solution is in contact with the gas in the absorber and the H_2S is directly oxidized to sulfur. The solution is regenerated at low pressure by a stream of air. The overall reaction can be expressed as $H_2S + 1/2\ O_2 = S + H_2O$. The sulfur is recovered as a solid. Examples of licensed processes are Sulferox™ and Lo-Cat®.

Physical adsorption typically refers to the use of molecular sieves for the removal of acid gases. This process is not suitable for the removal of high quantities of acid gases. It can be considered only to remove traces (in terms of ppm) of H_2S and/or CO_2. The process is similar to the process for gas drying and needs a specific adsorbent. Physical adsorption deploys molecular sieves for the removal of acid gases, and the adsorbent beds need to be regenerated at intervals and operate like a typical pressure swing adsorption (PSA) process. A scheme is shown in Figure 5.1.

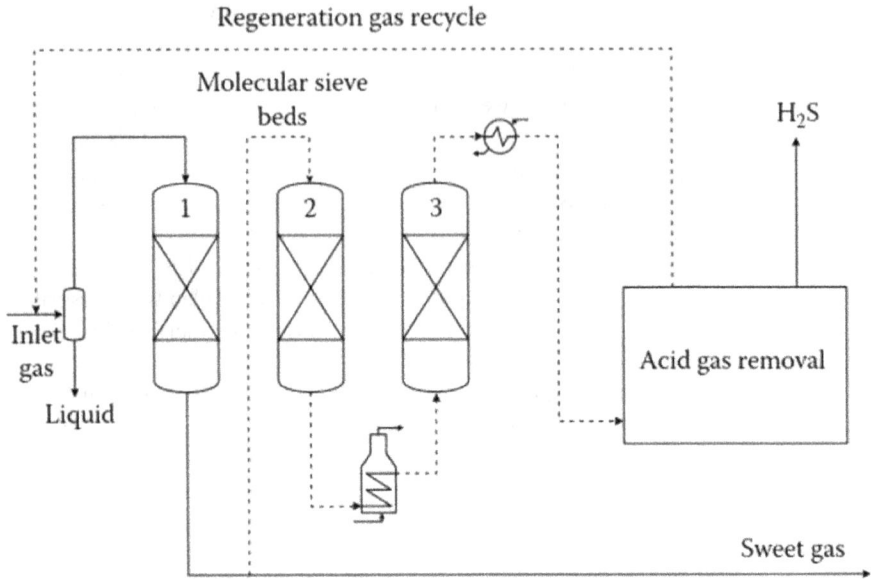

FIGURE 5.1   A scheme for physical adsorption with molecular sieve. (From Kidnay, Parrish, McCartney, *Fundamentals of Natural Gas Processing*, 2nd ed., 2011.)

Often the fixed-bed polishing is used to remove H2S deploying metal oxides like zinc oxide or copper zinc oxides and deployed as H2S guards.

Ethanol amines (e.g., MEA, DEA, MDEA, DGA) and hot potassium carbonate are chemical solvent processes that rely on chemical reactions to remove acid gas constituents from sour gas streams. The regeneration of chemical solvents is achieved by the application of heat.

Unlike chemical solvents, physical solvents are non-corrosive, requiring only carbon steel construction. At low partial pressures of acid gases, physical solvents are impractical because the compression of the gas for physical absorption is expensive. However, if the gas is available at high pressure, physical solvents might be a better choice than chemical solvents. Physical solvents can often be stripped off impurities by reducing the pressure generally without the application of heat. The heat requirement in solvent regeneration of a physical solvent can be low due to the relatively low heat of desorption and potentially due to lower solvent circulation rates. As a result, physical solvents can be competitive over chemical solvents (however, may need an external mechanical refrigeration and the additional energy requirement can offset the advantage of lower heat duty for regeneration).

Physical solvents tend to be favored over chemical solvents when the concentration of acid gases is very high. The concentration of heavy hydrocarbons in the feed gas also affects the choice of gas treating solvent. If the concentration of heavy hydrocarbons is high, a physical solvent may not be the best option due to higher co-absorption of hydrocarbons, particularly pentanes and heavier components. This is a major disadvantage since the hydrocarbons gets stripped along with acid gas during solvent regeneration and fed to Sulfur recovery plant (Claus unit), located downstream and can lead to problems in the Claus unit. Synthesis gases do not contain appreciable quantities of hydrocarbons. This makes physical solvents particularly applicable to synthesis gas treating.

Principles of removal of acid gases with solvents are described next:

- Chemical solvents are aqueous bases that undergo reversible acid-base reactions with the H2S and CO2 after they hydrolyze in water to weak acids. The absorption processes typically operate at room temperature. Examples are monoethanolamine (MEA) and methyldiethanolamine (MDEA).
- Physical solvents are polar molecules that have positive and negatively charged portions that "attract" the polar H2S and CO2 molecules (no chemical reaction; only attached by polar bonding). The absorption processes typically operate at low temperatures and require cooling or refrigeration. Solvent regeneration (stripping) is usually accomplished by flashing at lower pressure, then desorption with moderate heat input . Example processes are Selexol™, Rectisol®, Purisol®, and Morphysorb®.
- Physical-chemical or mixed solvents are mixtures of chemical and physical solvents that combine the features of the solvents. Cooling or refrigeration may be required for the absorption processes. The example processes are Sulfinol® and Amisol™. The Sulfinol-D process is a physico-chemical process based on the use of a mixture of diisopropanolamine (DIPA) and sulfolane in aqueous solution.

Only acid gas removal with solvents and membranes are covered further in upcoming sections.

5.3   COMMON PHYSICAL SOLVENTS FOR ACID GAS REMOVAL

All the processes based on physical absorption with different solvents are from licensors. So, the names of the licensors are mentioned along with the proprietary solvents used in the processes. Some processes that find applications in natural gas sweetening and synthesis gas treatment include:

- Selexol (dimethyl ether of polyethylene glycol) from UOP
- Rectisol (methanol) from Lurgi
- Purisol (N-methyl-pyrrolidone, NMP) from Lurgi
- Fluor Solvent (polypropylene carbonate) from Fluor

The flow diagram and required equipment of a dedicated process from a licensor can vary according to the feed quality acid gas removal level and operating conditions adopted. In many cases, the regeneration of the solvent is not only carried out through flash separation but also practiced with a regenerator with a reboiler and condenser.

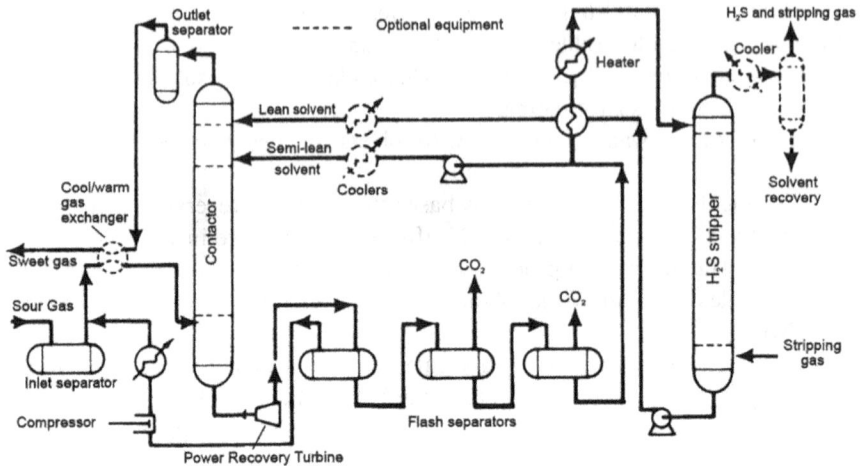

FIGURE 5.2 Typical physical solvent process. (From *GPSA Engineering Data Book*.)

A number of physical solvents are available for use in acid gas treating processes. A comprehensive list of common physical solvents can be found in a gas purification handbook by NETL, US Department of Energy[5].

In a typical physical solvent process, the solvent can remove H2S, CO2 and organic sulfur, and steam is used for regeneration. A solvent chiller is required for achieving low CO2 in the treated gas. Flash regeneration is used for CO2 removal from rich solvent. A simplified scheme is presented in Figure 5.2.

The four of the solvents discussed here are dimethyl ether of polyethylene glycol (DEPG), propylene carbonate (PC), N-methyl-2-pyrrolidone (NMP), and methanol (MeOH). Experts' opinions and suggestions on use of these physical solvents are presented in the following sections.

5.3.1 MeOH (Methanol)

There are a number of methanol processes for acid gas removal including the Rectisol process (licensed by Lurgi AG) and Ifpexol® (Prosernat). The Rectisol process was the earliest commercial process based on an organic physical solvent and is widely used for synthesis gas applications. The process operates at a very low temperature and is complex compared to other physical solvent processes. The main application for the Rectisol process is the purification of synthesis gases derived from the gasification of heavy oil and coal rather than natural gas treating applications. The Rectisol process can also be configured to address the separation of synthesis gas into various components depending on the required final products.

Methanol has a relatively high vapor pressure under normal process conditions, so deep refrigeration or special recovery methods are required to prevent high solvent losses. Water washing of effluent streams is often used to recover the methanol.

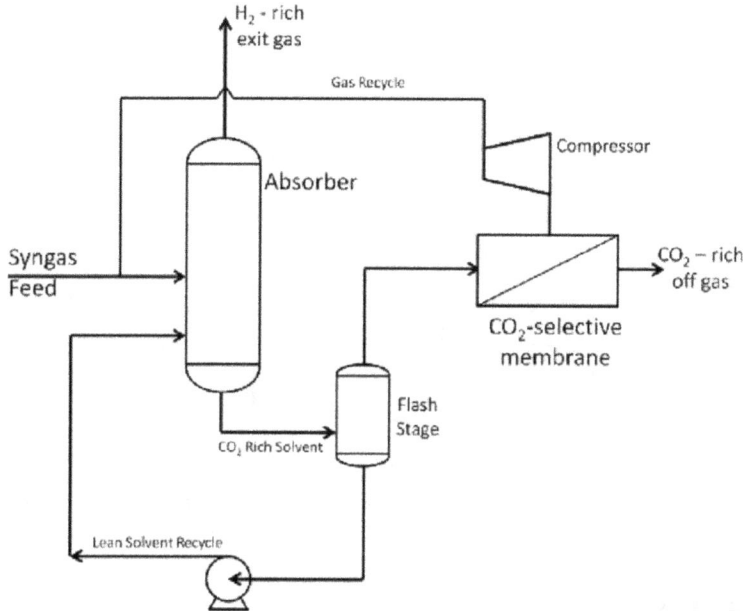

FIGURE 5.3 Simple process flow diagram for Rectisol process with CO2 selective membrane for CO2 recovery.

The Rectisol process typically operates below 32°F (0°C). Due to low temperatures, approximately 5% of the material in a Rectisol plant is stainless steel. High selectivity of methanol for H2S over CO2, combined with the ability to remove COS, is the primary advantage of the process. A simplified scheme of the Rectisol process with membrane is shown in Figure 5.3.

5.3.2 DEPG (DIMETHYL ETHER OF POLYETHYLENE GLYCOL)

DEPG is a mixture of dimethyl ethers of polyethylene glycol (CH3O(C2H4O) n CH3 (n is between 3 and 9) used to physically absorb H2S, CO2, and mercaptans from gas streams. Solvents containing DEPG are licensed and/or manufactured by several companies including Coastal Chemical Company (as Coastal AGR), Dow (Selexol), and UOP (Selexol). Other process suppliers such as Clariant GmbH of Germany offer similar solvents. Clariant solvents are a family of di-alkyl ethers of polyethylene glycol under the Genosorb® name. The Seloxol process uses dimethyl ether of polyethylene.

- Can be selective towards H2S and can bring down up to 4 ppm.
- CO2 can be brought down to very low level (2% to 0.1%).
- Water dew point can be controlled.
- Organic sulfur compounds like mercaptans, disulfides, and COS can be removed.

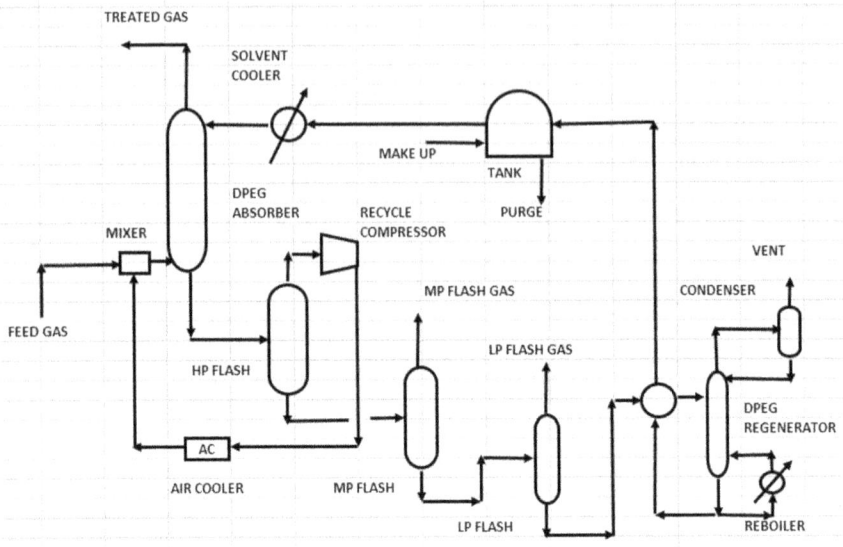

FIGURE 5.4 Simplified scheme of acid gas removal with DEPG.

- Siloxanes are removed from landfill gas.
- Metal carbonyls are removed from gas produced from gasifiers.

DEPG can be used for selective H2S removal that requires stripping and vacuum stripping. The process can be configured to yield both a rich H2S feed to the Claus unit as well as bulk CO2 removal. Selective H2S removal with deep CO2 removal usually requires a two-stage process with two absorbers and one regeneration column. H2S is selectively removed in the first column by a H2S lean solvent that has been thoroughly stripped with steam, while CO2 is removed in the second absorber. The second-stage solvent can be regenerated with air or nitrogen for deep CO2 removal, or use a series of flashes if bulk CO2 removal is required. Like the other solvents, DEPG also dehydrates the gas and removes HCN. Compared to the other solvents, DEPG has a higher viscosity that reduces mass transfer rates and tray efficiencies and increases packing or tray requirements, especially at reduced temperatures. Since it is sometimes necessary to reduce temperature to increase acid gas solubility and reduce circulation rate, this could be a disadvantage. DEPG requires no water wash to recover solvent due to very low vapor pressure. DEPG is suitable for operation at temperatures up to 347°F (175°C). The minimum operating temperature is usually 0°F (–18°C). A simplified sketch is presented in Figure 5.4.

5.3.3   NMP (N-METHYL-2-PYRROLIDONE)

The Purisol process, which uses NMP, is licensed by Lurgi AG. The flow schemes used for this solvent are similar to those used for DEPG. The process can be

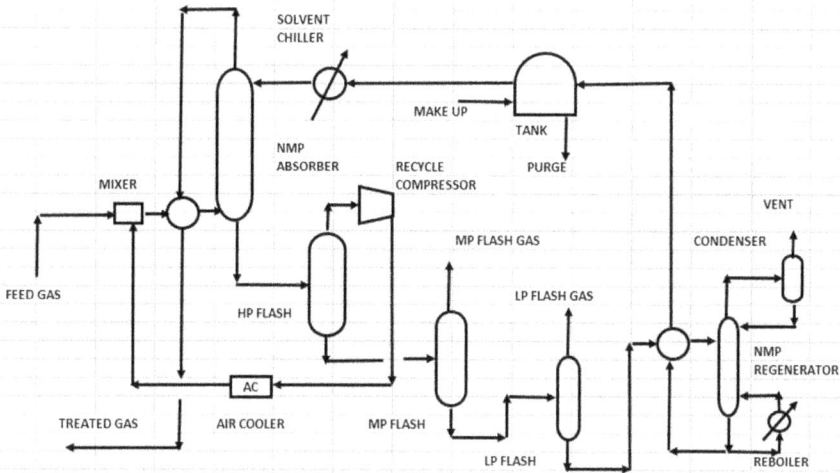

FIGURE 5.5 Simplified scheme of acid gas removal with chilled NMP.

operated either at ambient temperature or with refrigeration down to about 5°F (–15°C). NMP has a relatively higher vapor pressure compared to DEPG or PC, and the licensor usually recommends water washing of both the treated gas and the rejected acid gases for solvent recovery. NMP recovery with water is not necessary if the Purisol process is operated at subambient. NMP has the highest selectivity of all the physical solvents considered here for H2S over CO2. COS is not as soluble as H2S, but it is known to hydrolyze by the NMP solvent. *The Purisol process is particularly well suited for the purification of high-pressure, high CO2 synthesis gas for gas turbine integrated gasification combined cycle (IGCC) systems because of the high selectivity for H2S.* A simplified scheme is presented later in Figure 5.5.

5.3.4   PC (PROPYLENE CARBONATE)

The Fluor solvent process, which uses PC, is licensed by Fluor Daniel Inc. and has been in use since the late 1950s. PC is is particularly *advantageous in treating syngas.* PC has an advantage over the other solvents when little or no H2S is present and CO2 removal is important. PC has lower solubilities for light hydrocarbons in natural gas and hydrogen in synthesis gas. This lower solubility results in lower recycle gas compression requirements for the gas flashed from the rich solvent at intermediate pressures, and lower hydrocarbon losses in the CO2 vent gas stream. It is primarily intended for the removal of CO2 from high-pressure gas streams. The *CO2 removal process with Fluor solvent is known to improve synthesis gas plant reliability* (American Chemical Society, April 1974). Fluor solvent usually becomes competitive when the CO2 partial pressure in the feed is at least 75 psi (5.2 bar). A simplified scheme is presented later in Figure 5.6.

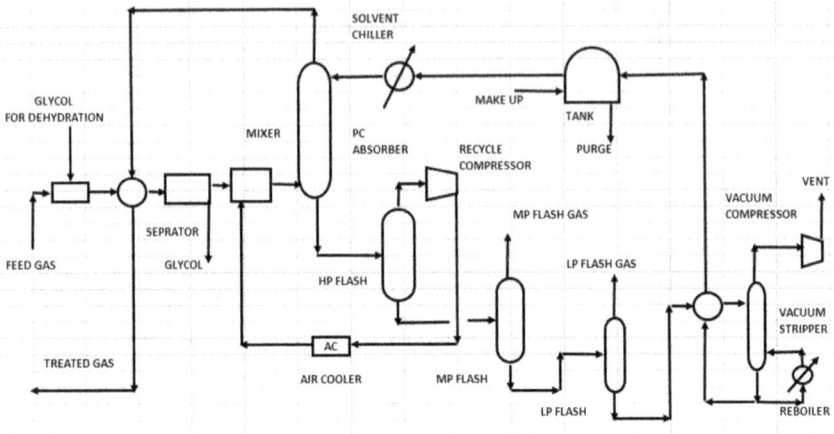

FIGURE 5.6 Simplified scheme of acid gas removal with chilled PC.

5.3.5 COMPARISON OF PHYSICAL PROPERTIES OF SOLVENT AND GAS SOLUBILITIES

The physical solvents considered here are non-corrosive, relatively non-toxic, and require only carbon steel construction. The selection of a physical solvent process depends on process objectives and characteristics of the solvents. Some of those characteristics include selectivity for H2S, COS, HCN, etc.; Table 5.1 compares some of the physical properties for the four physical solvents.

All of these physical solvents are more selective for acid gas than for other constituents (e.g. hydrogen, carbon monoxide, methane) and are presented in Table 5.2.

TABLE 5.1
Physical Properties of Solvents

Solvent	DEPG	PC	NMP	MEOH
Process Name	Selexol/Coastal AGR	Fluor Solvent	Purisol	Rectisol
Molecular weight	280	102	99	32
Specific gravity at 25°C (kg/m³)	1030	1195	1027	785
Viscosity at 25°C (c P)	5.8	3.0	1.65	0.6
Vapor pressure at 25°C (mmHg)	0.00073	0.085	0.4	125
Freezing point (°C)	−28	−48	−24	−92
Boiling point at 760 mm Hg (°C)	275	240	202	65
Specific heat at 25°C	0.49	0.339	0.40	0.566
Thermal conductivity (BTU/Hr) (ft) (°C)	0.11	0.12	0.095	0.122
Maximum operating temperature (°C)	175	65	—	—
CO2 solubility (ft³/gallon) at 25°C	0.485	0.455	0.477	0.425

TABLE 5.2

Solubilities of Various Gases in Physical Solvents Relative to CO2

Gas Component	DEPG at 25°C	PC at 25°C	NMP at 25°C	MeOH at −25°C
Hydrogen	0.013	0.0078	0.0064	0.0054
Nitrogen	0.020	0.0084	—	0.012
Oxygen		0.026	0.035	0.012
Carbon monoxide	0.028	0.021	0.021	0.020
Methane	0.066	0.038	0.072	0.051
Ethane	0.42	0.17	0.38	0.42
Ethylene	0.47	0.35	0.55	0.46
Carbon dioxide	1.0	1.0	1.0	1.0
Propane	1.01	0.51	1.07	2.35
I-Butane	1.84	1.13	2.21	—
N-Butane	2.37	1.75	2.21	—
Carbonyl sulfide	2.30	1.88	2.72	3.92
I-Pentane	4.47	3.5	—	—
Acetylene				
Ammonia	4.8	—	—	23.2
N-Pentane	5.46	5.0	—	—
Hydrogen sulfide	8.82	3.29	10.2	7.06
N-Hexane	11	13.5	42.7	—
Methyl mercaptan	22.4	27.2	34	—
N-Heptane	23.7	29.2	50	—
Carbon disulfide	2.3	1.88	2.72	3.92
Ethyl mercaptan	—	—	78.8	—
Sulfur dioxide	92.1	68.6	—	—
Dimethyl sulfide	—	—	91.9	—
Benzene	250	200	—	—
Thiophene	540	—	—	—
Water	730	300	4000	—
Hydrogen cyanide	1200	—	—	—

(Due to the relatively high volatility of Methanol as compared to others, the solubilities for Methanol are presented at −25°C.)

From Table 5.2 it can be seen that the minor gas impurities such as carbonyl sulfide, carbon disulfide, and mercaptans are quite soluble in most organic solvents. These compounds are removed to a large extent along with the acid gases. The solubility of hydrocarbons in organic solvents increases with the molecular weight of the hydrocarbon. Thus, hydrocarbons above ethane are also removed to a large extent and flashed from the solvent along with the acid gas. Physical solvent processes are generally not economical for the treatment of hydrocarbon streams that contain substantial quantities of pentane-plus hydrocarbons as needs a stripper for solvent regeneration. The capacity of solvent for absorbing acid gases increases as the temperature is decreased and reduces operating cost due to lower solvent circulation rate. The solubility of CH4, H2, and CO undergoes little change with temperature, increasing the selectivity of absorption of acid gascomponents.

5.3.6 PHYSICAL SOLVENT REGENERATION

The simplest version of a physical solvent process involves absorption followed by regeneration of the solvent by successive flashing to atmospheric pressure or vacuum, or by inert gas stripping in a stripper. If H2S is present at only very low concentrations or is entirely absent, this flow scheme is usually applicable since CO2 concentrations as high as 2% or 3% can often be tolerated in the product gas. When H2S is present in significant amounts, thermal regeneration is usually necessary to accomplish the thorough stripping of the lean solvent needed to reach stringent H2S purity of the treated gas required. As previously noted, PC cannot be thermally regenerated since it is unstable at the high temperature required to completely strip H2S from the rich solvent. Heat requirements in solvent regenerators are usually much less for physical solvents than for chemical solvents such as amines since the heat of desorption of the acid gas for the physical solvent is only a fraction of that for chemical solvents. The circulation rate of the physical solvent may also be less, particularly when the acid gas partial pressure is high.

5.3.7 SIMPLIFIED FLOW SCHEME OF FEW PHYSICAL SOLVENT PROCESSES

The flow configuration for the DEPG and NMP cases with no chilling are very similar. The DEPG flow configuration is shown in Figure 5.4.

The flow configuration for chilled DEPG and NMP cases is the same. The NMP chilled solvent configuration is shown in Figure 5.5.

For the PC case, the flow configuration is different from that of DEPG and NMP in the feed section as it needs chilling. PC requires ethylene glycol (EG) to be injected upstream of the feed gas cooler (to absorb moisture from the gas to arrest the formation of gas hydrates and ice), andseparator is provided downstream to separate the glycol from the chilled feed and is sent to glycol regenerator for regeneration of the EG and recycled. Since the PC operating temperature cannot exceed 65°C, a 2 psia vacuum stripping column utilizing medium pressure flash gas is used instead of the reboiler used for the other solvents. A simplified scheme is shown in Figure 5.6.

5.4   PHYSICAL-CHEMICAL SOLVENTS AND PROCESSES

The principle of this process combines the high absorption potential of alkanolamine (chemical absorption) and the low regeneration energy requirement of the physical solvent (physical absorption) for solvent regeneration. The Sulfinol process is the sole process with large industrial references. Shell is the licensor of the Sulfinol process. The Sulfinol process is based on the use of an aqueous mixture of alkanol amine (diisopropanolamine or methyldiethanolamine) and sulfolane

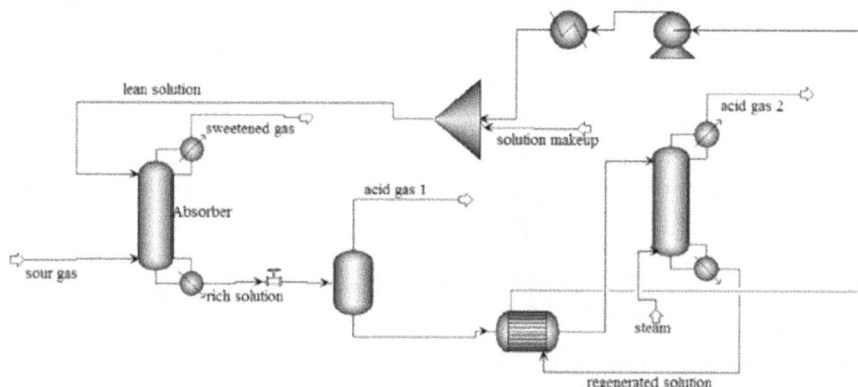

FIGURE 5.7 Simplified Sulfinol-M process.

(tetrahydrothiophene dioxide). The process is called Sulfinol-D when DIPA is used. The process is known as Sulfinol-M when MDEA is used. Sulfinol-D generally is a mixture (45 wt.% DIPA, 40 wt.% sulfolane, and 15 wt.% water). It combines the chemical absorption effect of the amine and the physical absorption effect of the sulfolane. Figure 5.7 shows the typical process flow diagram of a Sulfinol process. it is very similar to the classical alkanol-amine processes with an absorption section and a regeneration section working on the thermal swing principle.

Like in amine processes, the feed gas enters the bottom part of the absorber and is in counter-current contact with the descending solvent fed from the top of the column. The treated gas comes through the top of the absorber, and the rich solvent solution leaves the bottom and is flashed at a medium pressure. The flashed solution collects in the flash drum. The gas from the flash drum can be routed to the fuel gas pool. An amine wash is typically provided in the flash gas outlet to absorb H2S, if present.

The rich solution leaving the flash drum is preheated in a lean solution/rich solution heat exchanger and is let down to the regenerator operating pressure. The regenerator top product (acid gas and water vapor) is condensed (by water or air coolers) and collected in the reflux drum. The overhead product from the reflux drum is the wet acid gas which can be routed either to an acid gas flare or to a sulfur recovery unit. Condensed water from the reflux drum is pumped back to the top of the column as total reflux. A potential problem can occur when feeding an acid gas stream from a Sulfinol unit to a sulfur recovery unit (Claus process for instance) due to the presence of hydrocarbons (which can cause problems in sulfur quality and specification). The Sulfinol solution can solidify at temperatures around –2°C (special care is needed for storage of solution). Degradation of the solution due to COS is negligible. However, a reclaimer could be needed on DIPA-based units treating large quantities of CO2 (DIPA quality can be degraded with CO2).

Like amine processes, Sulfinol can be subject to foaming. Possible causes of foaming in the absorber are similar to those in case of amine units. Table 5.3

TABLE 5.3
Comparative Merits of Physical Solvent Processes[4]

Process	Advantages	Disadvantages
Selexol™	• Non-thermal solvent regeneration • Non-corrosive solvent • Dry gas leaves from the absorber	Most efficient at elevated pressures
Rectisol™ Ipexol-2™	• Non-foaming solvent • High chemical and thermal stability • Non-corrosive solvent	• High refrigeration costs • High capital costs • Amalgams formation at low temperature
Fluor™	• High CO2 solubility • Non-thermal regeneration • Simple operation • Non-corrosive solvent	• High solvent circulation rates • Expensive solvent
Purisol™	• Non-foaming solvent • High chemical and thermal stability • Non-corrosive solvent • Low volatility	• High compression cost • Most efficient at high pressure
Sulfinol™	• High capacity • Low solvent circulation rate	• Foaming issues • Corrosive solvent • Thermal regeneration

shows a comparison of the relative merits for all the common physical solvent processes.

Preferred application areas of the processes are presented next (opinion of experts):

- Physical solvent process Selexol is widely used in NG treating both for on shore and off shore applications.
- Rectisol process is usually deployed for treating synthesis gases produced from gasification of heavy oils, coal, petroleum coke rather than natural gas.
- The Flour process (using PC as solvent) is also known to be advantageous for treating synthesis gas.
- The Purisol process using NMP as solvent is particularly well suited for the purification of high-pressure, high CO2 synthesis gas for gas turbine integrated gasification combined cycle (IGCC) systems because of the high selectivity for H2S
- DPEG can additionally remove metal carbonyls from gases produced from gasification of heavy oil / petroleum coke and siloxanes from landfill gases.
- All the physical solvents remove H2S, CO2, COS, CS2 and mercaptans in varying degrees and are also able to remove water from sour gas streams. However, if water wash is provided in treated gas for solvent recovery the treated gas will be wet.

- Methanol process requires the lowest circulation rate and lowest net power but requires the maximum number of equipment.
- Stripping gas and a vacuum compressor are required for the chilled PC because thermal regeneration (i.e. reboiler) is not possible due to of degradation of solvent at higher temperature. Furthermore, PC cannot be used for selective H2S removal.
- All of the physical solvents can be used successfully for bulk removal of CO2. A detailed analysis must be performed to determine the most economical choice of solvent based on the product.

5.5   CHEMICAL SOLVENTS AND PROCESSES

5.5.1   CONVENTIONAL AMINE-BASED SOLVENTS USED FOR ACID GAS REMOVAL AND CARBON CAPTURE AND STORAGE (CCS)

The amine-based chemical absorption process has been used for CO2 and H2S removal (acid gas removal) in gas-treating plants since the 1950s and is considered to be by far the most developed CO2 capture process. Amine-based solutions are traditionally used for treating natural gas to remove acid gases to acceptable limits of H2S to 4 ppm v and CO2 to mutually agreed limits. A simplified sketch of an amine absorber and regenerator is presented in Figure 5.8.

This acid gas reaction is reversible and the acid gas can be released by heating the rich solution in a separate stripping column.

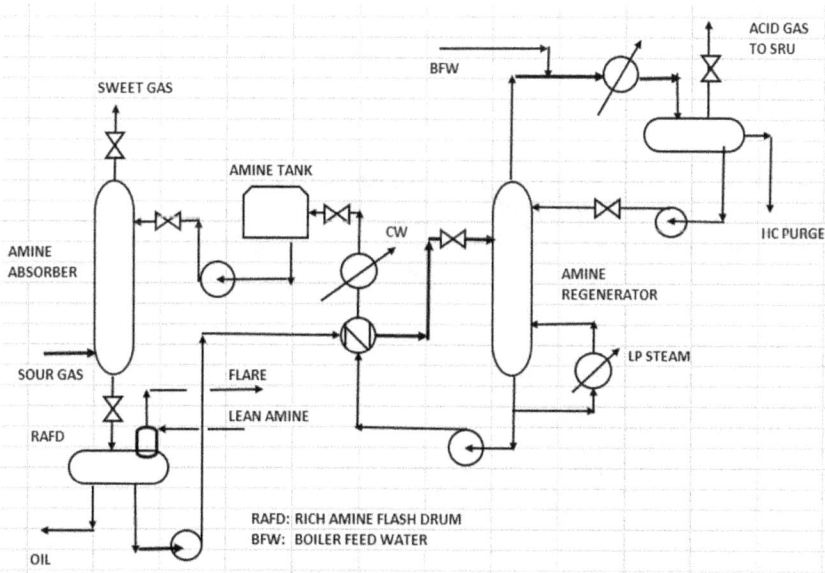

FIGURE 5.8   Typical amine absorber and regenerator.

Recently, amine-based chemical absorption came up as a potential technology that can be applied to reduce carbon dioxide emissions in industrial processes such as fossil fuels power plants, cement production, and iron and steel manufacturing. Post-combustion carbon capture using amines is developed for use in industrial applications.

The alkanol amines are volatile, cheap, and safe-to-handle compounds and are commonly classified by the degree of substitution of the central nitrogen atom. A single substitution denotes a primary amine, a double substitution denotes a second-ary amine, and a triple substitution denotes a tertiary amine. Each of these alkanol amines has at least one hydroxyl group and one amino (amine) group. The molecular structures of primary and secondary amines have hydrogen atoms attached with the central nitrogen, whereas the tertiary amines have no hydrogen atom attached to nitrogen. This structural characteristic plays an important role in the acid gas removal capabilities of the various treating solvents. Figure 5.9 shows the most widely used amines (primary amine MEA) for the treatment of acid gases, which are described in detail.

Primary alkanol amines, such as monoethanolamine (MEA) and diglycolamine (DGA), provide high chemical reactivity, favored kinetics, medium-to-low absorp-tion capacity, and acceptable stability. MEA, the first-generation and the most well-known amine-based absorbent, is highlighted by its high chemical reactivity with CO2 and low cost. These properties can reduce the absorber height and ensure a feasible operation. MEA-based scrubbing technology is suitable for acid gas removal and, in particular, post-combustion capture from the flue gas. However, needs high energy for stripping, due to high heat of reaction, higher solvent circulation rate due to lower concentration in aqueous solution and lower acid gas loading practiced (to arrest corrosion in CS metallurgy). Further, it is susceptible to, oxidative and thermal degradation and can cause corrosion.

DGA presents similar properties like MEA. in many areas, It has a low vapor pressure and can be used at higher concentrations, typically between 40% and 60% wt. in aqueous solution. This is considered the biggest advantage in reducing circula-tion rate and regenerator reboiler duty.

Secondary alkanol amines, such as diethanolamine (DEA) and diisopropanol-amine (DIPA), which have a hydrogen atom directly bonded to the nitrogen, show intermediate properties compared to primary amines and they are considered as an alternative to MEA. DEA is more resistant to degradation and shows lower corrosive

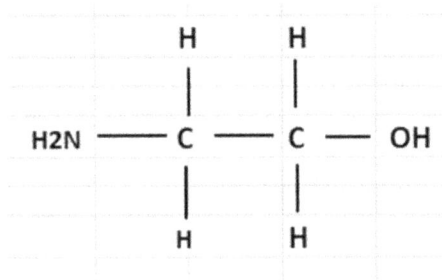

FIGURE 5.9 Structural general formula of a primary amine (MEA).

behavior and needs lower energy for regeneration than MEA. DIPA also needs a lower energy for solvent regeneration than MEA.

Finally, *tertiary amines*, such as triethanolamine (TEA) or methyldiethanolamine (MDEA), are characterized by having a high equivalent weight, which causes a low absorption capacity, low reactivity, but high stability. There are a few differences in the performance of primary and secondary amines as they are compared to tertiary amines for the CO2 separation process. Primary and secondary amines are very reactive; they form carbamate by direct reaction with CO2. On the other hand, tertiary amines can only form a bicarbonate ion by reaction of CO2 with water in amine solution (hydration of CO2) as they do not have a hydrogen ion attached to nitrogen. Hydration is a slower process than the direct reaction by carbamate formation. Hence, tertiary amines show low CO2 absorption rates. In general, the main issues of amine-based chemical absorption includes:

- High energy consumption during solvent regeneration.
- Lean amine concentration in amine solution is maintained within limit tolerated by carbon steel metallurgy.
- Corrosion requires the use of both inhibitors and resistant materials in their application or limiting the acid gas loading of rich amine for carbon steel metallurgy.
- Degradation in the presence of O2, SOx, and other impurities, such as particles, HCl.

Table 5.4 shows the typical amine solution concentrations, rich amine loading, and heat of reactions for components of acid gas.

5.5.2 STERICALLY HINDERED AMINE SOLVENTS

Sterically hindered amines are considered a type of amine that can improve CO2 absorption rates in comparison with the common primary and secondary amines. They are usually amino alcohols. A sterically hindered amine is formed by a primary or secondary amine in which the amino group is attached to a tertiary carbon

TABLE 5.4
Different Amines and Their Strength, Loadings, and Heat of Reactions

Attributes	MEA	DEA	DGA	MDEA
Maximum lean amine concentration, wt.%, In CS metallurgy	15%–20%	25%–30%	50%–65%	40%–50%
Rich amine maximum loading as moles of acid gas per mole of amine in CS metallurgy	0.2	0.3	0.35	0.5
		Heat of Reaction		
BTU per pound H2S	610	555	674	530
BTU per pound CO2	825	730	850	610

$$CH3$$
$$|$$
$$H2N \text{ ——— } C \text{ ——— } CH2 \text{ ——— } OH$$
$$|$$
$$CH3$$

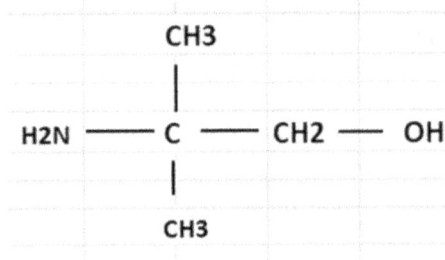

FIGURE 5.10   Sterically hindered amine, amino methyl propanol (AMP).

atom, as shown in Figure 5.10. The molecular structure of primary amines (MEA) is shown in Figure 5.9 and a sterically hindered amine (2-amino-2-methyl-1-propanol or AMP) is shown in Figure 5.10 to illustrate the difference.

These amines are characterized by the capability of forming carbamates of lower stability. The lower stability of the carbamate formed by CO2-amine leads to low energy consumption to release CO2 than the common primary and secondary amines. The use of this type of amine is known to reduce the energy requirement for regeneration by up to 20% compared to conventional MEA-based scrubbing, due to the formation of weak bonds.

Sterically hindered amine requires only one amine molecule to capture one molecule of CO2. Based on this assumption, the maximum CO2 loading using sterically hindered amines (AMP) is higher than for unhindered, primary or secondary amines.

5.5.3   NON-AMINE-BASED SOLVENTS

Non-amine-based solvents generaly use *potassium carbonate* as a solvent. Potassium carbonate (K2CO3) is an alkaline salt that is used for the removal of acid gas. Hot potassium carbonate (hot pot) is traditionally used for removing H2S and CO2 from producer gas/synthesis gas. One such process is the Benfield process. This uses an absorber for removing acid gas components and regenerates the rich solution to remove the acid gases for solvent reuse. The process contains a gas absorption step and a carbonate regeneration step. K2CO3 is the alkali in the absorption solvent. K2CO3 reacts with CO2 and H2S in the absorber tower forming bicarbonate (HCO3^{-1}) and bisulfide (HS^{-1}). The chemical reactions between the acid gases and K2CO3 are well known:

The spent solvent is steam stripped of the absorbed CO2 and H2S in the regeneration tower. The regenerated solvent is recycled back to the absorber after heat recovery using heat exchangers. At ambient condition, potassium carbonate is used in aqueous solution. In order to keep the carbonate dissolved in the water, the solution must be kept at a temperature (more than 60°C for a 30 wt.% carbonate solution). Unlike amine solutions, potassium carbonate solution is not susceptible to oxygen contamination. The product is available on the market, but processes based on its use are being developed by licensors. The major licensor is UOP with the Benfield processes, which have the largest number of industrial references in natural gas sweetening. The CATACARB process is also based on potassium carbonate but with less industrial reference. Sodium carbonate can be also a good solvent for CO2 absorption, but many commercial references are not available.

Potassium carbonate is used in aqueous solutions with concentrations of around 30 wt.% (molecular weight of potassium carbonate is 138.2).

5.5.4 CHEMICAL REACTIONS IN THE ABSORBER AND REGENERATOR

The reactions can be written as $K_2CO_3 + CO_2 + H_2O = 2KHCO_3$ and $K_2CO_3 + H_2S = 2KHS + KHCO_3$ for absorption. These reactions are reversible. They proceed from left to right in the absorber unit and from right to left in the regeneration unit. The absorption reaction is exothermic. The heat of reaction of K_2CO_3 with H_2S is 22 BTU/SCF (819 kJ/Sm3 of H_2S). The potassium bisulfide (KHS) is very difficult to regenerate, so the hot pot process is only suitable for sweetening gas mixtures having no H_2S or a very high CO_2/H_2S ratio.

The absorption reaction with CO_2 is very exothermic. The heat of reaction of K_2CO_3 with CO_2 is 32 BTU/SCF of CO_2 (1191 kJ/Sm3 of CO_2). When H_2S is present with CO_2 in the feed gas, sulfur salt resulting from the chemical reaction with H_2S is difficult to regenerate and thus loads unnecessarily the solution, which then loses its efficiency. *Therefore, this process is not suitable for selective removal of CO2 when H2S is present in considerable quantities in the feed gas.* To enhance the absorption of acid gas, an additive (amine) can be added to the potassium carbonate. COS and CS2 are removed by the hot pot process. Mercaptans can be removed to a certain degree (depending on their molecular weight) by the potassium carbonate solution. Typical sketches of the scheme are given in Figure 5.11 and Figure 5.12.

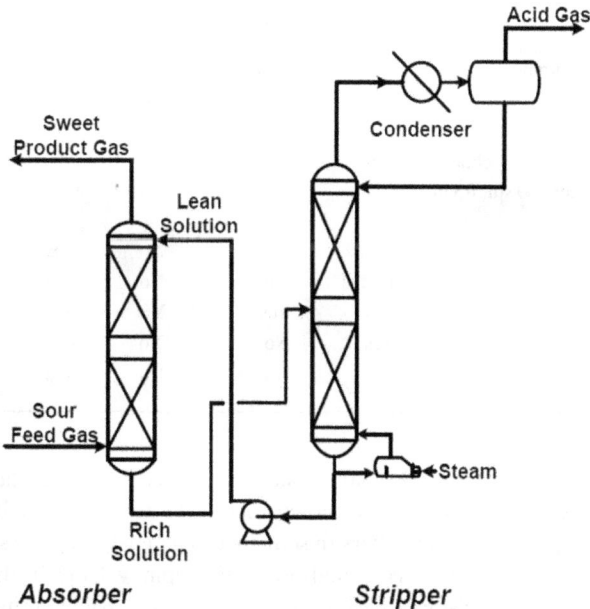

FIGURE 5.11   Benfield process.

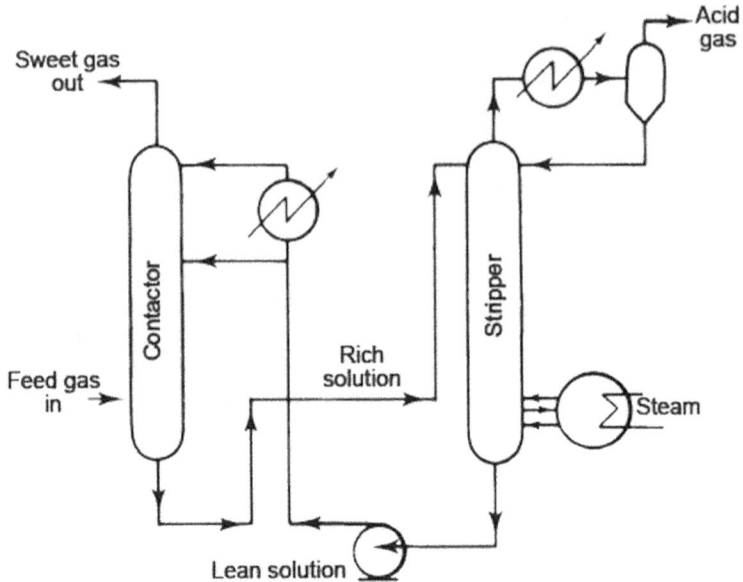

FIGURE 5.12   Split flow variant. (Source: *GPSA Engineering Data Book.*)

TABLE 5.5
Comparison of Attributes of Some Solvents

Attribute	MEA	DEA	K2CO3	Sulfinol	DGA
Lean solution concentration wt.%	15%	25%	30%	80%	60%
Stripping stream rate LB/Gal. Solution	1.2	1.2	0.6	0.8	1.35
Sweet gas specifications (pipeline grade of ¼ grain per 100 std. ft3 in treated gas)	Yes	Yes	No	Yes	Yes
HC (C3+) absorption	No	No	No	Yes	Little
Removal of RSH	Partial	Partial	No	Yes	Partial
Removal of COS	Yes	Yes	Yes	Yes	No
Non-regen product due to COS	Yes	No	No	No	No
Reclaimer required	Yes	No	No	Yes	No

A more complex process (Hi-Pure process) has been developed (also known to be a UOP process). The process is known to involve treatment by by amine and potassium carbonate in two absorbers in sequence. The Hi-Pure process can achieve LNG specifications (4 ppm V H2S and 50 to 100 ppm V CO2 in the sweet gas). Table 5.5 shows a comparison of performances between classical amines, potassium carbonate, and Sulfinol-D. A rigorous simulation may be performed to actually evaluate the performance.

5.6 RECENT TRENDS AND ADVANCES

5.6.1 SOLVENT BLENDS

Solvent regeneration is an energy-intensive process. Amine blends can offer potential improvements in CO_2 chemical absorption to reduce the regular reboiler duty and the solvent circulation rates. The addition of small amount of tertiary amines (MDEA and TEA) in aqueous solutions of primary or secondary amines (MEA and DEA) can constitute a solvent blend that enhances the overall behavior of the solvent in terms of energy requirements for solvent regeneration and higher resistance to solvent degradation. For this reason, different researchers are studying novel solvent formulations and blends, involving fast kinetic solvents such as MEA with other slow kinetic solvents like TEA, benzylamine (BZA), and MDEA.

MEA is a fast solvent, but it is known to be almost 50 times slower than PZ. The CO_2 absorption rate of MEA can be significantly improved by adding small amounts of PZ as a promoter. This blend can improve the individual CO_2 absorption rate. Potassium carbonate promoted with PZ is also considered as a promising solvent, along with the PZ and AMP blends. A summary of the most promising amines blends is given in Table 5.6.

Some advantages of blending these amines are

- Improved thermodynamic efficiency.
- Reduction in issues relating to degradation and corrosion.
- Flexibility is made available to tailor and optimize the composition of the solvent to achieve the highest absorption efficiency.
- Energy requirement for solvent regeneration can be reduced.

PZ is also a fluid used for CO_2 and H2S scrubbing in association with MDEA. Amine blends that are activated by concentrated piperazine are used extensively in

TABLE 5.6
Promising Amine Blends

Solvent blends	Abbreviation
Piperazine and potassium carbonate	$PZ + K_2CO_3$
2-amino-2-methyl-1-propanol and piperazine	AMP + PZ
2-amino-2-methyl-1-propanol and 1,2-ethanediamine	AMP + EDA
3-methylamino propylamine and dimethylmonoethanolamine	MAPA + DMMEA

commercial CO2 removal for CCS because piperazine advantageously allows for protection from significant thermal and oxidative degradation at typical coal flue gas conditions. The thermal degradation rates for MDEA and PZ are negligible, and PZ protects MDEA from oxidative degradation.

Piperazine's solubility in water is low, so it is often used in relatively small amounts in an amine solvent. Use of piperazine in solvent can remove a much higher proportion of CO2 from the sour (flue) gas stream per unit mass of solvent.

Piperazine can be thermally regenerated through multistage flash distillation and other methods after being heated to temperatures up to 150°C and recycled back into the absorption process, providing for higher overall energy performance in amine gas treating processes.

5.6.2 CHEMISTRY OF REACTION OF PIPERAZINE

The amine groups of piperazine react readily with carbon dioxide to produce PZ carbamate at a low loading (mol. CO2/equiv. PZ) range and PZ bicarbamate at an operating range of 0.31–0.41 mol. CO2/equiv. PZ, enhancing the rate of overall CO2 absorbed under operating conditions. Due to these reactions, there is limited free piperazine present in the solvent, resulting in its low volatility and rates of precipitation as PZ-6H2O. The reaction mechanism is shown in Figure 5.13.

Propietory simulation softwares provide flexibility to blend amines and piperazine for use in acid gas absorption/ regeneration studies and design and are extensively used.

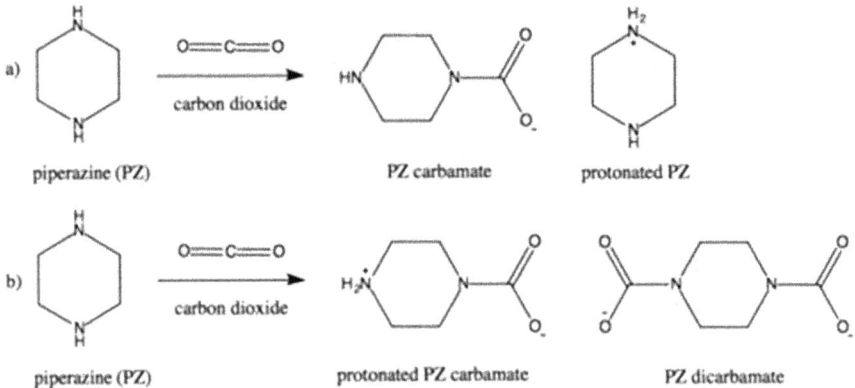

a) at low loading
b) at operating range of concentrated piperazine process

FIGURE 5.13 Chemistry of piperazine reactions.

5.6.3 INTRODUCTION TO IONIC LIQUIDS

A novel generation of solvents came up recently as an alternative for traditional amine-based solvents, namely ionic liquids (ILs). The IL solvents are discussed only to introduce the reader to the emerging new generation of solvents. Ionic liquids (ILs) are salts comprising cations and anions, and usually liquids at or below 100°C (having low melting points). The salts that are liquids at room temperature are generally called as room-temperature ionic liquids (RTILs). They have high boiling points and thus have low vapor pressure, which and can selectively absorb acid gases such as CO2 and SO2, needing relatively low regeneration energy. ILs are typically formed with the combination of a large organic cation like imidazolium, pyridinium, or phosphonium cation with either an inorganic anion such as Cl^-, BF_4^-, and PF_6^-, or an organic anion like RCO_2^- and $CF_3SO_3^-$.

Conventional ILs interact with CO2 as a physical solvent. Recently, researchers have been focusing on the development of ILs as a promising solvent for CCS based on their properties as a solvent for CO2 capture.

A major disadvantage of the application of ILs is known to be the relatively high viscosity, limiting their mass transfer capabilities. Further, they can become excessively viscous once CO_2 is absorbed, producing solvent pumping issues and mass transfer, and are also relatively more expensive than common amine-based solvents.

The next generation of ILs is known to be a combination of conventional ILs known with a functionalized amine group. Based on this configuration, amine-functionalized IL, also called task-specific IL, reacts with CO2 by chemisorption showing further improvements on the CO2 capture. The development of this type of IL is known to enhance the performance of IL in both biogas and natural gas treatment and CCS. Table 5.7 summarizes the application of a few ILs with reported best performance.

In addition, there are several other water free solvents like amino silicones, non-aqueous organic blends those may find wider applications. Amines with super base promoters, biphasic solvents, and TETA/ethanol blends also promise a greater application in future.

TABLE 5.7
Some IL Names and Possible Areas of Application

IL formulation	Abbreviated name	Areas of Application
1-butylpyridinium tetrafluoroborate	[Bpy] [BF4]	Post-combustion
Try-hexyl (tetradecyl)-phosphonium imidazole	—	Post-combustion
1-butyl-3-methyl-imidazolium hexafluorophosphate	[bmim] [PF6]	Post-combustion
1-butyl-3-methyl-imidazolium acetate	[bmim] [Ac]	Biogas/natural gas upgrading

5.7   DIFFERENT TYPES OF MEMBRANE PROCESSES

Seperation through membrane is based on the principle that certain gas compounds dissolve and diffuse through the polymeric material at a faster rate than others. Carbon dioxide, hydrogen, helium, hydrogen sulfide, and water vapor are highly permeable (fast gases). Conversely, nitrogen, methane and heavier paraffin compounds are less permeable. Gases that permeate fast are known as permeate and the ones that permeate slowly are known as residues or retentate. Membrane is an attractive process to remove CO_2 (bulk removal) and traces of H_2S from natural gas. H_2S removal when present in high concentrations (in terms of mole %) is not practiced due to the detrimental effect of this compound on the membrane. This technology cannot produce sales gas specifications or LNG specifications (H_2S lower than 4 ppm V and CO_2 less than 90 ppm V). Its application can be for the bulk removal of CO_2 from nearly H_2S free natural gases.

The adsorption of the CO_2 is better at high pressure (high CO_2 partial pressures). This process is not suited for low pressure operations. The polymeric material to be selected for membrane construction must be permeable to CO_2 and selective (to avoid permeation of hydrocarbon gas components).

There are two types of membranes:

- Spiral wound membrane
- Hollow fiber membrane

5.7.1   SPIRAL WOUND MEMBRANE

The membrane (see Figure 5.14) typically comprises of an active layer of cellulosic acetate supported by a porous material to form a leaf or flat sheet. Several leaves (or flat sheets) of this kind are wound over a central coil, drilled with holes. These

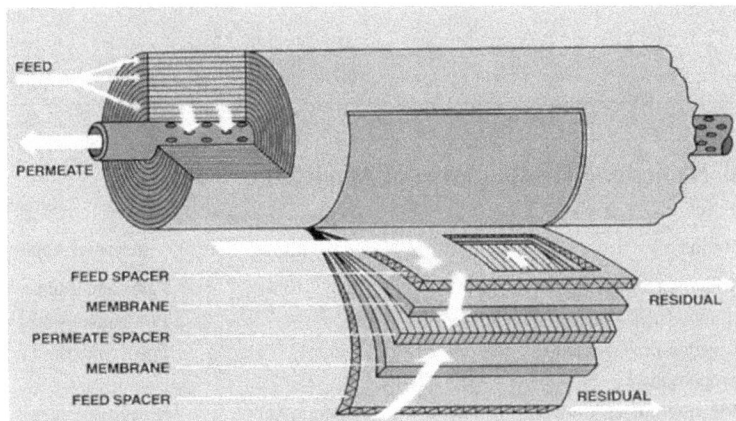

FIGURE 5.14   Spiral wound membrane.

leaves are separated by spacers for both residue and permeate gas (CO2 rich gas). The membrane element is then housed in a pressure tube with holes for collection of permeate. A tube can have several membrane elements. Feed gas is introduced to one side of the membrane element. It passes over the membrane and exits the membrane elements as a treated gas with essentially no pressure drop. The CO2-rich gas that permeates through the membrane flows through the holes drilled on the central tube and is recovered at lower pressure.

5.7.2 HOLLOW FIBER MEMBRANE

Approximately 70% of membrane systems used are hollow fibers. This technology utilizes small cylindrical fibers comprising an active layer supported by a porous material. These fibers are hollow in the center to carry the permeate stream. Each fiber generally has an external diameter of 600 microns and a bore less than 300 microns. Hollow fibers are arranged in a bundle around the tube. The fibers are tightened one against the other by an epoxy coating. Each membrane element is then fitted in a pressure tube. The raw gas enters the central tube and leaves the holes of the tube to contact the fibers. CO2 permeates through the fibers. Hollow fiber membrane (shown in Figure 5.15) generally cannot withstand high pressure (which is not the case for spiral wound membrane). Moreover, hollow fiber membrane generally has a lower permeability than spiral wound membrane. However, hollow fiber technology provides a better surface area per volume of membrane element (5 to 10 times more than spiral wound membrane).

The membrane process can be a single-stage process or a multistage (normally two-stage) process.

5.7.3 ONE-STAGE MEMBRANE PROCESS

Figure 5.16 shows a single-stage membrane process. The set of membranes handles the feed gas and removes the bulk of the CO2. However, light hydrocarbon losses (especially methane) cannot be avoided (light hydrocarbons permeate and come with the CO2 in the permeate stream). Methane concentration in the permeate stream can exceed 10 mole%.

The arrangement in Figure 5.16 involves no moving parts and is simple and reliable. But it results in low hydrocarbon recovery.

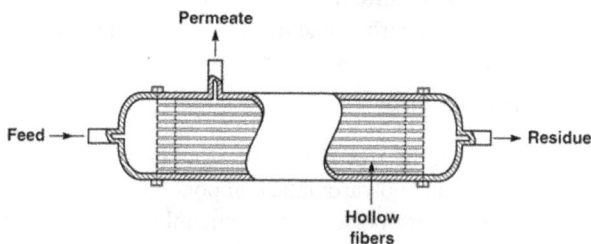

FIGURE 5.15 Hollow fiber membrane.

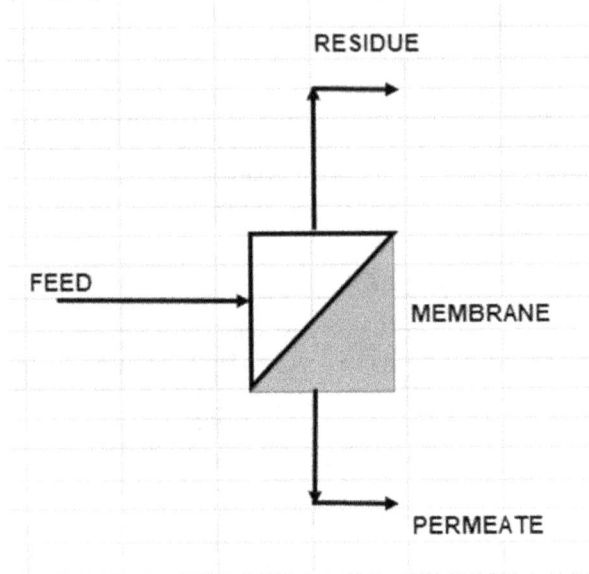

FIGURE 5.16 One-stage membrane process.

5.7.4 TWO-STAGE MEMBRANE PROCESS

A two-stage membrane process is shown in Figure 5.17. To mitigate the hydrocarbon losses pertaining to a single-stage system, a two-stage process can be considered. In a typical two-stage process, the permeate gas from the first stage of membranes (or first set), which still contains hydrocarbons, is sent (as a feed gas) to a second stage of membranes (second set), which ensures the separation between CO_2 and hydrocarbons (methane losses can be reduced to less than 2 vol.% of the feed gas methane content). With the permeate gas from the first stage being at low pressure, a compression is required to route this gas to the second stage. The methane-rich gas evolving from the second stage (second-stage residue gas) can be either recycled to the plant inlet at the relevant pressure or routed to the fuel gas system of the plant if required.

The scheme in Figure 5.17 allows higher CO_2 recovery and lower hydrocarbon losses, but needs a compressor and air cooler. A typical three-membrane stage arrangement is shown in Figure 5.18. It can handle feed with high CO_2 and can yield intermediate recovery of hydrocarbon and needs lower compression.

5.7.5 MEMBRANE PRETREATMENT

Membranes can get damaged by free liquids that can be carried over with the feed gas. Liquid water, aromatic and polyaromatic components (naphthalene for instance), glycol, methanol, and amine are poisons for the membrane. In order to protect and extend life of the membrane afeed gas pretreatment may be required at the upstream of the membrane package.

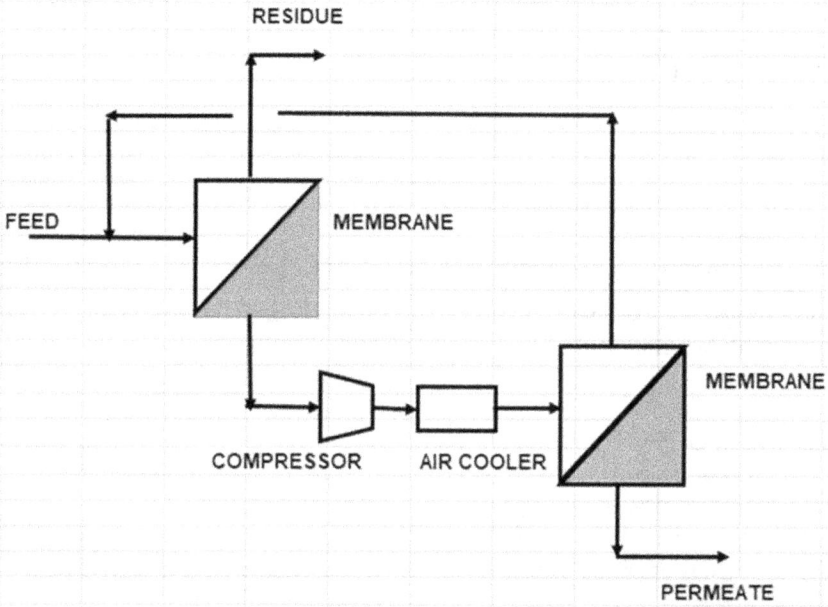

FIGURE 5.17 Two-stage membrane process.

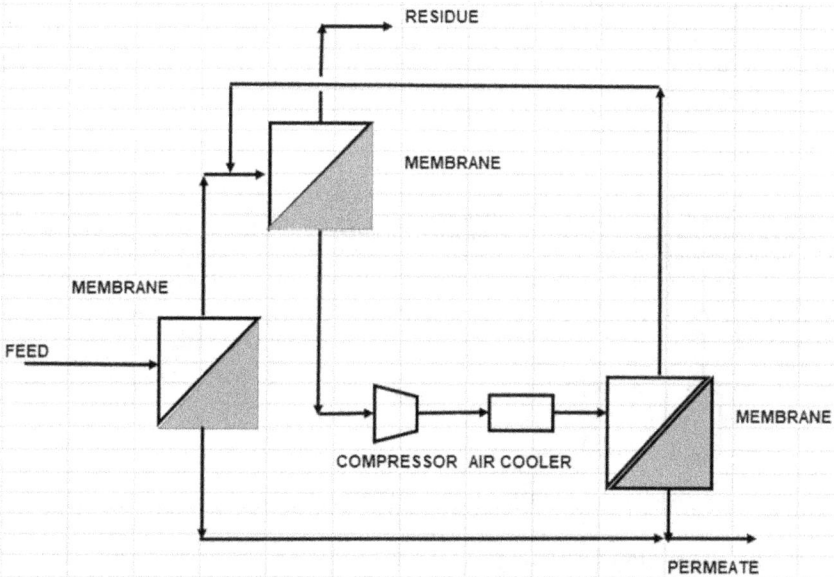

FIGURE 5.18 Three-stage membrane process.

5.7.6 Membrane Life

Even with a high feed gas quality, the life of membranes is from three to five years. After this period, it will be necessary to replace them with new membranes. However, a membrane system deployed for hydrogen separation from hydrocracker off-gasses (from medium pressure separator) lived far longer than the aforementioned period. Normally a very high extent of removal of acid gases by membrane is not practiced.

5.8 EXPERIENCES OF AMINE UNIT IN INDUSTRIES

Amine units are essentially comprised of several amine absorbers and usually with a common regenerator, as shown in Figure 5.19. It has an amine filtration section (clean up section), and storage and pumping facilities. The absorption of acid gases is a reversible exothermic process and the spent amine regeneration is an endothermic process needing heat for releasing the acid gas.

Continuous effort is made to optimize the heat required for amine regeneration by use of different types of amines, amine blends, configuration modification, use of additives, increasing amine strength, etc. Corrosion in an amine plant is another issue and needs attention. Other issues are amine losses, amine degradation, and amine foaming.

5.8.1 Effect of Amine Strength on Reboiler Condenser Duties of Amine Regenerator

An example is considered to compare the effect of amine strength on reboiler and condenser duties of an amine regenerator. In the first case, feed to the column (T1) is 30% wt. of amine (having MDEA 40 MT/Hr + 93.3 MT/Hr of water) having 4.56

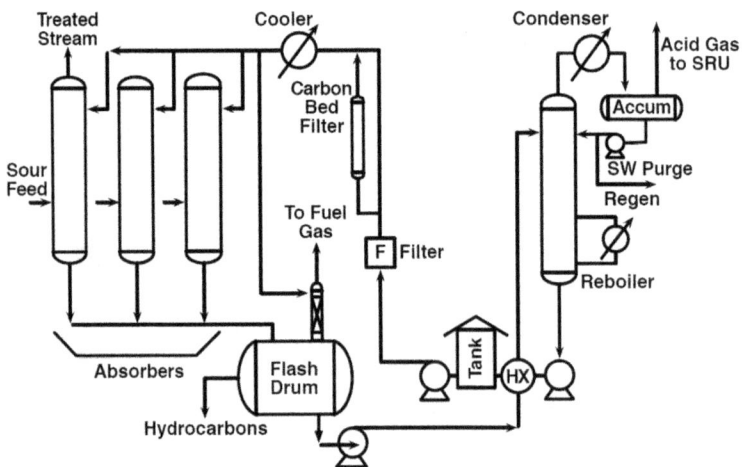

FIGURE 5.19 Flow scheme of amine absorption with regeneration unit.

MT of H2S absorbed in it. In the second case, feed to the column (T1_1) is 40% wt. of amine (having MDEA 40 MT/Hr + 60 MT/Hr of water) having the same quantity of 4.56 MT of H2S absorbed in it. Simulations are run to yield the same lean amine H2S loading of 0.0062 mole H2S/mole MDEA with the same feed preheat of 55°C and the regenerator operated at the same top pressure (1 kg/cm2 g). The resultant condenser and reboiler duties for both cases are computed and presented in Figure 5.20. They are further tabulated in Table 5.8.

The simulation results show a large reduction in reboiler and condenser duties when the amine strength is increased to 40% compared to corresponding duties in case of lower amine strength of 30%. This reduction is primarily attributable to the requirement of lower boil up in the regenerator to maintain the same lean amine H2S loading (0.0062 mole H2S/mole of MDEA) for a richer solution and in turn is due to a lower amine circulation rate.

In a few amines, rich amine flash drums (RAFD) are designed for higher pressure (5–6 Kg/cm2) operation and do not use a pump for feeding rich amine from the RAFD to the amine regenerator. In those units the hydrocarbon separation may be less efficient in RAFD compared to an arrangement with a lower operating pressure in the RAFD and a pump for feeding rich amine to the regenerator. This is primarily due to an increase of hydrocarbon solubility in amine at higher pressure. Solubility of hydrocarbons in heavier amines (like MDEA) is higher than amines with lower molecular weights (like DEA, MEA).

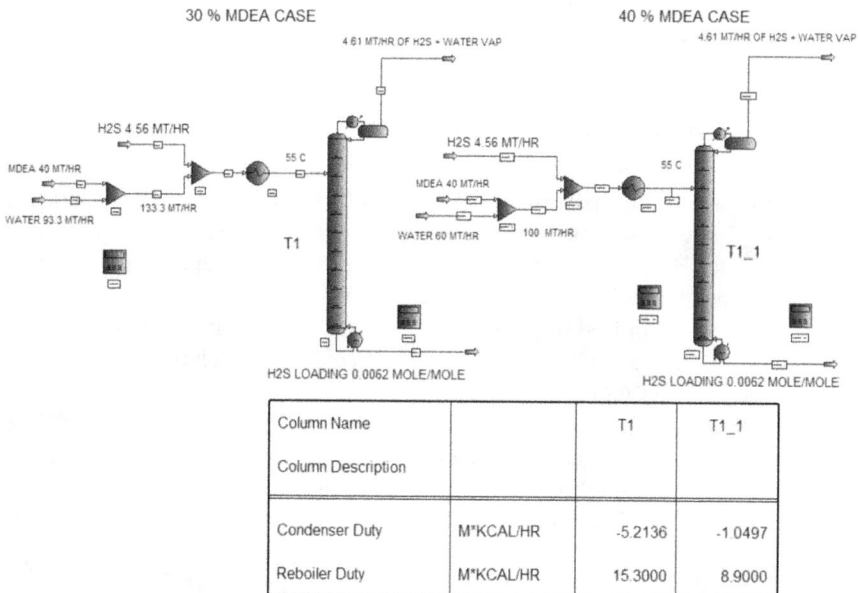

Column Name		T1	T1_1
Column Description			
Condenser Duty	M*KCAL/HR	-5.2136	-1.0497
Reboiler Duty	M*KCAL/HR	15.3000	8.9000

FIGURE 5.20 Reboiler and condenser duties with two amine strengths.

TABLE 5.8
Reboiler Condenser Duties with Different Strength of Amines (30% and 40%)

Item Name	Unit	T1 (30% Weight Amine in Water)	T1_1 (40% Weight Amine in Water)
H2S in rich amine solution	KG/Hr	4.56	4.56
Neat MDEA flow	MT/Hr	40	40
Flow of amine solution (amine circulation)	MT/Hr	133.3	100
Rich amine H2S loading	Mole of H2S/Mole of MDEA	0.399	0.399
Lean amine H2S loading	Mole of H2S/Mole of MDEA	0.0062	0.0062
Condenser duty (Qc)	MMKCAL/Hr	−5.2	−1.1
Reboiler duty (QR)	MMKCAL/Hr	15.3	8.9

5.8.2 COMMON PROBLEMS IN AMINE UNITS

- Corrosion in the unit
- Higher acid gas components in sweet gas
- Foaming and carryover
- Amine contamination
- Higher lean amine loading
- Amine degradation
- Carry over of amine to the reflux drum of the regenerator

5.8.2.1 Amine Degradation

Amine quality has a strong impact on the operation of an amine unit. Amine that is weak, dirty, or partially spent by the formation of heat stable salts (HSS) cannot do a good job of sweetening gas. Amine not bound by HSS, is the active ingredient for acid gas pick up. HSS in amine solution are those amine salts that do not dissociate at regenerator conditions. These salts tie up amine molecules, which then cannot react with acid gas, and can lead to fouling and corrosion. Acidic anions can come from the gases being treated (HCl, HCN, SO2, formic acid), and result in the formation of HSS. Oxygen can get into the amine from unblanketed tanks or from the treatment of vacuum column ejector off gases with amine solution. Formic acid can result from CO ingress in the alkaline amine solution and several other organic acids (including formic acid) are products of the hydrolysis of nitriles formed in cat crackers. Polymeric and other organic amine degradation products are often included in the HSS category.

5.8.2.2 Foaming

Amine solution becomes frothy when agitated by gas as on a tray in a gas absorber or regenerator. Fortunately, this froth quickly subsides in good quality amine solution and the column functions normally. However, this froth can be stabilized with changes in the surface properties of the amine solution or more commonly by the presence of a third phase in the column. The third phase can be solids such as iron sulfide or it can be a second liquid phase such as hydrocarbon. The froth does not break in the tray downcomer and flooding can occur. Pressure drop increases across the column, the operation becomes unstable, and as a result capacity is reduced.

5.8.2.3   Corrosion

In general, all amines are considered non-corrosive. Most amine units are built with carbon steel. Steel reacts with H2S to form a layer of FeS on the surface that protects the steel from further attack and corrosion. As long as the layer of FeS is intact, the overall corrosion rate is quite low. The iron sulfide layer is soft and friable. It can be worn away by high flow velocity or turbulence. High HSS content increases the viscosity and density of the amine solution and can destroy the protective layer. At the same time, if higher amine solution circulation rates are increased (required to circulate enough free amine to pick up the acid gas), solution velocities go up. This combination can result in erosion of the protective iron sulfide layer exposing bare steel to fresh attack by H2S and results in higher rates of corrosion. These are generally areas that are exposed to flashing of acid gas and areas subject to turbulence. Corrosion is usually observed in top and bottom of the absorber, areas where column shell comes in contact with rich amine. The rich lean amine exchanger and downsteam is also very susceptible to corrosion. The overhead of amine regenerator (air cooler/ water cooler is extremely corrosion prone area). The amine reboiler outlet and regenerator bottom section are also moderately corrosion prone location. Figure 5.21 shows the corrosion prone areas shown with dark lines and the most of the equipment provided in amine absorber, regenerator are shown in Figure 5.22.

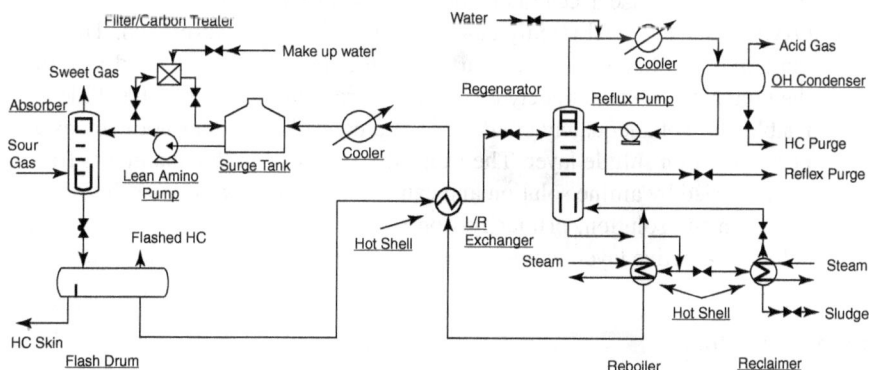

FIGURE 5.21   Corrosion prone locations in amine unit (shown with dark lines).

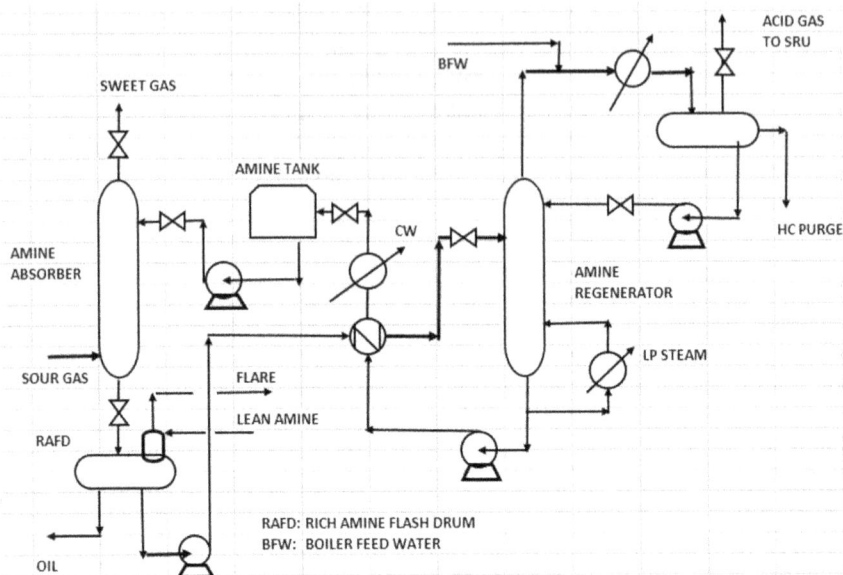

FIGURE 5.22   Typical amine absorber and regenerator.

Other contaminants that could accelerate corrosion include:

- Excessive ammonium bisulfide or cyanide in the regenerator overhead. Both compounds can penetrate the iron sulfide protective layer and attack steel.
- High levels of chloride ions in the amine solution can attack the stainless-steel alloys used in corrosion-prone areas of the amine unit.
- Excessive acid gas loading, above 0.5–0.6 mole /mole in the rich amine. Loadings above about 0.5 moles of acid gas per mole of amine forms the amine bisulfide salt. The bisulfide ion can penetrate the iron sulfide protective layer and cause accelerated corrosion of carbon steel.
- Oxygen and wet steel rapidly can result in scale or rust formation. This is most likely to happen during startup or turnaround. The oxidized iron in the pipe scale is immediately converted to iron sulfide. This iron sulfide is friable and not adequately bonded to the pipe wall, so it does not work as a protective iron sulfide layer. The iron sulfide formed in this manner is free to be carried by amine solution and can circulate along with the solution.
- Solids in the solution, primarily iron sulfide, can be abrasive to the iron sulfide protective layer, exposing the new steel surface to sulfidic attack.

5.8.2.4   Fouling

Another common problem amine systems experience is the accumulation of sludge or scale on column trays, and in stagnant or low velocity areas of the unit. Scale formation on trays leads to gradual choking of the disperser openarea. This can result in jet flooding and loss of gas capacity in the column. Areas with low flow velocity, such as the rich amine flash drum, often accumulate sludge. This sludge looks like black shoe polish or wax. It is thought to contain a mixture of contamination products—iron sulfide, heavy hydrocarbons, polymeric amine degradation products, and other solid material.

5.8.2.5 Poor Sweet Product Quality

If the level of contaminants increases to a point where they affect the good operation of the amine unit, sweet gas quality becomes inferior and may result in environmental issues. Poor treating in a natural gas plant can result in the sales gas failing to meet pipeline specification. Bigger upset in a sulfur unit results from the sudden entry of slug of hydrocarbon in the acid gas leading to fluctuations in air control. During the air control upset, high concentrations of SO_2 can be sent to the tail gas treating unit. If the SO_2 enters the amine system of tail gas treating unit can react with the amine to form heat stable salts.

5.8.2.6 Carry Over of Amine to Reflux Drum of the Regenerator

Often amine carry over takes place to the reflux drum of an amine regenerator. This can happen due to foaming and carry over and also due to poor performance of wash trays (typically 3 to 4 trays) located above the feed point.

5.8.3 Sources for Contamination

The sources of contamination can be through feed gas, make up water and oxygen ingress in amine tank and the in the system during turnarounds.

Feed gas can be the source of most of the contaminants to enter the amine system. Ammonia is a gas that is readily absorbed by the aqueous part of the amine solution in absorbers. Ammonia can cause corrosion in the amine unit and operability problems in the sulfur recovery unit (SRU). Organic acids like formic acid and mineral acids like HCl that can contribute to heat stable salt formation. Amine is an organic chemical so there is some natural affinity for hydrocarbon gas to dissolve in the amine solution. This is a physical phenomenon favored by the higher molecular weight amines such as MDEA or DGA, higher concentration amine solutions, and at higher pressures such as in gas plants or in high severity hydrotreaters in the refinery. Oxygen can contribute to amine degradation and corrosion.

Hydrocarbons can enter the amine system in the liquid phase. Feed gas to a scrubber is a hydrocarbon gas normally at its dew point. Thus, any cooling of the gas stream will result in condensation of the heavier components in the gas stream. A special case is liquefied petroleum gas (LPG) treating. LPG can be physically absorbed by the amine solution. It can also be carried from the LPG contactor in the rich amine stream exiting the LPG amine treater.

Make up water is used to prepare amine solution and compensate for water loss from the system. Fresh amine is usually delivered to the plant as 100% or 85%

amine and must be diluted to the desired solution strength. Practically, all amine systems lose water from the amine solution during normal operation. The amine solution loses water vapor with the treated gas in the absorbers, and some water is continually lost to the sulfur recovery unit along with the acid gas feed from the regenerator Water is frequently purged from the regenerator overhead reflux drum to control ammonia, cyanides in the reflux system. Therefore, water is routinely added to the amine system. Contaminants in the water will accumulate in the amine solution. Steam condensate is the best choice. Boiler feed water may contain sulfite, oxygen scavengers, phosphate corrosion inhibitors, and filming amines. City water may contain considerable calcium hardness, sodium, some chlorine, and may be high in chlorides.

Oxygen ingress in amines results from oxygen contamination is an amine surge tank, if the tank is breathing to atmosphere. In refineries, vapor recovery systems and flare gas recovery systems, vacuum off-gas treatment, and tail gas treatment of sulfur recovery can be common sources of oxygen ingress in amines.

Oxygen contamination and ingress in the system can also result during unit startup and maintenance turnarounds. Any pipe scale or rust, formed due to contact with oxygen is immediately converted to iron sulfide when exposed to H2S during operation. This iron sulfide formed does not remain bonded to the pipe surface and equipment walls, and is carried away by the circulating solution.

5.8.4   ACTIONS TO ARREST CONTAMINATION OF AMINE SYSTEMS

Arresting contaminants from getting in the amine solution is considered to be the first step to limit the buildup of contaminants in the system and reduce the operating problems caused by contaminants. Commonly, knockout drums, filter separators, or water wash drums are used for the gas streams.

A *knockout drum* at the inlet of an amine scrubber helps to arrest the entrained hydrocarbon liquids and occasional hydrocarbon slugs from the feed gas. Demister pads are commonly used to facilitate the separation of mist or liquid droplets from the gas stream. A typical filter separator is shown in Figure 5.23.

Proprietary filter coalescers are common in natural gas plants and are gaining increasing use in refineries. They can be used in place of knockout drums. Proprietary filter separators are available from several suppliers. The filter separators are typically two-stage devices with replaceable filter elements. The first stage removes solid material from the gas stream and protects the elements in the second stage coalescer from getting chocked by the solid fines. The second stage coalesces fine oil droplets into bigger ones and separates them from the gas stream. Proprietary filter coalescers can be very effective for both liquid and solid contaminants. A *water wash* facility is often provided to water wash the gas streams from a fluid catalytic cracking unit, coker, or vacuum unit that may contain cyanides, acids, ammonia, and particulates using a drum or a small column. Sour water stripper feed is commonly used as makeup water because the ammonia is a good buffer, and this also limits the sour water generation of the refinery. This is typically located downstream of the gas knockout drum and just at the inlet of the amine absorber.

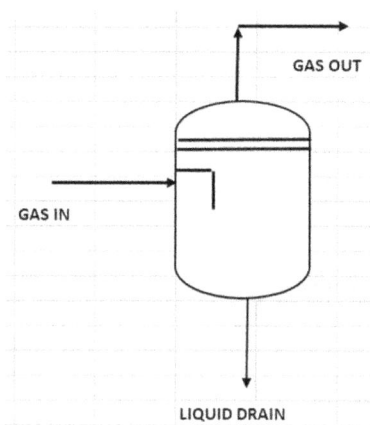

FIGURE 5.23   Knockout drums on feed gas.

Lean Amine Temperature is controlled always higher than the gas inlet temperature to avoid condensation of gas when contacted with lean amine. Condensed hydrocarbons can cause severe foaming in the absorber. Keeping the lean amine temperature 5°C (10°F) higher than the feed gas temperature helps to avoid condensation in the scrubber. However, the amine temperature may be kept lower for LPG or NGL wash so that NGL or LPG do not vaporize, in contact with hotter amine, under the pressure of the absorbers.

Iron sulfide formation cannot be prevented in a system that contains H2S. Good design practice involves maintaining low amine solution velocities in the rich amine piping (<1 meter/second) and lean amine line (<2 meter/second) to lower erosion of the sulfide layer. Alloys such as stainless steel are used in those areas prone to corrosion (where the likelihood exists for two-phase generation). These iron sulphide particles are removed only by the particulate filters located in the lean amine circuit (in the amine clean up section) and at the inlet of amine absorbers.

Blanketting of lean amine tank is needed to avoid oxygen contact with amine and is achieved by blanketing with nitrogen to avoid. Oxygen contamination can degrade the amine.

5.8.5   ACTIONS FOR LOWERING CONTAMINATIONS

Particulate filters are used to lower the solids and corrosion products in amine. Corrosion in the amine system produces fine iron sulfide particles. Excessive solids (>200–400 ppm wt.) can cause foaming in amine systems. Solids also tend to scour off the protective iron sulfide scale in piping and promote corrosion. Filtration of the lean amine is typically done to improve the quality of lean amine to absorbers Filtering the rich stream is also practiced and may be considered efficient. A variety of filtration systems are in use. Change of filter element can be based on a pressure drop across the filters. The rate of change of filter can be a rough measure of the rate of corrosion in the unit.

Carbon/charcoal bed filters are essentially provided in regenerated amine circuit to remove dissolved surface-active agents that may cause foaming. The dissolved contaminants are not removed by the particulate filters. The majority of amine systems use activated carbon treaters to remove surface active contaminants and traces of hydrocarbon. Particulate filters are provided upstream of the carbon filter to avoid plugging of the carbon bed with solids. Life of the carbon bed is monitored by visual inspection of the amine at the inlet and outlet of the carbon filter. There should be a perceptible color change and reduction of amine foaming tendency across the carbon filter. The typical carbon life is six months to a year. Carbon filters are generally not effective in controlling hydrocarbon content of the amine solution. The carbon filter is usually followed by another cartridge filter to arrest carbon particles slipping from the charcoal filter, if damaged.

Hydrocarbon skimming of the rich amine stream is performed in RAFD, at the inlet to amine regenerator. Rich amine from gas and liquid absorbers is saturated with hydrocarbons and can result in upset of the regenerator and the sulfur recovery unit. The rich amine from various absorbers is collected in a rich flash drum (RAFD) located at the inlet of amine regenerator, which removes the majority of the hydrocarbons. The drum should be designed with enough amine residence time to allow complete separation of hydrocarbons. Skimming facilities are provided to remove and recover the liquid hydrocarbons. Skimming can be done with a system of over-and-under baffles. The flash drum can operate at a pressure high enough to feed amine into the regenerator without pumps (3.2–4.6 bar G), or at a low pressure, and the amine can be fed to the amine regenerator by pumps. Observation of operating units suggests that the lower pressure option removed more hydrocarbon from amine and lead to fewer upsets downstream facilities.

Other skimming locations are also provided in the unit. Typically, the regenerator overhead accumulator drum is provided with an oil overflow baffle and provision of hydrocarbon skimming. Skimming connections are often provided in the amine scrubbers and at the lean amine surge tank.

Coalescer at inlet to the regenerator. are provided as an additional guard to arrest free oil ingress to the regenerator. A full-stream particulate filter should precede the coalescer on the rich amine stream. This arrangement (Figure 5.24) protects the

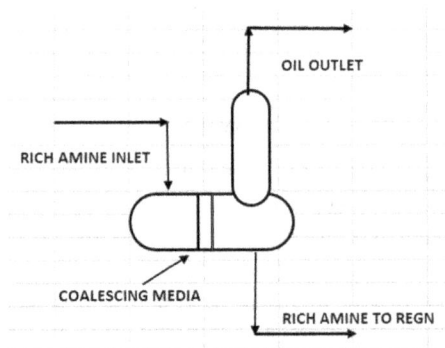

FIGURE 5.24   Oil amine coalescer on rich amine feed.

amine regenerator from fouling and is effective in limiting hydrocarbon ingress to the sulfur plant along with acid gas.

Kidney units are often installed to remove HSS from the amine system. Kidney units are small, permanently installed ion exchange systems that continuously treat a small slip stream of the circulating amine solutions to control HSS buildup in the circulating amine. Portable *Kidney* units can be also hired from amine suppliers to lower HSS. A *purge water* from amine regenerator overhead reflux drum is often provided to sour water stripper to arrest the build up of ammonia, cyanides in amine system. This is to limit the corrosion potential of ammonium bisulfide at the regenerator top and contain the reboiler and condenser duties of the regenerator as accumulation of ammonia can increase both the regenerator and condenser duties. Often amine carry over takes place to the reflux drum of an amine regenerator. This can happen due to foaming and carry over and also due to poor performance of wash trays (typically 3 to 4 trays) located above the feed tray. Higher amine content in reflux drum water does not permit bleed from the reflux drum water to sour water strippers as it would result in amine loss. Consequently, can result in accumulation of contaminants in amine system particularly ammonia. The wash trays in an amine regenerator have low liquid loading and normally operate in spray regime and likely to weep and adversely affect vapor liquid contact and can result in poor washing. Often splash baffles are used to arrest the liquid from the trays being flung directly to the down-comer and direct them to trays to ensure liquid level on the tray and good contact with vapor. If the liquid load is too low (lower than 0.1 GPM per linear inch of weir length) experts suggest gasketing of the tray segments or seal welding to avoid liquid weeping. Author has suggested gasketing / seal welding/ stitch welding of the sections of the wash trays in amine regenerator top to reduce weeping in several units. The implementation of the aforemention suggestion almost completely arrested the amine carry over to regenerator reflux drum and permitted the water purge from the drum to reduce contaminant build up. The heat of reaction for $NH_3 + H_2O \rightarrow NH_4OH$ is 87 Kcal / g-mole (highly exothermic). The forward reaction occur in top condenser while the reverse occurs in the reboiler. Thus, the purging of water improved the performance of the amine regenerators (reduced the lean amine loading with almost same reboiler duty.).

5.9   CASE STUDIES

5.9.1   HIGHER HYDROCARBON CONTENT IN ACID
GAS FROM AMINE REGENERATOR

A rich amine flash drum (RAFD) can be at low pressure as well as higher pressure. When a RAFD operates at low pressure, an amine pump is required to feed the rich amine through the rich amine/lean amine heat exchanger to the amine regenerator. Often the RAFD is operated at 5–6 kg/cm2 (g) pressure with N2 makeup to the RAFD, and no feed pump is provided for feeding rich amine to the regenerator. Amine is an organic chemical and is likely to dissolve a considerable quantity of hydrocarbons at higher pressure. Thus, the acid gas will contain a higher proportion

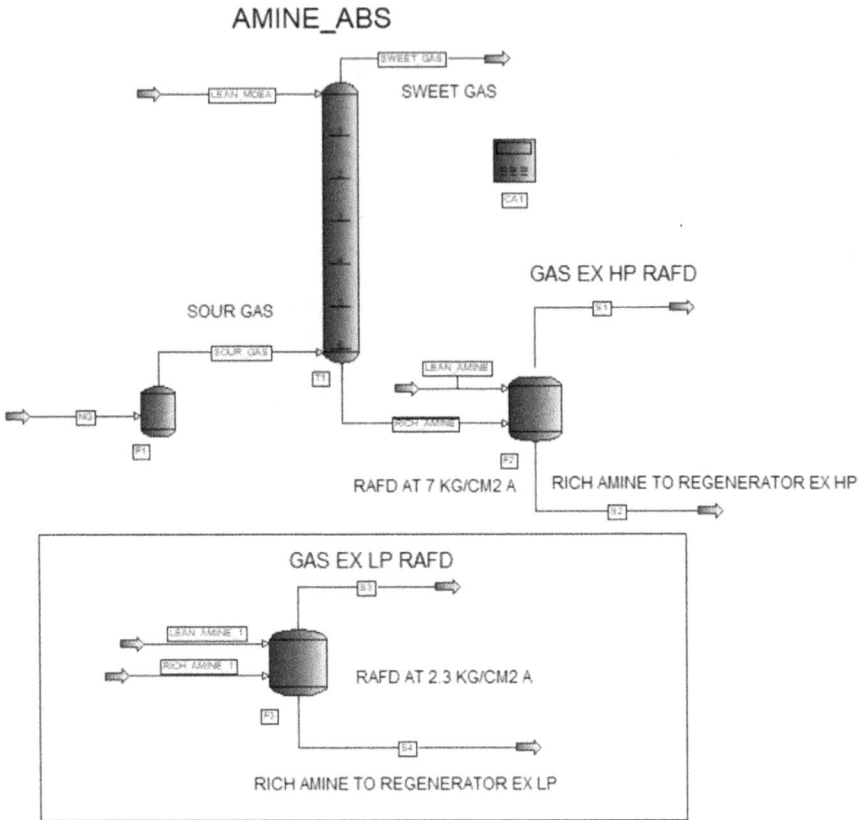

FIGURE 5.25   Simulation of RAFDs at two pressures.

of hydrocarbon. In the simulation Figure 5.25, sour gas is washed with lean amine. The absorber pressure is considered 18 kg/cm2 a. The rich amine loading at the bottom of the RAFD is 0.3482 mole H2S per mole of MDEA. In the RAFD top, a small stream of lean amine is introduced to capture some H2S. Typically it is done in an absorber having two to three stages. For simplicity, a single-stage is considered in the simulation shown in Figure 5.25. Results of the simulation are shown in Table 5.9.

The stream S1 and S2 denote flash gas and rich amine from the high-pressure (HP) RAFD at 7 kg/cm2 (a), and S3 and S4 flash gas and rich amine from low-pressure (LP) RAFD at 2.3 kg/cm2 (a). It can be seen that from the LP RAFD, almost double the quantity of hydrocarbon is released compared to the HP RAFD. This clearly shows that lower pressure operation of the RAFD may be better for reduction of hydrocarbons from rich amine and consequently from acid gas produced from the amine regenerator going to sulfur unit. The retention of heavier hydrocarbons at higher pressure is higher compared to lighter hydrocarbons. Further, the heavier amines have higher affinity for hydrocarbons. Thus, a low pressure RAFD

TABLE 5.9
Results of Simulation

Stream Name		Rich_Amine	S1	S3	S2	S4
Stream Description						
Stream Phase		Liquid	Vapor	Vapor	Liquid	Liquid
Total molar rate	KG-MOL/Hr	566.543	0.218	0.417	683.692	683.493
Total mass rate	KG/Hr	15583.272	5.007	10.212	18578.264	1857.3060
Temperature	°C	65.438	61.981	61.908	61.981	61.908
Pressure	KG/cm2	18.200	7.000	2.300	7.000	2.300
Total molecular weight		27.506	22.990	24.507	27.173	27.174
Total weight comp. rates	KG/Hr					
Water		8976.7746	0.1107	0.6429	10926.6639	10926.1317
H2S		597.5069	0.3371	1.9474	612.1698	610.5595
C02		0.0000	0.0000	0.0000	0.0000	0.0000
MDEA		5999.9997	0.0001	0.0004	7034.9997	7034.9994
Methane		4.2645	2.4675	3.7868	1.7970	0.4776
Ethane		1.3922	0.6272	1.1452	0.7650	0.2470
Propane		1.2269	0.5544	1.0071	0.6725	0.2198
IButane		0.5273	0.2630	0.4457	0.2643	0.0816
Butane		0.8352	0.3084	0.6276	0.5268	0.2076
IPentane		0.3892	0.1736	0.3163	0.2156	0.0729
Pentane		0.3551	0.1653	0.2926	0.1898	0.0625

(LP RAFD) is preferred in a unit operating with heavier amines like MDEA. Grass root amine regenerators were designed by author and team using LP RAFD. The author suggested switching to low pressure RAFD in couple of amine regenerators that changed the amine from DEA to MDEA to provide relief in the heavily loaded claus units located downstream.

5.9.2 SWEET GAS SPECIFICATION NOT ACHIEVABLE

It is often observed that acid gas content of sweet gas is not achievable. This can be attributed to higher lean amine H2S/CO2 loading leading to a lean end pinch. But mostly the absorber rich end pinch limits the achievable sweet gas H2S/CO2. This can be analyzed as follows.

A sour gas containing around 15% by mole CO2 and 0.5% by mole of H2S (in tail gas treating of sulfur unit) was to be treated at nearly atmospheric pressure (1.2 Kg/Cm2 a) to produce a sweet gas containing around 170 ppm of H2S. The lean amine (MDEA of around 40% wt. solution) flow maintained was 11000 kg/Hr. The desired sweet gas H2S content of 165 ppm mole could not be achieved. The lean amine loading was 0.007 mole H2S/mole MDEA. The case is presented in the simulation in Figure 5.26 and stream results are presented in Table 5.10.

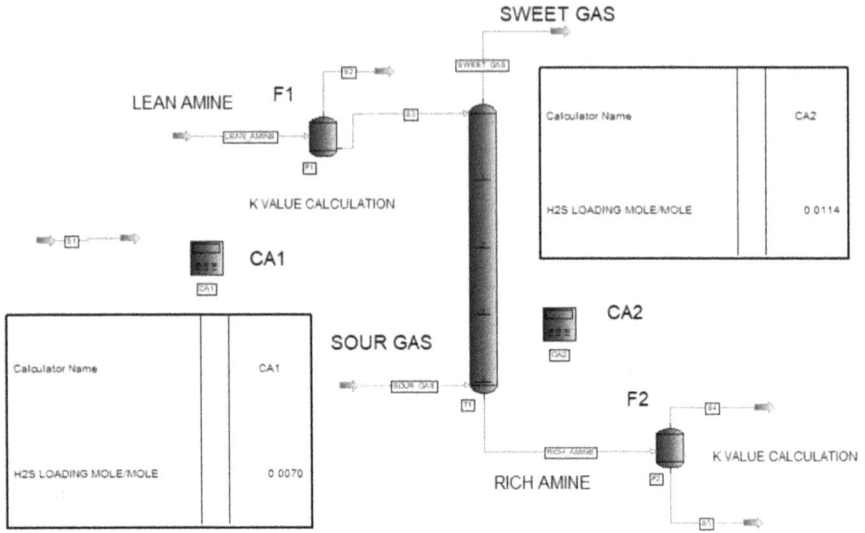

FIGURE 5.26 Simulation of amine absorber.

TABLE 5.10
Results of Simulation Showing Sweet Gas H2S Composition

Stream Name		Sour Gas	Lean Amine	Sweet Gas	Rich Amine
Stream phase		Vapor	Liquid	Vapor	Liquid
Temperature	°C	33.000	36.000	41.761	40.564
Pressure	KG/cm2	1.200	3.000	1.050	1.055
Total molar rate	KG-MOL/Hr	50.000	406.643	47.031	409.612
Total mass rate	KG/Hr	1670.382	11000.000	1449.580	11220.802
Total Molar Comp. (Percent)					
CO2		15.0000	0.0000	5.8975	1.1539
SO2		8.5000	0.0000	7.1098	0.2212
N2		75.0000	0.0000	79.6961	0.0044
H2S		0.5000	0.0625	0.1929	0.1009
H20		1.0000	91.0145	7.1034	89.6613
MDEA		0.0000	8.9230	0.0004	8.8583

The refinery operating team doubted a lean end pinch and was thinking of increasing the amine regenerator reboiler duty to lower the lean amine loading further. In the simulation in Figure 5.26 the sour gas amine absorber is configured with the flow rate and the composition of the sour gas as given (and also presented in Table 5.10) with a lean amine flow of 11000 Kg/H. A flash vessel F1 is used to generate the vapor–liquid composition and K value at the lean end, and a flash vessel F2 is used to generate the vapor–liquid composition and K value at the rich end. CA1 and CA2 are calculation blocks to compute lean amine and rich amine H2S loading in moles of H2S/mole of MDEA. The H2S loading computed at the lean and rich end, respectively, is 0.007 and 0.0114 Mole of H2S per mole of MDEA.

The vapor–liquid composition and K (equilibrium constant) value analysis at the lean end of the absorber was carried out and are presented as Table 5.11. It can be seen from the table that a sweet gas composition having H2S content of 55 ppm by volume can be achieved corresponding to the liquid molar composition of H2S of 0.0006246 in lean amine solution (which corresponds to lean amine H2S loading of 0.007 mole of H2S/mole of MDEA) at the lean end of the absorber. A pinch below the lean end (toward the rich end) may be responsible for higher H2S in sweet gas and can be resolved by an increase in flow of lean amine in the absorber.

Further, an X–Y plot was developed in all four stages of the absorber using the simulation of Figure 5.26 and is presented in Table 5.12. Practically, very little reduction of H2S loading is observed in sweet gas through stages (shown by Y values in stages from the fourth to first stage) indicating a near pinch situation in the column and suggesting an increase of flow of lean amine to the absorber.

Accordingly, in the same simulation as in Figure 5.26, lean amine flow is increased to 22 MT/Hr in the simulation and the sweet gas H2S ppm is found to get lower than 165 ppm (162 ppm) volume (or mole), as shown in Table 5.13. The refinery staff was convinced after examination of the analysis and agreed with the proposition to carry out necessary changes in the system to enable an increase flow of lean amine in the absorber and consequently to the amine regenerator. Similar situations are very common in amine absorbers where either the absorber lean end or rich end limits the achievable sweet gas H2S composition. The problem is more common with the rich end attaining equilibrium and can be resolved by increase of flow of lean amine. The lean end limit can only be resolved by lowering the H2S loading of lean amine from the regenerator by application of higher reboiler duty and sometimes by increase in amine strength.

TABLE 5.11

Vapor–Liquid Composition and K Values at the Lean End of the Absorber

	COMPONENT	VAPOR (1)	LIQUID	K-VALUE
1	C02	0.00000	0.00000	6.5016E+00
2	S02	0.00000	0.00000	3.1068E+01
3	N2	0.00000	0.00000	1.4738E+04
4	H2S	5.5187E–05	6.2461E–04	8.8B54E–02
5	H20	0.04601	0.91015	5.0555E–02I
6	MDEA	2.1835E–06	0.08923	2.4471E–05

(1) SUBCOOLED LIQUID, VAPOR SUM NOT EQUAL 1.0.

TABLE 5.12
X–Y Plot in Stages of the Absorber

Tray Molar Compositions

Component		Tray 1		Tray 2	
		X	Y	X	Y
1	CO2	0.00772	0.07298	0.00969	0.11772
2	SO2	0.00188	0.07263	0.00202	0.07947
3	N2	4.6642E-05	0.77934	4.3488E-05	0.72255
4	H2S	7.5348E-04	0.00216	8.1014E-04	0.00289
5	H2O	0.90138	0.07288	0.89938	0.07737
6	MDEA	0.08822	4.0229E-06	0.08805	4.3629E-06
Rate, KG-MOL/HR		336.51	48.10	337.17	51.90

		Tray 3		Tray 4	
Component		X	Y	X	Y
1	CO2	0.01040	0.12894	0.01192	0.13438
2	SO2	0.00206	0.07943	0.00226	0.08021
3	N2	4.B290E-05	0.71341	4.4803E-05	0.71847
4	H2S	8.8118E-04	0.00322	0.00106	0.003701
5	H2O	0.89847	0.07500	0.89599	0.06323
6	MDEA	0.08814	4.1047E-06	0.08872	3.0128E-06
Rate, KG-MOL/Hr		336.80	52.56	334.61	52.19

TABLE 5.13
Stream Summary with Higher Lean Amine Flow of 22000 Kg/Hr in the Absorber

Stream Name		Sour Gas	Lean Amine	Sweet Gas	Rich Amine
Stream phase		Vapor	Liquid	Vapor	Liquid
Temperature	°C	33.000	36.000	36.822	41.460
Pressure	KG/cm2	1.200	3.000	1.050	1.055
Total molar rate	KG-MOL/Hr	50.000	813.285	42.150	821.135
Total mass rate	KG/Hr	1670.382	22000.000	1242.883	22427.500
Total Molar Comp. (Percent)					
CO2		15.0000	0.0000	0.0130	0.9127
SO2		85.000	0.0000	5.6013	0.2301
N2		75.0000	0.0000	88.8794	0.0046
H2S		0.5000	0.0625	0.0162	0.0915
H2O		1.0000	91.0145	5.4899	89.9235
MDEA		0.0000	8.9230	0.0003	8.8377

5.9.3 Lower H2S Recovery in Tail Gas Treatment Unit Amine Absorber

In a tail gas amine absorber in a tail gas treatment unit (TGTU) the H2S recovery was not achievable. The lean amine (MDEA) temperature was kept higher than the water quenched gas temperature to avoid formation of hydrocarbon condensate and consequent foaming. The temperature of gas was 45°C–48°C. The operating team was maintaining the amine temperature around 55°C (as is practiced when the gas is saturated) to avoid condensate formation with contact with colder amine. It was not possible to bring down the treated gas H2S to the required level (150 ppm). It may be noted that the tail gas is not a gas saturated with hydrocarbons (but saturated with water) and not likely to produce hydrocarbon condensate (unless accidental hydrocarbon carryover beyond unit main combustion chamber (MCC) of Claus unit takes place). They were advised to lower the amine temperature almost to the same level as gas or lower. H2S recovery increased to near the required level. The problem was resolved.

5.10 CARBON CAPTURE

Current applications for CO2 include use in the food and beverage industry (higher relevance in Europe), water treatment, enhanced oil recovery (particularly in the United States), agriculture, fire extinguishers and inert agents, the metals fabrication industry, and in the production of urea and methanol. Several sources of CO2 and their CO2 concentration are given in Table 5.14.

The capture of CO2 can be divided in two main categories:

- Pre-combustion
- Post-combustion

In this case, the capture of CO2 from a hydrogen unit is of the first category and can be realized from the syngas stream (inlet of PSA) or from the tail gas stream (outlet purge gas/tail gas of PSA).

TABLE 5.14
Common Sources of CO2 and Concentration

Source	CO2 Purity in Mole %
Hydrogen generation unit	15%–20%
• Reformer flue gas	30%–40%
• Synthesis gas	
Power plant flue gas	15%–20%
Bioethanol unit	>80%
Natural source	>90%
Limekilns	Around 40%

The major processes available can be grouped as follows:

- Absorption processes
- Adsorption processes
- Membrane separation
- Cryogenic separation

To remove the CO2 from the streams aforementioned, all of the process options are viable. However, the capture of CO2 from syngas should be done with absorption processes, preferably chemical type (amine MDEA), and in tail gas mostly by cryogenic separation. There are a variety of amines that can be utilized in the treatment of acid gases. To remove of CO2 at low concentrations, the most widely used amines are MEA, DEA, and MDEA.

Comparison between these amines reveals that MDEA is the preferred one. MDEA results in a weak bond between MDEA and CO2 and the MDEA can be regenerated by letdown of pressure and the regeneration energy requirement can be minimized. For increasing the absorption of CO2, a higher residence time of MDEA in trays is used typically with increased weir height, or activated MDEA can be used (activation with piperazine is normal). The majority of CO2 is recovered by flashing at lower pressure. Balance can be recovered in a regenerator with application of minimum heat duty in the reboiler. In this process, CO2 is captured at higher pressure around 22–25 Kg/cm2 g. Hydrogen is obtained at around 20 Kg/cm2 g (Figure 5.27).

Alternatively, CO2 can be captured from tail gas of PSA (Figure 5.28) at low pressure. A wide range of technologies currently exist for separation and capture of

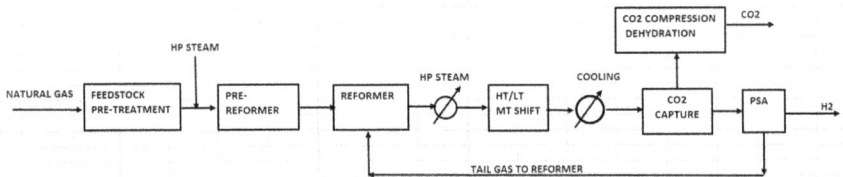

FIGURE 5.27   CO2 capture with amine at upstream of pressure swing adsorption (PSA).

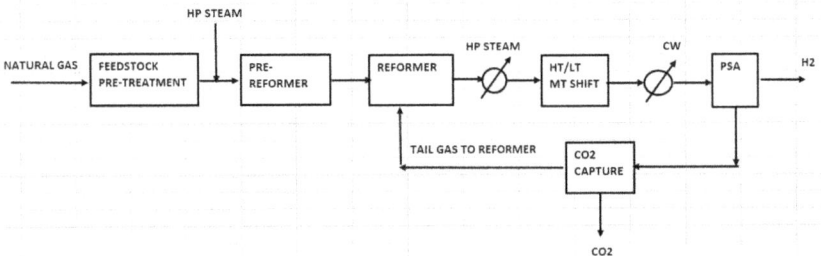

FIGURE 5.28   CO2 capture from low-pressure tail gas.

CO2 from gas streams. They are based on different physical and chemical processes including absorption, adsorption, membranes, and cryogenics. Normally cryogenic and membrane are applied for recovery. CO2 can be recovered with proprietary solvents. The overall scheme is shown in Figure 5.28.

Reformer furnace flue gas is normally very rich in CO2 as the PSA tail gas is also used as fuel in the reformer to the extent of around 70% or higher. A scheme of CO2 capture from reformer flue gas in a methanol plant is shown in Figure 5.29.

CO2 recovery from flue gas can be done by proprietary solvent and the scheme is shown in Figure 5.30. This may use a primary amine like MEA or DGA, or proprietary amine-based solvent for optimizing the reboiler duty in the regenerator. An MEA based scheme for CO2 recovery is shown in Figure 5.31. A scheme for recovery of CO2 from the coal gasifier is shown in Figure 5.32. After synthesis gas cooling and water separation, MDEA can be used to selectively absorb H2S. In the downstream CO2 capture section, MDEA or activated MDEA can be used with higher residence time in the tray to enable efficient capture of CO2.

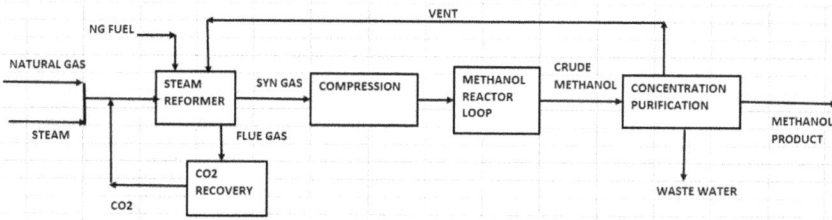

FIGURE 5.29   CO2 capture from reformer flue gas in methanol plant.

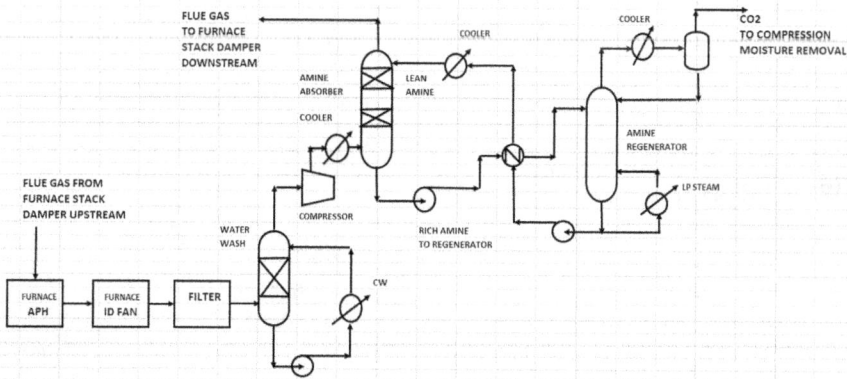

FIGURE 5.30   CO2 recovery from flue gas of furnace (licensed process).

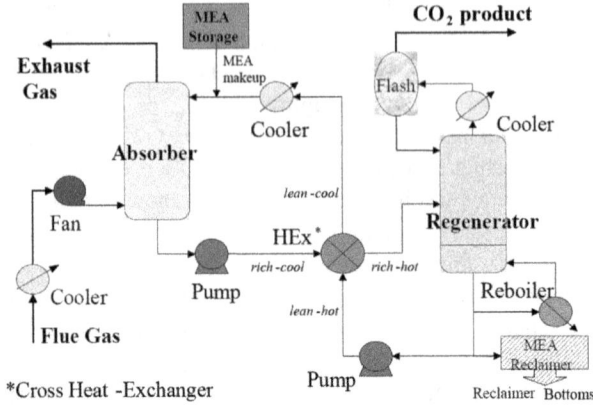

FIGURE 5.31 Flow sheet for CO2 capture from flue gases using amine-based system.

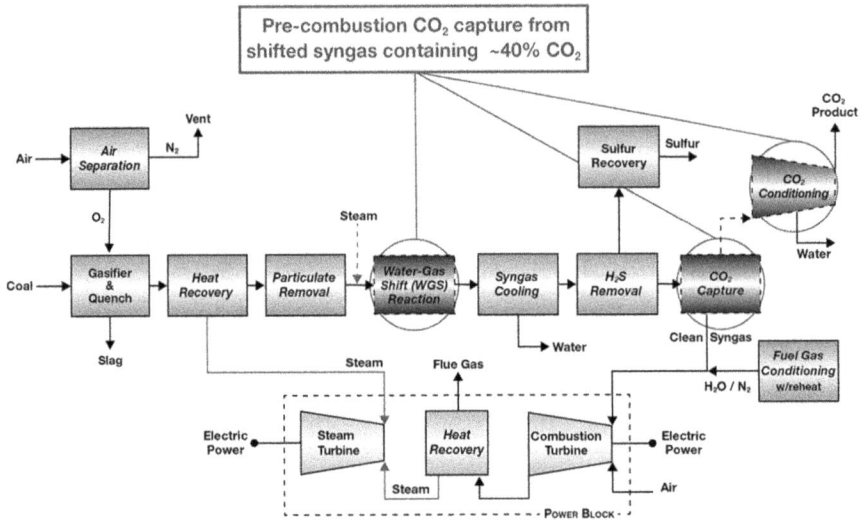

FIGURE 5.32 Pre-combustion recovery of CO2 from gases from the gasifier.

BIBLIOGRAPHY

Art Kidnay. *Acid Gas Treating*, Chapter 10, Presentation. Colorado School of Mines, copyright John Jechura, 2019.

Arthur Kohl and Richard Nielsen. *Gas Purification*. 5th edition, Houston, TX: Gulf Publishing Company, 1997.

Barry Burr and Lili Lyddon. A comparison of physical solvents for acid gas removal. Presented at *Gas Processors Association Convention*. Bryan, TX: Bryan Research & Engineering, Inc., 2008.

Fernando Vega, Mercedes Cano, Sara Camino, Luz M. Gallego Fernández, Esmeralda
    Portillo, and Benito Navarrete. Solvents for carbon dioxide capture. 2018. https://doi
    .org/10.5772/intechopen.71443.
GPSA Engineering Data Book. 14th edition, Tulsa, OK: GPSA, 2017.
Louis Beke. *Contamination in Amine Systems*. BSDT seminar, September 2010.
National Energy Technology Laboratory (NETL). *Acid Gas Removal (AGR)*. Morgantown,
    WV: US Department of Energy.
National Energy Technology Laboratory (NETL). *Carbon Dioxide Capture Handbook*.
    Morgantown, WV: US Department of Energy, August 2015.

6 Hydrogen Generation Units

6.1 INTRODUCTION

Hydrogen is generated as a byproduct from catalytic reforming units during the production of high-octane reformate, used in the production of gasoline. Earlier semi-regenerative units were converted to continuous regeneration types (which operate at lower pressures with regenerated catalysts at higher temperature) further increasing the production of hydrogen per unit weight of feed.

For higher needs of hydrogen for hydrotreating, hydrocracking, etc., hydrogen is manufactured by steam reforming of natural gas, methane, or naphtha. In some scenarios partial oxidation is used with heavy oil as feed where it is available at lower cost or as a means for disposal for heavy high-sulfur streams like residues and coke. However, the process needs oxygen plants and much capital. Normally, it is economically attractive if accompanied with electricity generation, where electricity is the primary product and hydrogen is produced as a byproduct. Wherever lighter feedstocks are available, steam reforming is preferred. Hydrogen generation has become a necessity for a refinery with higher emphasis on production of clean fuel and introduction of residue upgradation units demanding a considerable amount of hydrogen (1.8%–2.8% on feed). Hydrogen generation from various process units are presented in Table 6.1.

A steam reformer is traditionally used by a refinery for production of hydrogen.

6.2 DESCRIPTION OF A STEAM REFORMER

A brief description of the unit is necessary for appreciation of the typical problems experienced in a steam reformer unit. A steam reformer can consist of several sections:

- Pre-desulfurization section (PDS)
- Final desulfurization section (FDS) or feed purification section
- Pre-reformer
- Tubular reformer furnace
- Reformed gas cooling section
- Shift section
- Final cooling section
- Synthetic gas purification section to yield pure hydrogen (pressure swing adsorption, PSA)

Typical schemes for different configurations of reformers, tubular reformer furnace, and PSA are presented in Figures 6.1 to 6.5.

DOI: 10.1201/9781003268246-6

TABLE 6.1

Typical Hydrogen Production from Different Units

Process Unit	Hydrogen Generation wt.% on Feed
Semi-regenerative catalytic reformers	1.6–2.0
Continuous catalytic reformers	3.0–3.5
Residue gasification	20–25
Catalytic cracker (FCCU)	0.1–0.15
Gas/naphtha steam cracker	1.0–1.2
Steam reformer	30–33

6.2.1 PROCESS DESCRIPTION

If feed sulfur is high, the naphtha is treated in a conventional naphtha hydrotreating unit to typically bring down the sulfur to about 5–10 ppm wt. The predesulfurisation section (PDS) deploys a hydrogenation reactor (usually with Co/Mo catalyst) to convert sulfur compounds to H2S and strip the H2S in a stripper to reduce the sulfur content of the feed to 5–10 ppm.

In the final desulfurization section FDS feed is vaporized mixed with hydrogen and heated again and routed to the hydrogenation reactor to convert the balance sulfur compounds to H2S. The generated H2S is absorbed in zinc oxide (ZnO) beds provided in the FDS, and the feed sulfur to the pre-reformer is brought down below 1 ppb. Normally, two ZnO beds are placed in series in a lead lag fashion. The freshly loaded one is kept at the back as a guard bed. Once the first one gets exhausted, confirmed by detection of H2S at the reactor outlet, the reactor is taken out and fresh catalyst is loaded and placed as the second absorber.

Pre-reforming is used for the gas/naphtha feed and is typically used to increase the capacity of the existing unit by reducing the load on the tubular reformer. The pre-reforming reactions convert the heavier hydrocarbons (propane and butane and higher hydrocarbons) to methane ahead of the heat-intensive reforming reactions, essentially shifting part of the load of the reformer heater. In the process, feed from the purification section or FDS is further heated in the flue gas section of the tubular reformer, mixed with superheated steam in a fixed ratio with feed flow, and introduced in the fixed bed pre-reformer. In the pre-reformer, heavier hydrocarbons are converted to methane with the help of the following set of reactions:

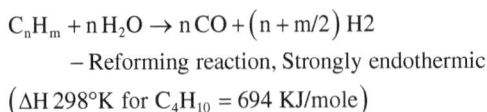

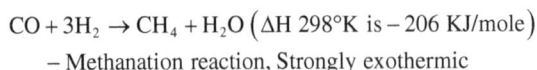

$$C_nH_m + n\,H_2O \rightarrow n\,CO + (n + m/2)\,H2$$

– Reforming reaction, Strongly endothermic

$$(\Delta H\ 298°K\ \text{for}\ C_4H_{10} = 694\ KJ/mole)$$

$$CO + 3H_2 \rightarrow CH_4 + H_2O\ (\Delta H\ 298°K\ \text{is}\ -206\ KJ/mole)$$

– Methanation reaction, Strongly exothermic

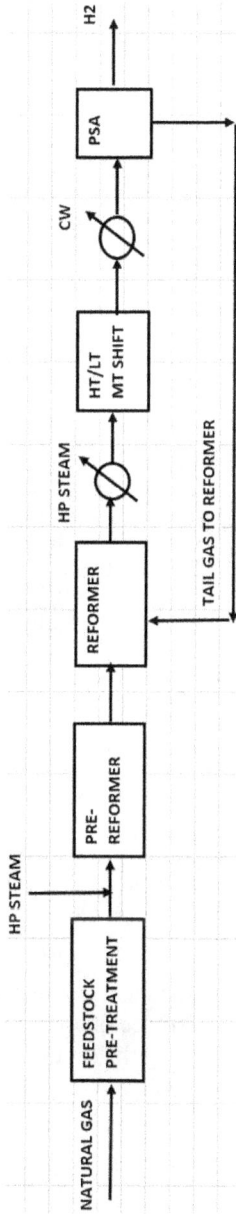

FIGURE 6.1 A typical block diagram of a steam reformer.

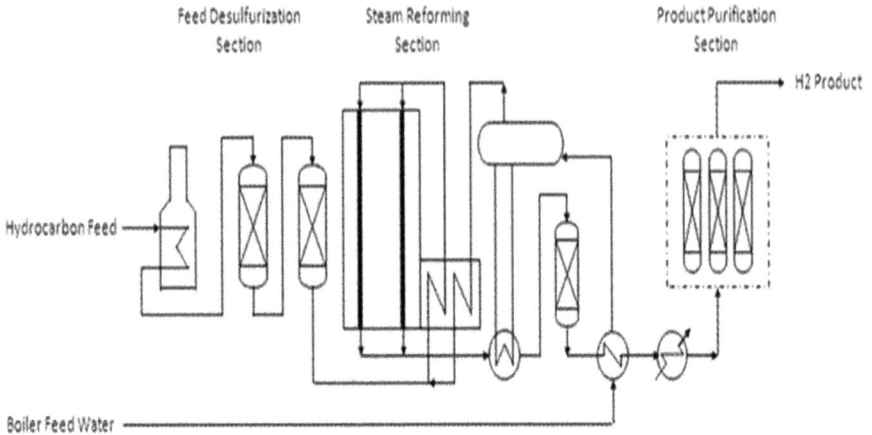

FIGURE 6.2 Typical scheme of a reformer (horizontal convection section).

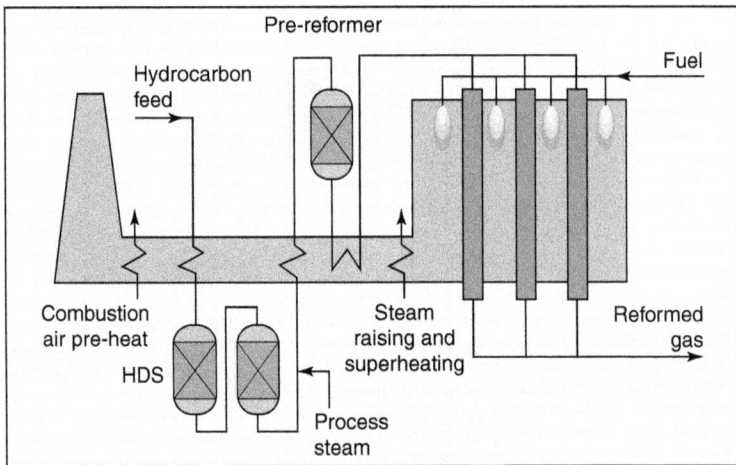

FIGURE 6.3 Typical scheme of a pre-reformer/reformer.

Thus, the temperature profile along the height of the performer catalyst shows first a drop in temperature suggesting occurrence of an endothermic reforming reaction, then an increase in temperature suggesting the exothermic methanation reaction. No heavy hydrocarbon should break through the pre-reformer. The addition of the pre-reformer as a retrofit to an existing facility may have two problems:

• Absence of space and in convection—Physical space constraint may not allow adding a feed reheat coil within the convection section for preheating pre-reformer feed.
• Also, the metallurgy of the inlet pigtails may not be able to handle the higher feed temperature at the inlet to the reformer.

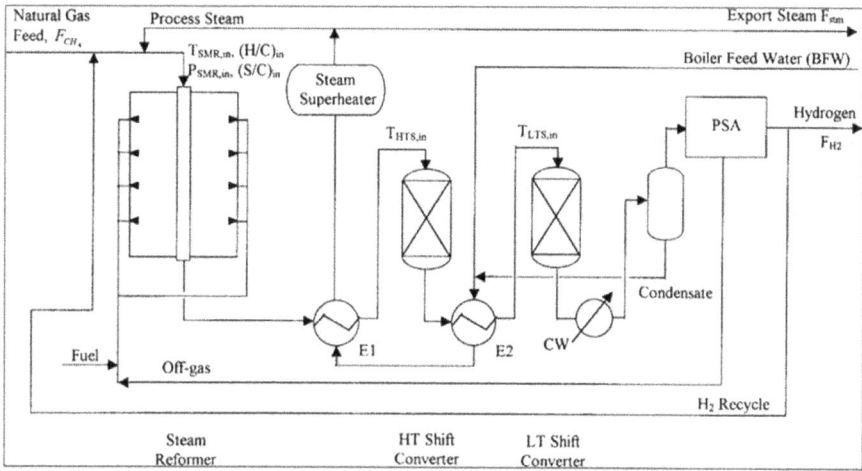

FIGURE 6.4 Typical scheme of a reformer with combination of HT and LT shift reactors.

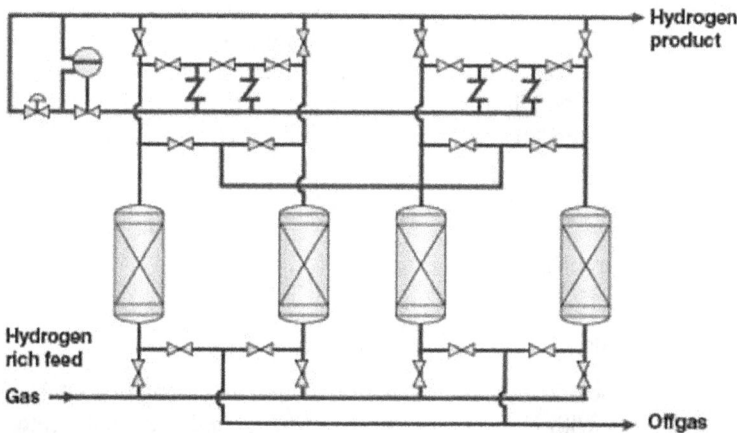

FIGURE 6.5 Typical arrangement of a reformer PSA with interconnections.

The preformed feed is then further heated to around 600°C in the flue gas section of the reformer and introduced in the reformer tubes heated by burners to supply the heat required for reforming. In the reformer tubes, methane reacts with steam and produces mainly hydrogen and CO. The typical reformer outlet temperature is around 835°C. The typical flue gas outlet temperature of the reformer furnace is around 930°C, which is used to heat pre-reformer feed, heat reformer feed, generate high-pressure (HP) steam, superheat the HP steam generated, BFW preheating, etc. The reformed gas from the tubular reformer is cooled to around 130°C–300°C by HP steam generation (in the reformed gas cooler), BFW preheating, etc., and fed to a medium-pressure (MP) shift reactor or a combination of high-temperature (HT) and low-temperature (LT) shift reactor (with inter-reactor cooling) to convert the majority of CO to CO_2 and hydrogen. This further increases the production of hydrogen by the shift reaction. The following shift reaction is an exothermic reaction. So, the reaction is favored at lower temperatures to achieve higher equilibrium conversion. Thus, when MT shift is deployed, a higher proportion of hydrogen is produced and the CO proportion in the purge gas, produced from PSA, reduces and results in lower calorific value of the purge gas (tail gas). Thus, a higher amount of supplemental fuel in the tubular reformer furnace is needed.

The shift reaction is represented as $CO + H2 = CO_2 + H2$. This reaction is equilibrium controlled and moderately exothermic (ΔH 298°K is −41 KJ/mole). The outlet of the shift reactor is further cooled in the air cooler and excess stream is condensed followed by the water cooler to condense/cool the synthesis gas to around 40°C.

The cooled synthesis gas is typically fed to pressure swing adsorption beds to recover hydrogen of purity around 99.99% volume. The beds are automatically regenerated by letting down the pressure of the beds. The tail gas, having impurities like CO, CO2, and CH4 along with some hydrogen, is rejected and routed to purge gas drum and is subsequently routed to reformer furnace as the main fuel. The purge gas hydrogen content is typically around 20% by volume. The low calorific value purge gas (tail gas) typically supplies 70%–75% of fuel to the tubular reformer, and balance is supplied by supplementary fuel. The reformer furnace firing arrangement is different for technologies from different technology suppliers like top-fired burners, side-fired burners, and terrace-mounted burners. Higher firing is required at the top section (at the entry to supply the heat for the endothermic reforming reaction in the section).

6.3   COMMON PROCESS PROBLEMS

Common process problems experienced are described next.

6.3.1   SULFUR SLIPPAGE TO PRE-REFORMER/REFORMER

The problems can be of the slippage of sulfur components from the FDS hydrotreating reactor (reactor with Co/Mo catalyst) to downstream due to inadequate conversion of sulfur species to H2S in the reactor. This may be due to hydrotreater catalyst deactivation, more sulfur in feed than in design, low H2 recycle rate, low inlet temperature to the HDS reactor than necessary, etc. It may be noted that ZnO-absorbent beds located downstream can only remove H2S but no other sulfur species. Thus, if sulfur is not converted to H2S in the HDS reactor, then it will pass downstream to the pre-reformer or tubular reformer. The slippage of H2S from ZnO absorbers can also lead to damage of pre-reformer/reformer catalysts. The sulfur slippage may poison pre-reformer catalysts and lead to ingress of hydrocarbons heavier than methane into the tubular reformer, and can coke catalysts in tubes, resulting in a higher pressure drop across in reformer tubes, and can restrict the flow of feed and steam mixture through tubes. The lower flow through tubes will further increase the tube temperature and coke, and may entirely block the flow. This can lead to tube rupture. This is a grave emergency.

Breakthrough of H2S from the ZnO bed can deactivate pre-reformer catalyst and in turn may lead to heavier hydrocarbon ingress to tubular reformer and coking. The reformer catalyst is also susceptible to poisoning by sulfur, and pre-reformer catalyst is more sensitive to sulfur poisoning. So, the units that do not have a pre-reformer ahead of the reformer, the reformer can get catalyst damage due to sulfur slippage from the purification section. Typically, the inlet temperature of the tubular reformer is lower at around 400°C if the tubular reformer is not preceded by a pre-reformer. The inlet temperature of the tubular reformer is maintained much higher around 600°C if preceded by a pre-reformer. The pre-reformer converts the heavier hydrocarbons to methane and provides flexibility to maintain a higher temperature at the inlet of the tubular reformer. In units with the pre-reformer upstream of the tubular reformer, slippage of the sulfur from upstream will first damage the catalyst of the pre-reformer and will eventually damage the reformer catalyst by coking.

6.3.2   SLIPPAGE OF HIGHER HYDROCARBONS FROM PRE-REFORMER TO TUBULAR REFORMER

Hydrocarbon heavier than methane can slip to the tubular reformer in case of deactivation of pre-reformer catalyst or catalyst getting exhausted. The pre-reformer catalyst temperature profile is continuously monitored to mark the location of inversion of temperature due to shift of endothermic reforming reaction to exothermic methanation reaction. If that inflection point reaches near 90% of bed height from the top, the unit is brought down to replace pre-reformer catalyst.

6.3.3   OPERATION WITH LOWER STEAM-TO-CARBON RATIO

Steam-to-carbon ratio is the most important parameter of a reformer. Typically, a steam-to-feed carbon mole ratio of around three is maintained in reformers at the inlet of the pre-reformer (or at the inlet to tubular reformer). Higher steam drives reforming reaction forward and also prevents carbon deposition (Boudouard carbon) due to the reaction of $2\,CO = C + CO_2$ (Boudouard reaction). Higher stream is always good for reforming, unless too much flow of steam results in feed flow restriction and higher pressure drop across the tubes of the reformer. Higher steam also consumes more fuel. The steam to carbon flow ratio also depends on the type of catalyst used in the shift section. Steam is also required as a reactant for the shift reaction and it helps to propagate the exothermic shift reaction $CO + H_2O = CO_2 + H2$ in the forward direction. An HT shift deploying iron-based catalysts needs more steam flow (steam-to-dry gas ratio). If steam flow is low, the iron catalyst deployed in HT shift may initiate a highly exothermic Fischer–Tropsch reaction (methanation reaction). In the past, some runaway of temperature was reported in HT shift reactors and it is known to have been caused by a low steam-to-dry gas ratio in HT shift reactors. Normally, an HT and LT shift combination or a medium-temperature shift reactor is used to maximize hydrogen yield. The medium-temperature shift catalysts are known to be copper-based catalysts that can operate with a lower steam-to-dry gas ratio, enabling the reformer to operate at a lower steam-to-carbon ratio.

6.3.4 OVERHEATING OF TUBES OF TUBULAR REFORMER

Overheating of tubes is the most common problem experienced in tubular reformers. Visual inspection of tubes by experienced operators are very common to detect overheating. The tube metal temperature is periodically monitored by measurement of temperature of the tubes by optical pyrometers. The overheating of tubes can result from higher localized firing, flame impingement, or lower flow of feed in some tubes. Before starting the unit, a back pressure measurement of all the tubes is carried out before and after loading of catalyst to ensure an almost similar flow path resistance in all the tubes. This is to ensure that flow of feed and steam is almost uniform through all the tubes. Some tubes may offer more resistance to flow due to catalyst breakage and higher packing after a prolonged run and can get overheated. Carbon makes catalyst break and pack the tubes causing low feed flow and tube overheating. Flame impingement may overheat tubes and catalyst, and may lead to formation of coke on catalyst leading to catalyst disintegration that eventually leads to flow resistance and can overheat the tubes.

6.3.5 QUALITY OF STEAM FOR REACTION

Quality of steam is very important with respect to silica content. Suspended solids (SS) and total dissolved solids (TDS) are also very critically monitored in the steam drum. Typically, the steam generation system of a reformer is not combined with other process units in the complex. Only steam is exported during a normal run and imported during startup.

6.3.6 SLIPPAGE OF CHLORIDES

Organic chlorides can be present in naphtha feed stock and that can get hydrogenated to form HCL. HCL can react with ZnO and deactivate the catalyst. Further, chloride is also a poison for pre-reformer catalyst. Chloride guard catalyst is placed on the top of ZnO beds to arrest chloride slippage downstream of chloride beds.

6.4 MECHANISM FOR DEACTIVATION OF CATALYSTS

Deactivation of catalysts in different reactors and guard beds are described next.

6.4.1 DEACTIVATION OF HDS CATALYST

HDS catalyst is active in sulfide form. The deactivation of HDS catalyst can take place if pre-sulfiding is not proper or sulfur gets stripped off while processing low sulfur feed stocks. Operators are often reluctant to add sulfur (DMDS) to the hydrocarbon feed to arrest sulfur strip off. This is due to the apprehension that the lifespan of ZnO catalyst may get reduced and it also increases the operating cost. This is particularly true for natural gas as feed that has low sulfur. The author has experience of at least two cases where sulfur stripped off from HDS catalyst and deactivated the catalyst. Organic sulfur passed downstream and suddenly deactivated the pre-reformer catalyst. Unnecessary higher inlet temperature can also result in coke laydown and decline in HDS catalyst activity.

6.4.2   Deactivation of Zinc Oxide (ZnO) Catalyst

The reaction of H2S absorption in ZnO is represented by the equation $ZnO + H_2S = ZnS + H_2O$. The ZnO placed in the ZnO adsorbers may get saturated over a period of time in operation and can result in slippage H_2S to the pre-reformer or reformer.

Feed to the hydrogen generation unit (HGU) may contain organic chlorides. The organic chlorides are converted to HCL in the hydrogenation reactor in the presence of recycled hydrogen. The HCL thus formed is removed in purification section. Normally, chlorine guard catalyst is placed at the upstream (loaded on the top) of ZnO beds to protect ZnO from reaction with HCL. Some catalysts used as chlorine guards contain potassium carbonate to trap HCl and with the formation of potassium chloride following the reaction $K_2CO_3 + HCL = 2KCL + CO_2 + H2O$. The breakthrough of HCL from chlorine guard may lead to the reaction of ZnO with HCL and form $ZnCl_2$. The $ZnCl_2$ has a low melting point and will be at liquid state at the prevailing temperature and also have high vapor pressure, and may land up to the reformer or pre-reformer and get hydrolyzed by reaction with steam, and may reconvert into zinc oxide and HCl. This may adversely affect the shift catalyst. Thus, slippage of HCl from chlorine guard may result in damage to ZnO beds and also have adverse effects downstream.

6.4.3   Deactivation of Pre-Reformer Catalyst

Pre-reformer catalyst can get oxidized during steaming without hydrogen during startup and emergencies. Unlike reformer, if pre-reformer catalyst gets oxidized, it cannot be brought to its reduced state with circulation of hydrogen along with steam before introduction of feed. Emergency hydrogen should continue along with steam to avoid oxidation and loss of activity of catalyst during startup and emergencies.

Sulfur/H2S slippage from the upstream desulfurization section can also result in poisoning and deactivation of pre-reformer catalyst. This is the most common reason of deactivation of pre-reformer catalyst. Ingress of poisons like silica can also damage pre-reformer catalyst. A lower steam-to-HC ratio can also lead to more carbon deposit and loss of activity.

6.4.4   Deactivation of Reformer Catalyst

The deactivation of reformer catalyst can result from formation of coke on the catalyst and is typically reflected in higher methane content at the outlet of the tubular reformer. This may be due to a lower steam-to-carbon ratio being maintained. The carbon deposit can also result in a tubular reformer if the pre-reformer upstream is unable to convert all the heavier hydrocarbons to methane and hydrocarbon heavier than methane is exposed to a higher inlet temperature (around 600°C) at the inlet to tubular reformer. It may be noted that the units without a pre-reformer operate at a relatively lower temperature of around 400°C at the inlet to tubular reformer. The most common reason for this problem is formation of carbon on catalyst (through Boudouard reaction or reverse water gas reaction). The carbon deposited on catalyst weakens the catalyst and often disintegrates the catalysts, and results in resistance in flow path and consequent tube overheating. Low flow through few tubes and flame impingement can also result in high tube and catalyst temperatures and may partially reduce activity of the catalyst.

6.4.5 DEACTIVATION OF SHIFT CATALYST

The inlet temperature of shift converters is maintained around 10°C–20°C above the dew point of the feed vapor. Three alternative reactor configurations are typically practiced in the shift section:

- A high-temperature (HT) shift reactor
- A combination of high-temperature (HT) shift and low-temperature (LT) shift reactors
- A medium-temperature shift (MT) shift reactor

In the first option only, a HT shift reactor is used. The second option deploys an HT and LT shift combination where the HT shift is placed ahead of the LT shift reactor. HT shift catalyst is not active typically below a temperature of 330°C, and HT shift deploys iron oxide, chromium, etc., in the catalyst. Operating with a lower steam-to-carbon (steam-to-dry gas) ratio can result in iron carbide formation. Iron carbide will weaken the strength of catalyst and can cause breakage of catalyst and result in higher pressure drop across the bed. Further, iron carbide is a Fischer–Tropsch catalyst and can lead to several undesired reactions, consume hydrogen, and cause problems to the downstream section. HT shift needs a higher steam-to-dry gas ratio and in turn requires a higher steam-to-carbon ratio to be maintained in the reformer. A low steam-to-dry gas ratio can result in a big exotherm in HT shift and can partly or fully deactivate the catalyst. If HT shift loses much activity and passes more CO to the LT shift located downstream, a large exotherm can result across the LT shift, and the catalyst containing copper/Zn/Al can get sintered and lose activity. The LT shift reactor temperature should not go above 250°C as it can cause an increase in a sintering rate of Cu present in the catalyst. Lowering of inlet temperature to control exotherm is also not possible, limited by dew point of the feed.

The MT shift catalyst typically operates between 190°C and 330°C. This catalyst does not contain iron oxide. Thus, operation is possible at a relatively low steam-to-carbon ratio. The catalyst is known to have resistance towards condensing water and relatively higher resistance toward chlorine and silica poisoning. Problems in MT shift can be high pressure drop due to

- Improper loading of the catalyst support grid and hold down resulting in movement of catalyst inside the reactor
- Carryover/deposition of solids from the steam generation system
- Deposit of dust from refractory material, foreign material, upstream catalyst, etc.
- Catalyst breakage

Loss of activity can be due to the following reasons:

* Channeling.
* Operation at high temperature (higher than 330°C).
* Operation at too low a temperature, below the dew point of the gas. Some activity can be temporarily lost due to condensate blocking the pores of the catalyst. Inlet temperature to shift reactor may be maintained10°C–20°C higher than the water dew point temperature.
* Catalyst poisons like sulfur, silica, and chlorine can also lower the activity.

6.5 CASE STUDIES

6.5.1 CASE STUDY 1

In one hydrogen generation unit having low sulfur naphtha as feed (without pre-reformer) all of a sudden the pressure drop increased across the reformer during startup. A couple of tubes ruptured discharging hot gases with flames and hitting the refractory wall. Feed to the reformer was immediately stopped and firing gradually reduced and stopped. Steam continued to purge the tubes and to prevent carbon lay down on catalyst. The reformer cooled. The pressure drop was checked in all the tubes with the back pressure measuring device. About ten to twelve tubes showed a very high pressure drop when back pressure was checked with the special device. Catalysts from all the tubes were unloaded. About three to four tubes could not be unloaded with application of vacuum. The tubes were later cut and removed. New tubes were welded.

In the choked tubes catalyst was found mostly disintegrated and blocking the flow path of the feed vapor and steam. All the tubes were reloaded with recovered good catalyst and partly with new catalyst. Back pressure of all the tubes were again noted and found similar and concluded in order. Catalyst of damaged tubes were analyzed and found to have a lot of coke that might have led to the disintegration of the catalyst, choked the flow path, overheated, and ruptured the tubes. Figure 6.6 shows the circuit for startup nitrogen circulation for heating the purification section and downstream prior to introducing steam in the tubular reformer and feed cut-in. It was concluded that there was some hydrocarbon in the section between the ZnO beds and reformer inlet shutoff valve (shown in Figure 6.6) that was not thoroughly purged and drained or the shutdown valve at the inlet of the reformer section had passed. This led to the entry of some hydrocarbon in the tubes during startup nitrogen circulation for heating before steam was introduced in the tubular reformer.

The startup procedure was modified to purge the section with longer time than earlier defined. The section between the ZnO beds and reformer was thoroughly purged with nitrogen and ensured free of hydrocarbon by repeated draining, flaring, etc. Each drain and vent point were checked with an explosive meter to ensure absence of hydrocarbon in the section. Then the nitrogen circulation started through the reformer tubes and the reformer furnace was fired, steam introduced at the prescribed temperature. Startup took off normally.

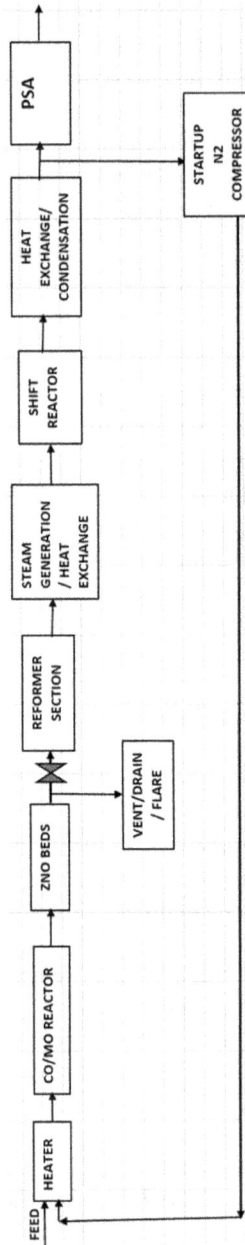

FIGURE 6.6 Startup N2 circulation circuit (long loop).

6.5.2   CASE STUDY 2

A similar incident was observed in another unit where the tubular reformer was located at some distance from ZnO beds. Thus, the holdup volume in the section was higher. However, provision was there to circulate N2 through the startup compressor only for the reformer section (Figure 6.7). The refinery had connected the purification section with the reformer section before introducing steam in the reformer. Some hydrocarbon content, left out in the section from the ZnO bed to the reformer, entered the reformer and coked the catalyst in absence of steam.

The startup procedure was modified. First, N2 circulation in the short loop was established in the tubular reformer. The reformer fired, heated, and steam was introduced in the reformer. The desulfurization or the purificationsection was under hydrocarbon removal/nitrogen purge. The shutoff valve isolating the reformer section from purification section was maintained in the closed position. Once thorough nitrogen purging is completed in the purification section, then N2 circulation started through the hydrotreater reactor, ZnO beds, and reformer, and the long loop circulation resumed. At this point, even if some hydrocarbon gets into the reformer, in the presence of steam the same will get reformed and no carbon deposition will result. Thus, this ensured a full-proof startup. Both units had a startup fired heater for heating the hydrogenation reactor and ZnO beds.

In some units it was found (particularly in units with pre-reformers) that licensors provide a block and bleed isolation, as shown in Figure 6.8 (having two shutoff control valves with a bleed line with control valve from the section between the isolation shutoff valves) at the inlet to the pre-reformer/reformer section, and the purification section is heated separately (using an HP steam heater) with a separate nitrogen blower, and the vapor outlet from ZnO right from the upstream of block and bleed assembly is recirculated to the feed drum. So, all the sections get purged and heated up to the block and bleed isolation with nitrogen. Provision is also extended to introduce feed in the purification section and the vapors from the ZnO outlet can be cooled/condensed and returned to the feed surge drum (with DMDS injection to arrest sulfur strip off from the desulfurization reactor catalyst), while the reformer section is being prepared for startup and remains isolated from the purification section by the block and bleed isolation system. A separate nitrogen startup compressor is used for circulation in the pre-reformer/reformer section and the reformer fired. Steam is introduced (with or without emergency hydrogen) and then feed from the purification section (ZnO outlet) is introduced in the pre-reformer/reformer section. So, the hydrocarbon content of the line can never enter the pre-reformer/reformer section in the absence of steam in the section. This arrangement, as in Figure 6.8, may be a safe scheme to avoid such eventualities.

6.5.3   CASE STUDY 3

In a HGU of very big capacity, frequent tube failures occurred. The reformer was a side-fired reformer having a pre-reformer preceding the tubular reformer and MT shift reactor following the reformer. The units used to operate with around a 2.8:1

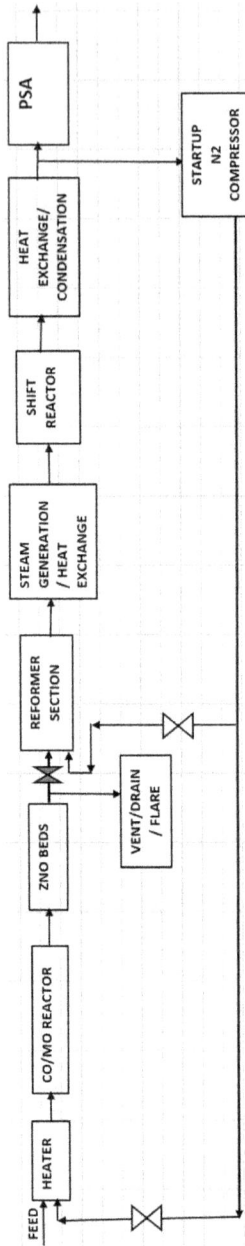

FIGURE 6.7 Startup N2 circulation circuit (showing short loop and long loop).

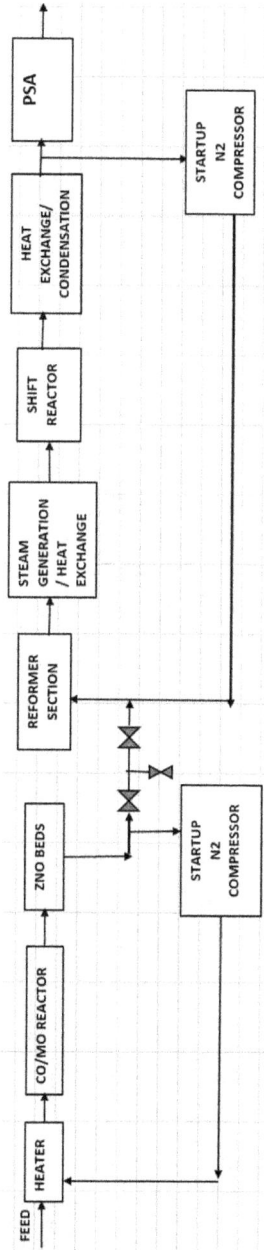

FIGURE 6.8 Startup N2 circulation separate circuits for HDS and pre-reformer/reformer section (showing block and bleed isolation).

steam-to-carbon ratio. Five to six tubes got ruptured at a time creating a very big emergency. New tubes were welded, catalyst reloaded, and units started. An inspection examined the failure and reported carburization of tubes resulting in the failure. The failures repeated at least twice. Units were restarted after repair. Experts thought the failure may be due to non-uniform localized hard firing and possible flame impingement and consequent overheating, etc. Later, the author and team revisited the problem. The feed to the units were natural gas (NG) with some quantity of hydrogen containing gases obtained from hydrotreating units. Hydrogen percentage by volume in the feed was around 30% vol. The licensor had earlier put a restriction of 40% (volume) maximum hydrogen in the feed, possibly to avoid stripping of sulfur from the hydrotreater (HDS) catalyst. It was learned that a few days ago the refinery had routed some additional hydrogen-rich gas from the MP separator of another hydrotreater to the hydrogen unit feed, but the hydrogen percent in gas was still around 32%–35% by volume. The steam flow meter and naphtha flow meters were checked and found alright. It was further analyzed and concluded that the reverse water gas reaction can take place to a greater extent with a higher proportion of hydrogen in the feed and deposit carbon following the reaction $CO + H2 = C + H_2O$ (reverse water gas reaction). So, it was advised to increase the steam-to-carbon ratio marginally to about 2.85:1 and operate normally. Higher steam was also expected to lower the likelihood of formation of carbon from the Boudouard reaction. The likelihood of carbon formation from reverse water gas reaction may reduce with higher steam and likely arrest deposition of carbon (due to the presence of higher proportion hydrogen in feed). This may in turn reduce and may stop reoccurrence of the tube carburization and failures. All the problems got resolved. The unit has run for more than three years without any problems. Further, the refinery has taken up a project of

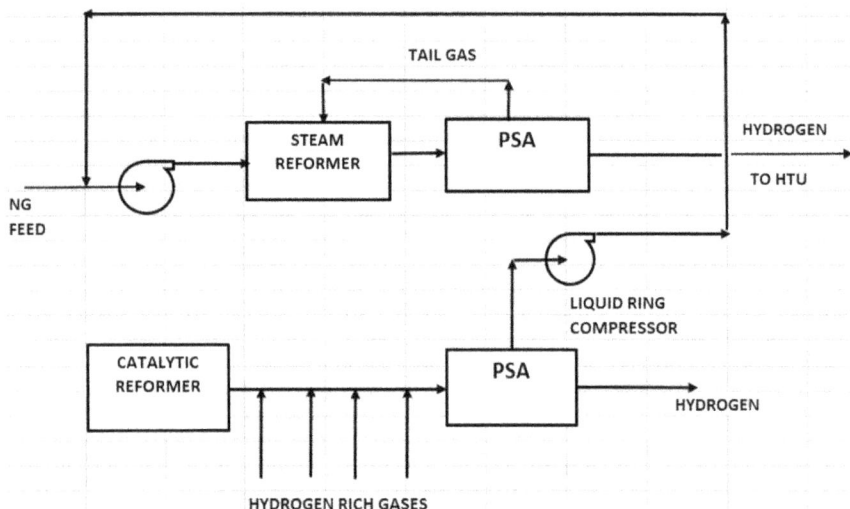

FIGURE 6.9 Hydrogen recovery through PSA and integration with HGU feed.

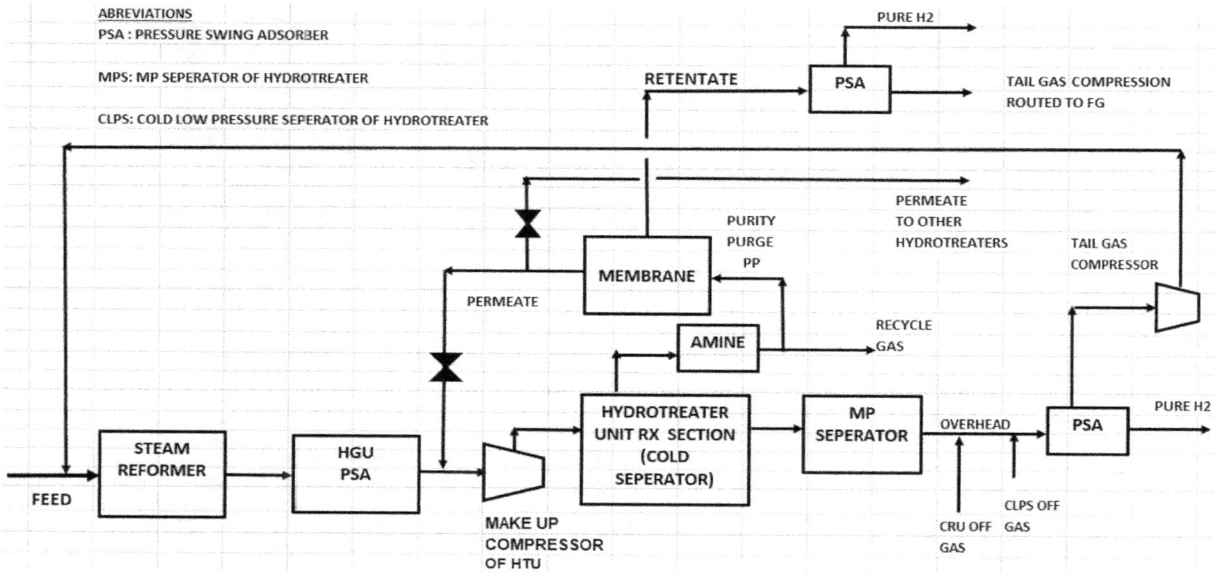

FIGURE 6.10   Hydrogen recovery scheme, including recovery from purity purge (PP) of hydrotreaters.

installation of PSA to recover hydrogen from hydrogen-rich gases and to reduce the proportion of hydrogen in feed to the hydrogen generation unit. It may be worth noting that a high hydrogen proportion in the feed can adversely affect tubular reformer tubes due to carburization of tubes.

6.6 HYDROGEN RECOVERY

Hydrogen is a very precious component and needs to be recovered from hydrogen-bearing gases to the extent possible. Pressure swing adsorption (PSA) units are installed to recover hydrogen from the catalytic reformer unit (CRU) off gases and other hydrogen-rich gases (after amine wash) from a refinery. The PSA off gases may still contain an appreciable amount of hydrogen at low pressure along with hydrocarbon light ends. Thus, this PSA tail gas is compressed and can be introduced as feed to the hydrogen unit along with normal feed for the recovery of hydrogen. The scheme is presented in Figure 6.9.

A scheme to recover hydrogen-rich purity purge from the cold separator of hydrotreaters and hydrocrackers (after a high-pressure amine wash) is shown in Figure 6.10, using a membrane and routing of the permeate hydrogen to make up compressor suction (or interstage suction) or to other hydrotreaters at lower pressures. The retentate is routed to another PSA to recover pure hydrogen. The tail gas produced from the PSA of retentate can be routed to fuel gas. The comprehensive scheme of hydrogen recovery is shown in Figure 6.10 along with recovery from other hydrogen-rich streams like CRU off gas and hydrogen-rich gases from hydrotreaters.

6.7 CAPACITY AUGMENTATION

Hydrogen is in big demand, particularly with the introduction of heavy high-sulfur crudes in crude mix and requirements for superior product qualities. An appreciable amount of hydrogen can be recovered from different recovery schemes described earlier. For further production of hydrogen, modification in the HGU may be required. In recent years, a fixed-bed reformer reactor is added downstream of the conventional tubular reformer. This uses catalysts usually placed in tubes in a fixed-bed reactor. The process gas reformer flows through the tubes packed with catalysts and is heated by hot flue gases of the tubular reformer from outside. This is often termed as the convection reformer (CR) or heat transfer exchanger reactor (HTER). The addition of HTER can generate around 20% additional hydrogen from the hydrogen generation unit (HGU).

BIBLIOGRAPHY

Jonna Benson and Andrew Celin. Recovering Hydrogen – and Profits – from Hydrogen-Rich Offgas. *CEP Magazine*, January, 2018, pp. 55–59.
Satish Reddy, Mukund Bhakta, John Gilmartin, and Joseph Yonkoski. Cost effective CO_2 capture from flue gas for increasing methanol plant production. *Energy Procedia*, 63, 2014, pp. 1407–1414.

7 Sulfur Recovery Unit

7.1 INTRODUCTION

Sour natural gas (NG) streams are treated with amines to bring down the H2S and CO2 levels in the gas to acceptable limits since decades. The rich amine is regenerated in amine regenerators (strippers) to strip the acid gas components and regenerate the amine for further absorption. The acid gases produced are fed to sulfur recovery units (SRU).

In refineries and petrochemicals, sour gas streams and liquified petroleum gas (LPG) are treated in amine absorbers, and spent amine is regenerated to produce acid gas rich in H2S. The gas produced is fed to sulfur recovery units. Most often sour gas from the sour water strippers join with the acid gases and they are fed together to the sulfur recovery unit for recovery of sulfur from the gases, primarily to reduce pollution.

Different sections of a sulfur recovery unit are shown in Figure 7.1. In the Claus section, 95%–97% of sulfur is recovered. After recovery of the majority of sulfur, the tail gas is fed typically to a tail gas treatment (TGT) unit for further recovery up to 99.5% (shown in Figures 7.5 to 7.8). Other processes, like the Super Claus, are often deployed for additional sulfur recovery.

The unreacted tail gas is fed to an incinerator to convert the balance of H2S to sulfur dioxide and discharge to the stack. The liquid sulfur flows to the sulfur pit through sulfur locks and is degassed to remove the H2S from the pool by blowing air, and the H2S-bearing gas is also fed to the incinerator. The liquid sulfur is pumped to storage, solidified, and dispatched. Often sulfur in the form of liquid is also dispatched.

7.2 PROCESS DESCRIPTION AND REACTION MECHANISM

The acid gas is first fed to a knockout pot to knock off liquid, if any. Then the acid gas is typically preheated to 110°C–200°C by high-pressure steam and fed to the SRU. Similarly, sour water gases are routed through the knockout drum and fed together with acid gas to the main combustion chamber (MCC). Sour gas from a sour water stripper is normally available at a temperature of 90°C–95°C and does not need further heating. Air is introduced in the MCC to promote the Claus reactions. Typically, air is also heated by high-pressure steam before feeding to increase the MCC temperature. A temperature above 1250°C and higher is preferred in the MCC to ensure complete conversion of ammonia in MCC. Otherwise ammonia can form salts and deposit in the reactors in downstream conversion section, and increase the pressure drop of the unit. After reaction, the hot gases are cooled by generation of high-pressure (HP) or medium-pressure (MP) steam and fed to the sulfur condenser

DOI: 10.1201/9781003268246-7

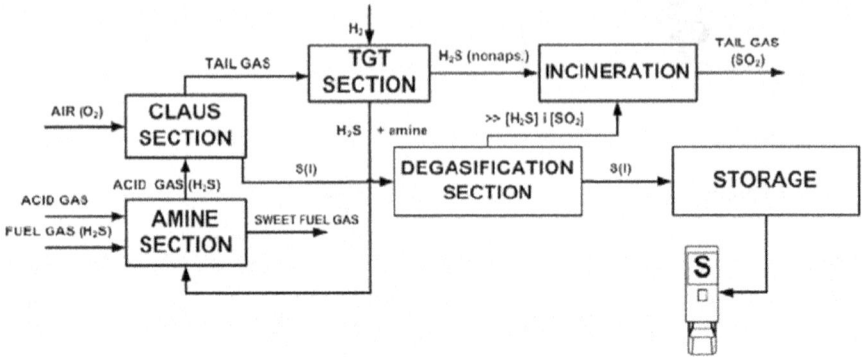

FIGURE 7.1 Different sections of sulfur unit.

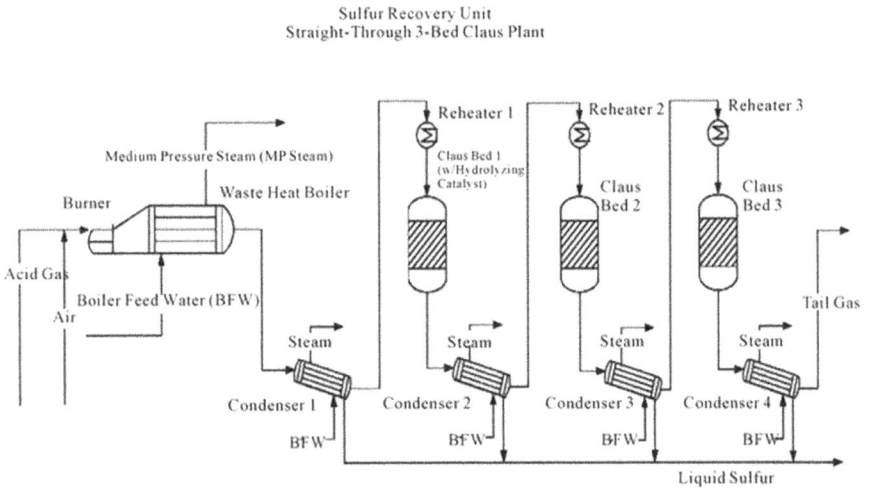

FIGURE 7.2 Claus sections of sulfur unit.

for further cooling while generating low-pressure (LP) steam. In the MCC, a conversion of around 65% is achieved and in the sulfur condenser, downstream of the MCC 65%–70% of total feed sulfur is recovered. The gas is further admitted into catalytic converter section. Typically, two to three catalytic reactors are provided. The reactors are typically preceded by HP steam heaters to heat the gas stream to 210°C–220°C prior to entry to the catalytic converters. The reacted gases from the catalytic reactor is routed to sulfur condensers for condensation of sulfur. The liquid sulfur flows out of the condensers through sulfur locks (provided to arrest blowing of gases through the liquid sulfur outlets) to sulfur pit. Two different flow schemes are shown for the catalytic conversion section in Figure 7.2 and Figure 7.3.

The SRU is based on the modified Claus process, which is a two-step reaction process. A portion of the total H2S is burned in the reaction furnace or main combustion

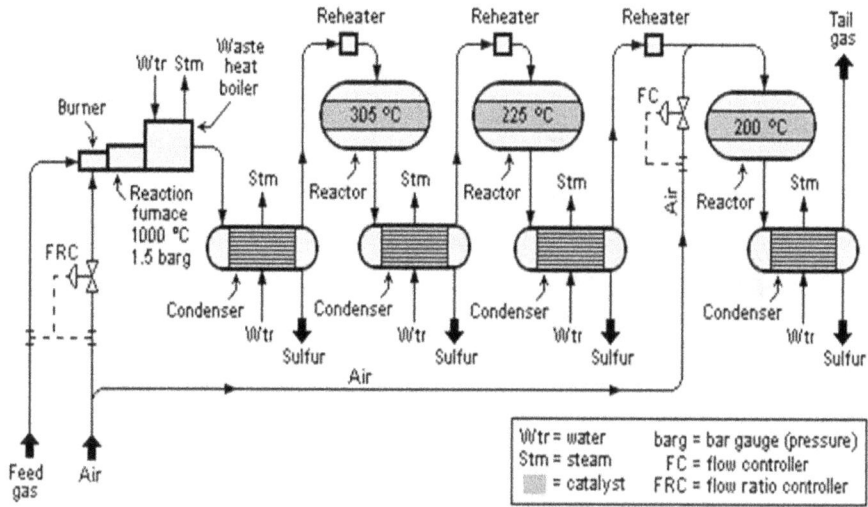

FIGURE 7.3 Different flow scheme of sulfur unit.

chamber(MCC) to form SO2. Then, the balance H2S reacts with the SO2 formed at an optimal ratio of 2:1 to form elemental sulfur (S_x) in the main combustion chamber and in the Claus reactors. After each catalytic stage, liquid sulfur is recovered in the Claus condensers. The remaining unreacted H2S and SO2 then proceed to the next stage, where the equilibrium-limited Claus reaction continues in the presence of Claus catalyst.

In general, gas processing plants and oil refineries are required to recover between 95% and 99.99% of the total sulfur introduced to the SRU. A conventional two-stage SRU, with two Claus reactors, is typically expected to recover 96%-plus. For a three-stage SRU, the general expectation is 98%-plus. If necessary, a tail gas cleanup unit (TGCU) is required to recover the balance. Emissions regulations are becoming more stringent and normally dictate the recovery of sulfur.

7.2.1 CLAUS THERMAL STAGE

The Claus furnace or main combustion chamber (MCC) plays a very important role in the sulfur recovery unit. The two main reactions occur in the Claus furnace (MCB) and are as follows:

$$H_2S + 3/2\ O_2 = SO_2 + H_2O \text{ with } \Delta Hr = -518\ KJ/g\text{-mole, exothermic}$$

$$2H_2S + SO_2 = 3/2\ S_2 + 2H_2O \text{ with } \Delta Hr = +47\ KJ/g\text{-mole, endothermic}$$

The allotrope of sulfur produced in the MCB is known to be S_2. The second reaction to produce sulfur is endothermic, thus a drop in temperature is expected due to this reaction. Often, a split feed arrangement is used where a part of the feed is

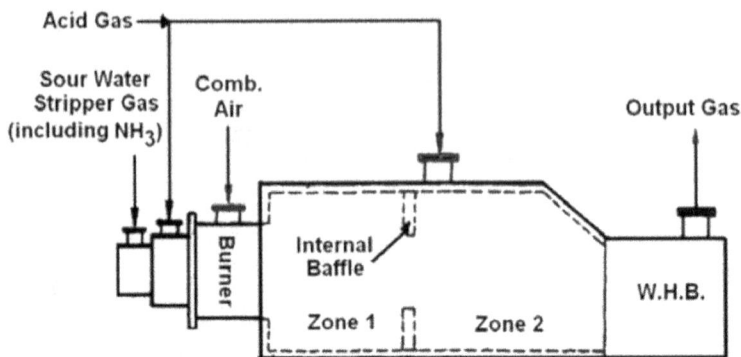

FIGURE 7.4 A split flow arrangement in the MCB to maintain MCB temperature.

introduced in the MCB near the inlet and the balance toward the outlet for obtaining a higher temperature in the MCC inlet to help full ammonia destruction, as in Figure 7.4. Many other side reactions also take place in the furnace and produce unwanted components that end up as ambient pollutants.

H2S is partially oxidized with air (one-third of H2S is converted to SO2) in the Claus furnace. The acid gas/air mixture goes into the furnace operating at a temperature of 1300–1700 K, where the reaction is given sufficient time to reach equilibrium. Depending upon the calorific value of the acid gas, various methods for stable burning are practiced. If lean gases (low in H2S content) with very low calorific value are involved, then auxiliary fuel, gas–air preheating, oxygen enrichment, and the split feed method need to be used to achieve high temperature in the furnace to destroy ammonia fed along with sour gas (sour water stripper off-gas).

7.2.2 CLAUS CATALYTIC STAGE

The remaining H2S from the Claus furnace is reacted with SO2 at a lower temperature (to increase equilibrium conversion) over alumina or titanium dioxide-based catalyst at 200°C–240°C to produce more sulfur. In the catalytic stage, allotrope S_8 is produced, which involves an exothermic reaction between H2S and SO2: $2H_2S + 1.5 SO_2 = 3/8 S_8 + H_2O$, $\Delta Hr = -108$ KJ/g-mole. Other allotropes of sulfur may also be produced in smaller quantities. The overall reaction for the entire process can be written as $3 H_2S + 1.5 O_2$ $3/n$ $Sn + 3 H_2O$, $\Delta Hr = -626$ KJ/g-mole.

Liquid sulfur polymerizes at a temperature of about 430 K (157°C) (viscosity of liquid S increases sharply with temperature above 160° C). Therefore, the temperature of the liquid sulfur needs to be monitored/controlled to prevent polymerization and clogging of lines. Lines are normally jacketed with LP steam to maintain temperature of liquid sulfur typically between 135 to 155° C. In conventional units liquid sulfur getting into the catalyst in converters is avoided. Several modes of heating the gas prior to feeding into catalytic converters are practiced.

7.2.2.1 Reheat Methods

The reheat methods include hot gas bypass, Steam re-heaters, acid gas fired line burners, and natural gas fired line burners. Different practices followed for heating the gas for reaction in catalytic converters include the following:

Hot gas bypass. This involves mixing of the hot gas bypassing the waste gas cooler upstream. It is very important to ensure there is adequate mixing of the streams prior to entry to catalytic converters.

Steam reheaters. Heating with HP steam generated in heat exchangers at the upstream of each catalytic converter.

Gas/gas exchangers. The cooled gas is heated by hot gas from catalytic converters in gas/gas heat exchangers at the upstream of each catalytic converter. This system is not much in use.

Direct fired heaters or line burners. Fired heaters are deployed utilizing fuel gas to substoichiometrically burn (to avoid oxygen breakthrough and damage of catalyst in the converters) and heat the gas at the upstream of each catalytic converter. The fuel preferred is natural gas having more hydrogen to lower formation of soot that can deposit on the catalyst of catalytic converters.

A *cold bed adsorption* (CBA) process, also known as sub dew point process is also practiced. In the CBA process the heterogenous catalytic reaction is allowed to take place at lower temperatures below the sulfur dew point, thus increasing the equilibrium sulfur conversion. The process is known to deliver a conversion of 97.5%–99.0%.

Some of the contaminants that enter the amine system get removed in the regenerator and then get into the sulfur recovery unit. The primary concern in the SRU is ammonia and hydrocarbons present in the acid gas feed. The sulfur plant uses oxygen of air that is introduced with an air blower into the unit to convert H2S to elemental sulfur. The overall reaction is $H_2S + \frac{1}{2} O_2 \rightarrow S + H_2O$. The amount of oxygen, generally introduced by an air blower into the sulfur unit, is controlled to maintain a ratio of 2:1 of H2S:SO2 by mole in the tail gas at the outlet of Claus section. On a volume basis, both NH3 and hydrocarbon consume more oxygen in the sulfur unit than required for the conversion of H2S to sulfur evident from reactions shown next.

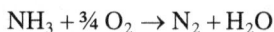

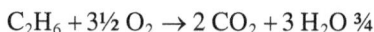

$$NH_3 + \tfrac{3}{4} O_2 \rightarrow N_2 + H_2O$$

$$C_2H_6 + 3\tfrac{1}{2} O_2 \rightarrow 2\, CO_2 + 3\, H_2O\, \tfrac{3}{4}$$

So, both ammonia and hydrocarbon rob the sulfur plant of capacity. Typically, the first limit in sulfur plant capacity is the volume of air the air blower can deliver and the second is the pressure drop caused by the higher flow of gases through the unit.

In refineries, ammonia is often routed to the sulfur unit primarily through sour water off-gases. If ammonia enters the sulfur plant it needs to be fully destroyed.

If ammonia does not get destroyed, it can lead to deposition of ammonium salts in cooler parts of the sulfur plant and can result in plugging, increasing pressure drop and localized corrosion issues.

The effluent gases from the Claus plant are typically vented to the atmosphere/ stack or are directed to a tail gas treatment system. But irrespective of the Claus plant having a tail gas unit or not, the final effluent gas is usually incinerated to oxidize any residual sulfur to sulfur dioxide. Air pollution control regulations in most industrialized countries prohibit the discharge of large amounts of sulfur compounds into the atmosphere. Therefore, Claus plants with tail gas treatment units are often mandatory. The sulfur produced from the Claus process are extremely good in quality and thus is a starting stock for production of valuable basic chemicals like sulfuric acid.

With growing air pollution concerns, sulfur recovery in Claus type units is increasing to the point where units that normally would not be considered economical are installed strictly for the purpose of air pollution control. In addition, the recovery efficiency of Claus-type plants is continuously being improved by better plant operation, better design methods, and developments in the process technology.

Since the inception of the process by Claus in 1883, it has undergone several modifications. As presently used, most process configurations are similar in their basic concept and differ only in the design and arrangement of the equipment. Oxygen enrichment technology is deployed to increase sulfur unit capacity and restrict the size of the unit by lowering the nitrogen content of the gas. But oxygen enrichment of air may result in a very high temperature inside the furnace and may need special refractory inside.

7.3   ENHANCED SULFUR RECOVERY

Typically, several methods are deployed for increasing the recovery of sulfur. A selective oxidation process, in which the tail gas is selectively oxidized by air in the presence of metal oxide catalyst, is also used and a known to have achieved a sulfur recovery of 99%.

The modified Claus process for tail gas cleanup unit or tail gas treatment unit (TGCU/TGTU) is used when very high conversion of H_2S is desired. The recovery achieved with TGTU is 99.8%–99.9%. The problem with TGTU is that it demands a high investment for increase in conversion by around only 2%.

Several processes are in practice for treatment of tail gas to increase the recovery of sulfur and achieve a recovery more than 99.5%. The Super Claus process flow scheme is shown in Figure 7.5 and the typical control scheme of the process is shown in Figure 7.6.

In the Shell Claus off-gas treatment (SCOT) process shown in Figures 7.7 and 7.8, the SO_2 and other sulfur compounds in tail gas is essentially reduced by hydrogen to convert the sulfur compounds to H_2S in a reactor. The hot effluent is cooled and the H_2S produced is absorbed by amine. The rich amine is regenerated and acid gas from regenerator is recycled to the inlet of Claus section. In locations where hydrogen is not available, it is generated by substochiometric combustion of tail gas with air and fuel gas to produce H_2 that reacts with SO_2 and other sulfur compounds

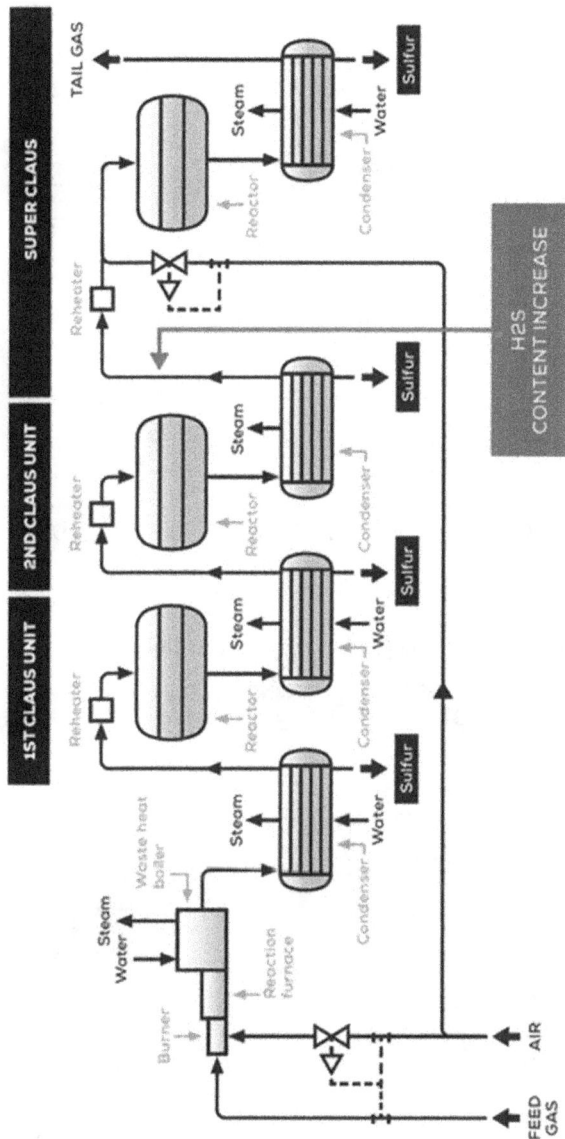

FIGURE 7.5 Scheme of Super Claus for TGTU.

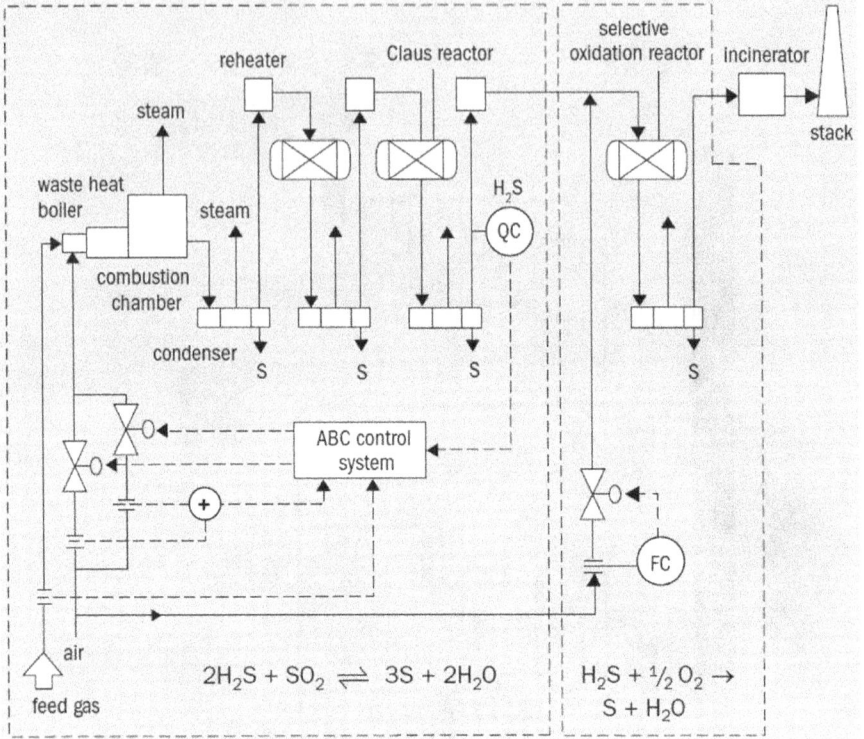

$$2H_2S + SO_2 \rightleftharpoons 3S + 2H_2O$$

$$H_2S + \frac{1}{2}O_2 \rightarrow S + H_2O$$

FIGURE 7.6 Arrangement of Super Claus with typical controls.

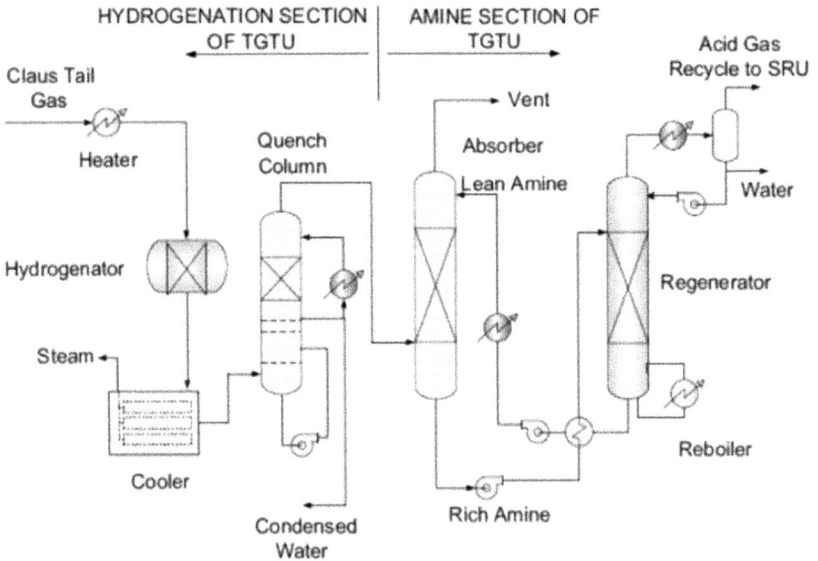

FIGURE 7.7 TGTU showing hydrogenation section and amine section.

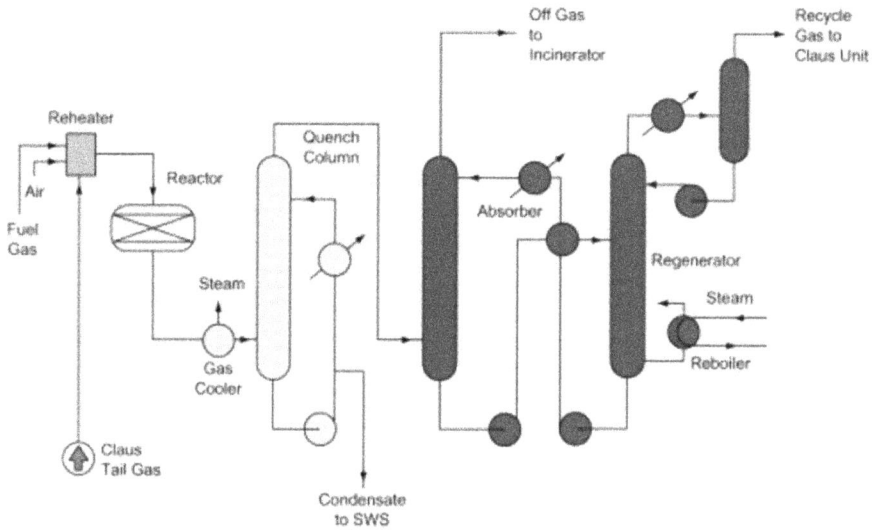

FIGURE 7.8   TGTU deploying SCOT reactor with hydrogenator (with in situ generation of hydrogen).

in tail gas to convert them into H2S. The arrangements are shown in Figure 7.7. In locations where hydrogen is available, the hydrogen is mixed with the tail gas, then heated and introduced in the hydrogenation reactor to convert sulfur compounds present in tail gas into H2S.

Typically, a separate amine regenerator is used to regenerate the rich amine. The lean amine loading of H2S is usually maintained low (around 0.005 moles of H2S/mole of amine) to enable the lean amine to absorb more H2S from the tail gas stream at near atmospheric pressure. A recovery of more than 99.7% is achievable in the process.

7.4   COMMON PROBLEMS IN SULFUR UNITS

Problems in sulfur units are listed next:

1. Lower temperature in MCC results in lower conversion in the Claus furnace. Typical conversion in the Claus furnace is 65% to 70%. A lower conversion loads the catalytic stages and may result in lower conversion to sulfur in the Claus furnace.

   Further, if the temperature of the MCC falls below 1200°C, ammonia destruction may not fully take place and ammonium salts may eventually deposit in catalytic converters and give rise to higher pressure at the inlet and may result in tripping of the unit. If the gas is lean in H2S, the MCC temperature may be lower. Typically, acid gas and combustion air are heated by HP steam and support fuel firing is kept on to attain the desired temperature in MCC. Often the split feed technique is also used to increase MCC temperature, as shown earlier in Figure 7.4.

2. Instrumentation issues in the MCC can lead to problems. A key to reliable operation of burner and furnace instrumentation is to adequately purge each instrument nozzle even when the SRU is shutdown. The process gas contains elemental sulfur, which will condense and solidify if allowed to enter the cool instrument nozzles. The purge rate should be measured with a rotameter to verify and ensure flow.

3. In case of too much auxiliary firing of fuel gas, carbon or soot may deposit in the catalytic converter beds and result in higher pressure at the inlet.

4. Hot spots in the MCC shell is a common problem due to damage/poor quality of refractory provided inside.

5. Solidification of sulfur inside lines and blocking the flow of liquid sulfur through pipes and sulfur locks can be a source of problems. The flow of steam in the steam jacketed lines may be checked and quality internal insulation may be ensured. *The equipment and lines in sulfur units are sloped toward the sulfur locks to avoid stagnant portions.* The slopes of the equipment and lines need to be ensured for trouble-free operation.

6. Steam coil failure in the sulfur pit can be a problem. Experts recommend using alloy piping for the steam coils, steam supply downcomers and condensate risers, and alloy for any internal components such as ladder rungs that will be alternately covered with liquid sulfur and then exposed to air as the pit level changes.

7. Sulfur transfer pump failure can be another source of problem. Usually dual-jacketed pumps are used to avoid failures.

8. Sulfur pit eductor failure can be a source of problems. Experts suggests using a fully steam jacketed steam eductor to continuously draw atmospheric air into the pit, sweeping vapor space to prevent the accumulation of H2S.

9. Breakage/dislodgement of refractory and chocking the liquid sulfur flow path is often observed.

7.5   CAPACITY AUGMENTATION

Traditionally, an increase in capacity of a sulfur unit is achieved by the addition of parallel train mostly with common TGTU. Oxygen enrichment is sometimes used but needs extra care and may result in high temperature in the main combustion chamber (MCC) needing special refractories. The injection of oxygen also needs a precise control strategy and should be only implemented with the support and guidance of experts. Reduction of hydrocarbon in acid gas and sour gas streams help reduce air requirements in sulfur unit and help in increasing capacity. Low pressure operation of rich amine flash drum (RAFD) of amine regenerator, can reduce hydrocarbon entry to the sulfur unit and may result in incremental gain in capacity.

BIBLIOGRAPHY

Hamid Reza Mahdipoor, Keyvan Khorsand, Reza Hayati, and Hooman Javaherizadeh. Effect of reaction furnace and converter temperatures on performance of sulfur recovery units (SRUs). *Journal of Petroleum Science Research (JPSR)*, 1, no. 1, April 2012, pp. 1–3.

Mohamed Sassi and Ashwani K. Gupta. Sulfur recovery from acid gas using the claus process and high temperature air combustion (HiTAC) technology. *American Journal of Environmental Sciences*, 4, no. 5, 2008, pp. 502–511.

Reza Rezazadeh and Sima Rezvantalab. Investigation of inlet gas streams effect on the modified claus reaction furnace. *Advances in Chemical Engineering and Science*, 3, 2013, pp. 6–14.

Steve Fenderson. Improving claus sulfur recovery unit reliability through engineering design. In *Brimstone Engineering Sulfur Recovery Symposium*, Goar, Allison, & Associates, Inc., September 15–18, 1998.

8 Plant Safety

8.1 INTRODUCTION

The equipment used in industry is protected by an overpressure relief system. When the pressure of equipment increases beyond the designed pressure or maximum allowable working pressure (MAWP), the valve mounted on the vessel opens and releases the content to a safe destination. The emergency relief system is devised to protect equipment, the environment, and operating personnel from abnormal conditions and catastrophic failure of equipment. The system may be comprised of a pressure relief valve or rupture disc, and piping to a disposal system like a flare or vent stack (API 521). The pressure relieving device may be classified as follows (API 521):

- *Relief valve.* A spring-loaded safety valve actuated by the static pressure upstream of the valve. The valve opens normally in proportion to the pressure over the opening pressure. A relief valve is used primarily for incompressible fluids.
- *Safety valve.* A spring-loaded pressure relief valve actuated by the static pressure upstream of the valve and characterized by rapid opening or pop action. The valve is normally used for compressible fluid.
- *Safety relief valve.* A spring-loaded pressure relief valve that may be used as either a relief valve or safety valve.
- *Rupture disc.* A non-reclosing differential relief device actuated by inlet static pressure and designed to function by bursting of a rupture disc.

Some of the contingencies that can lead to overpressure of equipment can be a closed outlet on vessels, inadvertent opening of valve from a higher-pressure source, utility failure, loss of cooling fans, reflux failure in columns, tube leak in exchangers, loss of instrument air or electric power, abnormal heat inputs, and plant fires. Relieving load or quantity is traditionally computed by the guidelines indicated in API 521.

8.2 METHODOLOGIES OF RELIEF LOAD COMPUTATION OF COLUMNS

Estimation of relief loads under extreme conditions is important for the correct sizing of relief valves and flare headers, and also for selection of the disposal system as it can be very costly to implement. Thus, relieving loads and relieving conditions need correct prediction.

Estimating accurate relief loads for different contingencies in distillation columns is quite complex. The computation methods for relief load of distillation columns are

DOI: 10.1201/9781003268246-8

not well documented in any publication; they are mostly available with design organizations and experienced professionals. The conventional methods need to be adopted with adequate care for a distillation column due to the compositional changes along the column height. The traditional method for estimating relief load for distillation column is the *unbalanced heat method or UBH method* described next.

8.2.1 Unbalanced Heat Method (UBH Method)

In the UBH method, the relieving load of column is computed as follows:

$$R = Q \text{ unbalanced} (\text{excess})/\lambda$$

$$Q \text{ unbalanced} = F\,h_F - B\,h_B - D\,h_D + Q_R - QC - (F - B - D)h_L$$

where

F=column feed rate at relief; h_F=specific enthalpy of feed at relief
B=column bottom rate at relief; h_B=specific enthalpy of bottom at relief
D=distillate rate at relief; h_D=specific enthalpy of distillate at relief
Q_R=reboiler heat input at relief
Q_C=condenser duty at relief (generally, the design duty can be considered)
h_L=specific enthalpy of top-tray liquid; λ=latent heat of vaporization of top-tray liquid
R=relief load in mass in units (kg/Hr)
Flowrates, specific enthalpies, and λ are in consistent units.

This method may be conservative and may lead to bigger relief valves and flare headers but is mostly practiced. With the advent of reliable simulation softwares, use of process simulation is becoming popular and being increasingly used for estimating relief load and properties of the relieving fluid. Steady-state simulation can also be used to estimate the relief load with few limitations and can address some of the assumptions and limitations involved in the conventional (UBH) method.

The conventional approach of tower relief load calculation, especially for grass-roots units, is to compute the unbalanced heat across the tower during an upset scenario. Although the unbalanced heat method has its limitations, it is one of the most trusted and most used methods for relief load estimation for distillation columns. One of the basic assumptions in the method is availability of an unlimited supply of liquid to the top tray, and the lighter liquid (having low molar latent heat of vaporization) is considered to vaporize from the top tray during a relieving scenario. This may result in a high relief load in mass units or actual cubic meter per hour (ACMH; owing to the usually low latent heat of vaporization and lower molecular weight) of the top-tray liquid at its bubble point.

It may be noted that the latent heat of vaporization for lighter compounds are always lower in in molar units (kcal/k-mole). This reduces either from Trouton's rule ($\Delta H_V/ T_B = K$), or by the Kistiakowsky equation ($\Delta H_V /T_B = 8.75 + 1.987\ln T_B$), where ΔH_V is the molar latent heat of vaporization in kcal/k-mole and T_B is the boiling point in kelvins, K is a constant (the value of constant K in Trouton's rule is close to 21 for many substances). Further, due to the lower molecular weight of lighter material in top trays the ACMH of the vapor relieved becomes higher and that primarily controls the seizing of relief valves and the connected lines and disposal system.

The major limitation of UBH method is in the prediction of an accurate relieving temperature. The liquid in the trays below the top tray are progressively heavier, and the bottom-tray liquid is the heaviest liquid in the column. In case of reflux failure, the top-tray liquid dries after a certain time and the liquid begins to vaporize from the lower trays, resulting in a higher relief temperature than that of the top-tray liquid. This effect was much more pronounced in small- to medium-sized towers where the top tray is likely to dry out quickly as the reflux fails and liquids heavier than the top-tray liquid are likely to vaporize and also contribute to the load. The relieving load may be lower owing to higher latent heat of vaporization, but there is likelihood of increase of relieving temperature.

The consideration of vaporization of top-tray liquid (considered in UBH method) would result in higher relieving load (in ACMH) but would indicate a lower relieving temperature and may lead to incorrect material selection for design of the column overhead, relief valve, and downstream disposal system in many cases. The effect on relieving temperature may be more pronounced in a column that has a large difference in boiling temperatures between the liquids in the top and bottom trays.

The total inventory of the system, mainly depend on the diameter of the column and the number of side draws and side strippers, also contributes to the scenario in question. If the reboiling/stripping is continued during a relief scenario, then the likelihood of column overheads being exposed to higher boiling fluids is more realistic for a small-diameter column with no or a limited number of side draws. This may render the overhead system vulnerable to high temperature exposure during the relief scenario.

8.2.2 STEADY-STATE SIMULATION METHOD

The concern of inaccurate relieving temperature in the UBH method can be adequately addressed by using a steady-state simulation method. Dynamic simulation provides an alternative method for determining relief load under abnormal conditions but involves more rigorous computations. A case study is presented to demonstrate the difference of steady-state relieving temperatures calculated by the UBH method and a simulation method, and its impact on design temperatures of the column overhead system and relief valve laterals for a benzene–toluene fractionating column (Figure 8.1).

Reboiler pinch may also be encountered during relieving. During relieving, the column pressure rises consequently. The bottom liquid temperature (at bubble point) also increases and the difference of temperature between the column bottom liquid temperature and reboiler heating fluid temperature reduces, usually leading to a lower heat input to the column reboiler. This is described as reboiler pinch. The credit of lower heat input to reboiler is often considered for evaluating relieving load of a distillation column. If there was a significant reduction in the reboiler duty

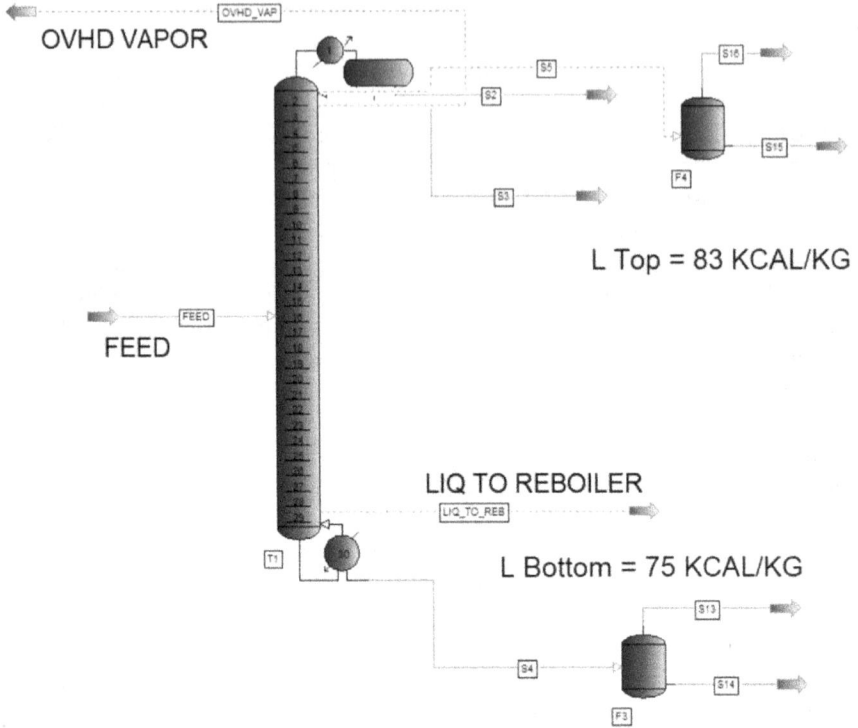

FIGURE 8.1   Simulation of the column for separation of benzene and toluene.

at relief, the lighter components would begin traveling toward the bottom, causing the duty to rise again. Many designers re-rate the reboiler with feed composition instead of bottom composition in these circumstances to maintain a more conservative/realistic reboiler duty at relief. A general guideline is not to consider any reduction in reboiler duty, if the reboiler is a once-through reboiler (i.e., the tray liquid is directly routed to reboiler without mixing with the liquid in the bottom spout). Reduction in reboiler heat input with an increase in bottom liquid temperature can be considered if the reboiler is a recirculation type (i.e., liquid from the bottommost tray mixes with the liquid in the bottom spout and routed to the reboiler with adequate inventory/residence time in the bottom spout). The majority of reboilers deployed in industries are recirculation types. It may advisable to re-rate the reboiler to evaluate the reboiler duty during relieving conditions. The important contingencies for evaluation of relief load of a distillation column as indicated in API 521 are referred to in the following section.

8.2.3   ILLUSTRATION/EXAMPLES FOR COMPUTATION BY THE METHODS

To illustrate the computation of relieving load an example of benzene–toluene separation column is considered with feed flow rate (F) of 100 MT/H at 70°C and 50:50

TABLE 8.1

Column Parameters at Relieving Pressure of 3.5 Kg/cm2(g)

Column Name		T1
Column condenser duty	M*KCAL/HR	−16.5985
Column reboiler duty	M*KCAL/HR	20.7252
Column reflux ratio		3.0000

weight percent of benzene and toluene in the feed. D and B are top and bottom products, respectively. The column has 30 stages producing benzene of purity of 99.5% wt. as the top product (D) operating with a reflux ratio of 3 at an operating pressure of 0.5 kg/cm2 (g). The required reboiler duty is 20.04 G-Cal/Hr at operating conditions. The top product (D) and bottom product (B) flow rates are 50 MT/Hr each. Relieving pressure is 3.5 Kg/ Cm2 (g). The simulation at 3.5 Kg/cm2 (g) delivers the enthalpy of all streams, reboiler condenser duties, and delivers F h_f = 3.08 Kcal/ Hr, B.h_B = 4.084 Kcal/Hr, h_L = 62.45 Kcal/Kg, latent heat of the top tray liquid λ = 83 Kcal/Kg. The required reboiler duty marginally increases to 20.72 G-Cal/Hr and condenser duty Qc = 16.6 G-Cal/Hr at relieving pressure of 3.5 Kg/cm2 (g). The simulation flow diagram is presented in Figure 8.1, and the column parameters in Table 8.1 and stream summary in Table 8.2. The overhead vapor (OVHD_VAP) in Table 8.2 represents the relieving duty of the column at 3.5 Kg/cm2 (g) pressure by simulation using a column.

8.2.3.1 Computation by UBH Method

The relieving loads are calculated for different contingencies using the UBH method.

8.2.3.1.1 Loss of Reflux

A loss of reflux may occur due to failure of the reflux pump/product pumps or failure of coolant in the column overhead condensers. The effects can be summarized as follows:

- Reflux stops immediately. The reflux drum and the condenser flood, restricting the overhead vapor path or coolant flow in the condenser stops, resulting in no condensation and pressurization of the column.
- The feed continues at the same flow rate during relief condition.
- Bottom product continues at the same rate. If a feed vs. column bottom heat exchanger is provided to preheat the feed, it will continue to deliver constant feed preheating.

$$Q \text{ unbalanced} = F h_F - B h_B - D h_D + Q_R - QC - (F - B - D) h_L$$

TABLE 8.2
Stream Summary of Column Simulation

Stream Name		Feed	S3	S4	OVHD_ VAP	LIQ_TO_ REB
Stream Phase		Liquid	Liquid	Liquid	Vapor	Liquid
Total Mass Rate	KG/HR	100000.000	50000.396	49999.604	200001.584	325428.240
Temperature	°C	70.000	139.047	177.028	139.103	176.817
Pressure	KG/CM2	6.000	4.500	4.800	4.500	4.789
Total molecular weight		84.549	78.144	92.099	78.144	92.069
Latent heat	KCAL/KG	n/a	n/a	n/a	n/a	n/a
Total sp. enthalpy	KCAL/KG	30.797	62.451	81.677	145.443	81.553
Liquid sp. enthalpy	KCAL/KG	30.797	62.451	81.677	n/a	81.553
Dry vapor sp. enthalpy	KCAL/KG	n/a	n/a	n/a	145.443	n/a
Total enthalpy	M*KCAL/ HR	3.080	3.123	4.084	29.089	26.540

Thus, in this scenario of reflux failure, Qc and D reduce to zero, Q unbalanced is computed from the preceding equation, and the relief load R is evaluated by dividing the Q unbalanced by the latent heat of vaporization of the top tray liquid. To obtain the property, particularly the latent heat of the top-tray liquid, the column may be simulated in a steady-state simulator with a dummy topmost tray liquid draw of 1 kg/Hr or a pseudo liquid stream may be drawn from the topmost tray. The latent heat of the topmost liquid can be used in calculating the relief load as (Q unbalanced/λ).

$$Q \text{ unbalanced} = 3.08 - 4.084 + 20.72 - 50 \times 62.45 = 16.58 \text{ G-Cal/Hr}$$

$\lambda = 83$ Kcal/Kg. Thus, R (relief load) $= 16.58 \times 1000/ 83 = $ **199.7 MT/Hr.**

8.2.3.1.2   Loss of Feed

- Feed stops immediately.
- After some time, when the column level drops, the bottom product decreases to maintain the column sump level, and finally reduces to zero.
- The column overhead vapor rate decreases, the reflux drum level drops, and the distillate rate decreases to maintain the condenser level and finally becomes zero.

Therefore, the terms F, D, and B reduce to zero. The Q unbalanced can be calculated equating the terms F, D, and B to zero and in turn the relief load can be estimated by division of the Q unbalanced by the latent heat of vaporization of the top liquid obtained earlier as Q unbalanced/λ.

$$Q \text{ unbalanced} = QR - Qc = 20.5 - 16.6 = 3.9 \text{ G-Cal/Hr}$$

$$R\,(\text{Relief load}) = 3.9 \times 1000/83 = \mathbf{46.98\ MT\,/\,Hr}$$

8.2.3.1.3   Site-Wide Power Failure (SWPF)

- In a SWPF, all electrical equipment fails, therefore, the feed pumps, the column bottom pumps, and the reflux pumps stop. Assuming all cooling water pumps are electrically driven, the condensing duty is also immediately lost.
- The reboiler hot fluid/steam is assumed to flow continuously to the reboiler. If the reboiler hot fluid flow discontinues, then the reboiler duty theoretically reduces to zero. If the bottom reboiler is a furnace, then flow through the furnace will stop due to loss of the circulation pump and firing will stop, but 20% to 30% of operating duty can be considered, imparted by the hot refractories of the furnace on the liquid content in the furnace coils or purging medium used to displace the liquid content of the furnace tubes (to avoid liquid degradation inside furnace tubes).

Therefore, Qc, F, D, and B reduce to zero and Q unbalanced is evaluated (reduces to Q_R). Relief load is also evaluated by dividing by the latent heat of vaporization of liquid in the topmost tray as (Q unbalanced/λ).

$$Q \text{ unbalanced} = 20.72 \text{ G-Cal/Hr}, R = 20.72 \times 1000/83 = \mathbf{249.6\ MT\,/\,Hr}$$

The comparison of relief loads evaluated for different contingencies with the UBH method are presented in Table 8.3 (controlling loads are highlighted).

TABLE 8.3
Relief Loads Evaluated for Different Contingencies with UBH Method

Attribute	Top Reflux Failure (UBH)	Site Power Failure (UBH)	Feed Failure (UBH)
Relief load (MT/Hr)	199.7	**249.6**	46.98
Relieving temperature (°C)	139.1	139.1	139.1
Molecular weight	78.1	78.1	78.1
Moles of relieving vapor (Kg-moles)	2559.41	3121.12	587.45
Relief load in ACMH	18199.66	23452.4	4414

8.2.3.2   Computation by Steady-State Method

The relief loads can be also computed using the steady-state simulation method. The method can more accurately determine the relieving temperature.

8.2.3.2.1   Top Reflux Failure

For columns with only reflux cooling and no side pumparound (PA) heat removal, failure of the top reflux leads to a total cooling loss to the column, and the overhead section of the column is exposed to the feed vapor temperature. In case of loss of reflux, all trays above the feed tray become dry after some time, which is the basis for the assumption here.

So, to simulate this, a stripper column without condenser can be configured with the same number of stages as the stripping section of the original column. The feed will enter at the topmost location of the column (now a stripper) and pressure will be considered as the relieving pressure. Alternatively, for easy and quick computation, a flash drum (as in Figure 8.2) can be modeled to simulate the cases having the feed entering the flash drum. The reboiler duty may be considered based on judgment, normally without any reduction in case of a once-through reboiler. However, stripping steam, if any, will be considered in the flash drum as a second feed.

The feed flow to the vessel is 100 MT/H at a pressure of 6 Kg/cm2 and at a temperature of 70°C, as in the illustration in Section 8.2.3. The flash drum pressure is taken as 3.5 Kg/cm2 (g) as relieving pressure, and a reboiler heat input of only 20.72 G-Cal/Hr is considered as an input to the flash drum.

Reflux flow in the original column may still need to be considered as an input to the column to account for the vaporization of accumulated liquid in trays for computation of the relieving load for reflux failure using a flash drum. Otherwise the relief load will be maximum equal to the total feed to column. Considering the reflux as input will make the relief quantity essentially the same as the column top vapor in column simulation. Stream S7 of Figure 8.2 will represent the relieving load.

REFLUX FAILURE

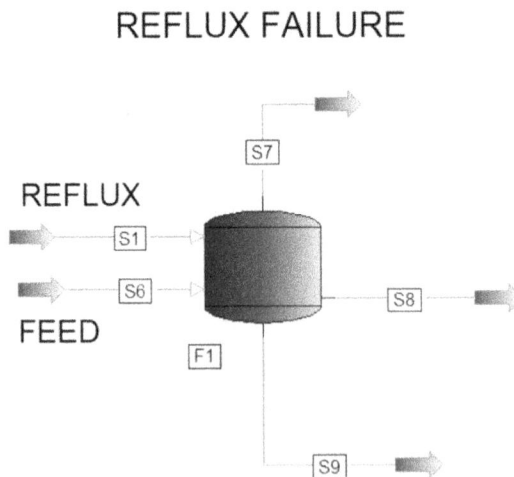

FIGURE 8.2   Load due to reflux failure evaluated using a flash drum.

TABLE 8.4
Relief Loads Evaluated by different methods for Reflux Failure

Attribute	Top Reflux Failure (Using UBH Method)	Top Reflux Failure (Using Column Simulation)	Top Reflux Failure (Using Flash Drum)
Relief load (MT/Hr)	199.6	200	204.3
Relieving temperature (°C)	139.1	139.1	146.5
Molecular weight	78.1	78.1	80.32
Moles of relieving vapor (Kg-moles)	2559.41	2559.4	2543.96
Relief load in ACMH	18199.66	18199.66	18400.6

Alternatively, the relieving duty in *reflux failure* case can be considered as the total overhead vapor (OVHD_VAP) in the steady state column in Table 8.2. This will be essentially the same as the relieving load computed through UBH method. The relieving duty in reflux failure case computed by three methods—UBH, steady state column, and simulation using a flash drum—is presented in Table 8.4.

The net vapor flash drum (S7 of figure 8.2) is to be relieved through the relief valve in case of reflux failure. An appropriate cushion (over the load calculated by this method) may be considered based on the computed latent heat (of vaporization) ratio of the top liquid and bottom liquid (up to 10%–15%).

8.2.3.2.2   Site Power Failure

In case of *site power failure* there will be no feed, the appropriate reboiler duty will con-tinue to be imparted to the bottom liquid that will generate vapor. The bottom spout of the column will have adequate hold up much higher than total tray hold up. This inven-tory is likely to vaporize with application of reboiler duty. For the purpose of modeling, the bottommost tray liquid (a pseudo stream as the original column bottom tray liquid) as feed entering the flash vessel can be considered having the appropriate reboiler duty or stripping steam, as in Figure 8.3. The pressure of the flash drum will be considered the same as relieving pressure of 3.5 Kg/cm2 (g). The model will be run, and vapor

SITE POWER FAILURE

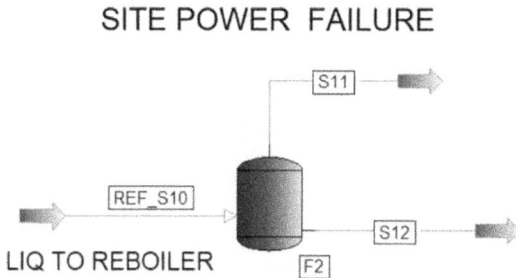

FIGURE 8.3   Relief load for site power failure contingency evaluated using a flash drum.

generated from the top (S11 of figure 8.3) will represent the relieving vapor load and temperature. The relieving load calculated by this steady-state simulation method may be lower than the one calculated by the UBH method, but the relieving temperature will be higher. So, some cushion on the relieving load can be considered for a conservative design based on the latent heat ratio indicated earlier. So, it may be imperative to choose the method by analysis and critical judgment. A conservative approach may be better. However, a simulation is always essential (for both methods).

8.2.3.2.3 Feed Failure

In case of feed failure, the relieving load can be evaluated by configuring a flash vessel at relieving pressure of 3.5 Kg/cm2 (g), as in Figure 8.4, with the bottom tray liquid as feed (pseudo stream going to reboiler). The net heat duty imparted to the drum is the reboiler duty less the condenser duty from the column simulation result (20.72 – 16.6 = 4.12 G-Cal/Hr) at relieving pressure. Streams S7, S11, and S11_1 represent the streams of relieving vapor in reflux failure, site power failure, and feed failure, respectively, and shown in Figure 8.2, Figure 8.3, and Figure 8.4, respectively. Table 8.5 and Table 8.6 present the relieving loads and others relevant properties for the three contingencies reflux failure (RF), total steam power failure(TSP) and feed failure (FF) using the steady-state simulation method (using a flash drum). The controlling loads are shown highlighted in Table 8.6.

Table 8.7 compares the controlling (highest) relief loads and other relevant properties arrived by both the UBH and steady-state simulation methods. Though the relief rate expressed in mass unit in the simulation method is higher (280.6 MT/Hr) than that of the UBH method (243.7 MT/Hr), the molar rate of UBH (3121.12 Kg-mole/Hr) is higher than that of the simulation method (3048.03 Kg-mole /Hr) due the lower molecular weight of relieving vapor in case of UBH method (78.1) as compared to molecular weight in simulation method (92.1). This molar load calculated by the UBH method is higher by (3121.12/3048.03) × 100 = 102.4%. The variation of molar loads governs the volume of vapor flow. The variation observed is 2.4% in this case. Thus, an increase of 3% to 5% over the relief load (in ACMH) computed through the simulation method is advisable as a conservative approach. A comparison of controlling loads determined through the UBH method and simulation method is presented in Table 8.7 (controlling loads highlighted).

FEED FAILURE

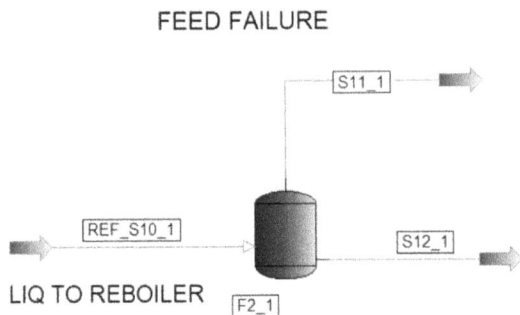

FIGURE 8.4 Load due to feed failure evaluated using a flash drum.

TABLE 8.5

Summary of Results of Simulation with Relief Load for contingencies of RF, TSP, and FF Using Flash Drum

Stream Name		S7	S11	S11_1
Stream phase		Vapor	Vapor	Vapor
Total mass rate	KG/HR	204333.923	280624.413	62184.728
Total molar rate	KG-MOL/HR	2543.960	3048.150	675.732
Vapor act. vol. rate	M3/HR	18400.5580	23227.0924	5148.5632
Temperature	°C	146.501	173.864	173.795
Pressure	KG/CM2	4.500	4.500	4.500
Total molecular weight		80.321	92.064	92.026
Vapor Z (from k)		0.92166	0.91377	0.91380
Vapor ideal CP/CV ratio		1.0719	1.0564	1.0565

TABLE 8.6

Relief Loads Evaluated for Different Contingencies Using Flash Drum

Attribute	Top Reflux Failure (Using Flash Drum)	Feed Failure Simulation (Using Flash Drum)	Site Power Failure Simulation (Using Flash Drum)
Relief load (MT/Hr)	204.3	62.18	**280.6**
Relieving temperature (°C)	146.5	173.86	173.86
Molecular weight	80.32	92.06	92.1
Moles of relieving vapor (Kg-moles)	2543.96	675.72	3048.15
Relief load in ACMH	18400.6	5148.48	23227.1

TABLE 8.7

Comparison of Controlling Loads by UBH and Simulation Method

Attribute	Site Power Failure (UBH)	Site Power Failure (Simulation) (Using Flash Drum)
Relief load (MT/Hr)	243.76	280.6
Relieving temperature (°C)	139.1	173.86
Molecular weight	78.1	92.1
Moles of relieving vapor (Kg-moles)	3121.12	3048.15
Relief load in ACMH	23452.4	23227.1

8.3   RELIEF LOAD CALCULATION OF CRUDE DISTILLATION UNIT

The determination of the relief load for a crude distillation unit requires a similar approach. A crude distillation unit is normally provided with a top reflux and two to three pumparound (PA) refluxes. There may be crude preheaters, air coolers, and water coolers in the column overhead circuit. Reflux failure usually results in the highest relief load of crude columns and dictates governing (highest load) case. In some columns, a top pumparound reflux is provided to maintain the top temperature of the column and top reflux is provided only to supplement (the shortfall of top pumparound), if any. Normal crude columns are provided with charge heaters that provide the heat duty for the required vaporization of the crude, and stripping steam is introduced at the bottom of the main column and side strippers.

To examine and illustrate the situation a crude distillation column with feed rate of 282 MT/H having three side draws with side strippers (namely, heavy naphtha of 120–140°C cut), kerosene and gas oil are considered with two pumparound refluxes (namely, kerosene pumparound and gas oil pumparound) drawn from respective draw trays. The charge heater heat duty and top condenser heat duty are found from the column simulation as 28.1 G-Cal/Hr and 15.6 G-Cal/Hr, respectively. Kerosene PA and diesel PA duties are 8 G-Cal/Hr and 12 G-Cal/Hr, respectively (as per simulation) at the relieving pressure of 3.5 kg/cm2 (g) or 4.5 kg/cm2 (a). The atmospheric distillation column profile (including the top condenser heat pumparound duties) from simulation is presented in Table 8.8.

8.3.1   TOP REFLUX FAILURE

To compute the relief load, temperature, and properties by the simulation method, a flash vessel is configured at the relieving pressure, where feed and stripping steams enter as feeds to the drum, and total pumparound heat removal (some people prefer removal of around 85% of total PA duty to be more conservative) can be considered from the flash vessel as there will be no condenser heat removal in case of reflux failure. The total pumparound heat duty of $(12 + 8) = 20$ G-Cal/Hr is withdrawn from the flash vessel. The net vapor from the top of flash vessel S 28 can be considered as the relief load and its temperature as the relieving vapor temperature for reflux failure, as shown in Figure 8.5. The net vapor properties can be used as relieving vapor properties. An increase by 3% to 5% over the relief load thus, evaluated (in ACMH), can be used to take care of any inaccuracies and to be conservative. The relieving vapor mass flow and temperature can be seen (from Figure 8.5) as 120.5 MT/Hr and 284.5°C, respectively.

8.3.2   PUMPAROUND FAILURE

A pumparound failure may also lead to overpressure of the column. The relief load can be estimated again by configuration of a flash vessel, as shown in Figure 8.6, which will have crude feed and stripping steam as inlet streams. Heat removal by the gas oil pumparound (original duty 12 G-Cal/Hr) is considered as zero. The other

TABLE 8.8
Atmospheric Column Vapor–Liquid Profile from Simulation

Column Summary

Net Flow Rates in Kg/Hr

Tray	Temperature (°C)	Pressure Kg/CM2	Liquid	Vapor	Feed	Product	Heater Duties M*KCAL/ Hr
1C*	45.0	4.20	74616.7			33680.0L	−15.6143
						5279.3W	
2	125.0	4.50	103906.2	113576.0			
3	141.7	4.52	111082.5	142865.5			
4	149.3	4.54	113720.6	150041.8	4083.1V		
5	154.6	4.57	106389.7	148596.8		8583.1L	
6	158.8	4.59	106638.4	149849.0			
7	162.5	4.61	106348.1	150097.7			
8	166.2	4.63	105397.0	149807.4			
9	170.5	4.65	103252.0	148856.3			
10	176.4	4.67	98789.8	146711.3			
11	185.8	4.70	426682.6	142249.1	28543.8V		−8.0000
					244954.9P		
12	198.9	4.72	109688.9	196643.2		244954.9P	
						87343.8L	
13	223.7	4.74	109383.2	211948.2			
14	239.3	4.76	109546.0	211642.5			
15	248.7	4.78	107726.6	211805.3			
16	255.5	4.80	102805.0	209985.9			
17	261.9	4.83	91095.4	205064.3			
18	271.6	4.85	520912.8	193354.7	23453.4V		−12.0000
					329372.6P		
19	290.5	4.87	95631.6	270346.1		329372.6P	
						96393.9L	
20	331.6	4.89	80721.0	270831.4			
21	352.8	4.91	76018.2	255920.8			
22	361.7	4.93	64937.5	251218.0			
23	367.9	4.96	135411.2	240137.3	282000.0M		
24	363.3	4.98	123333.9	28611.0			
25	355.2	5.00		16533.7	3500.0V	110300.2L	

SAFTY LOAD REFLUX FAILURE

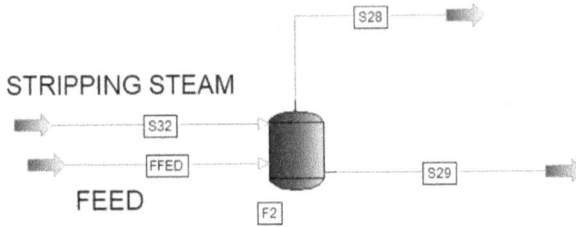

STRIPPING STEAM

FEED

Stream Name		FFED	S28
Stream Phase		Mixed	Vapor
Total Mass Rate	KG/HR	282000.000	120529.487
Total Molar Rate	KG-MOL/HR	1584.060	1261.627
Vapor Act. Vol. Rate	M3/HR	13471.8503	12695.0071
Temperature	C	372.000	284.487
Pressure	KG/CM2	5.000	4.500
Total Molecular Weight		178.024	95.535
Vapor Z (from k)		0.92893	0.94912
Vapor Ideal CP/CV Ratio		1.0209	1.0377

FIGURE 8.5   Relief load due to reflux failure of a crude column evaluated using a flash drum.

heat removals by the overhead condenser (15.6 G-Cal/Hr) and the other pumparound (8 G-Cal/Hr) need to be extracted from the flash vessel. So, total heat removal from the vessel becomes the top condenser duty plus the smaller pump around duty (15.6 + 8 = 23.6 G-Cal/Hr). The net overhead vapor S 37 at relieving pressure can be considered as the relief load and temperature of relief vapor (shown in Figure 8.6). The vapor quantity becomes 105.19 MT/Hr.

8.3.3   SITE POWER FAILURE

In case of site power failure, feed to furnace and column will stop. A crude feed rate of 15%–20% is assumed at a preheat temperature of 270°C. Heat input by the furnace is considered as 4.5 G-Cal/Hr based on 16% of earlier furnace duty (28.1 G-Cal/Hr) available to heat the crude. As furnace firing will stop and only radiation from hot refractories may continue to impart heat to the fluid inside furnace tubes, a coil purge medium is introduced (usually steam) to avoid product degradation in furnace tubes. Heat removal by pumparounds will discontinue. The reflux will stop so no heat removal is done by the condenser. Stripping steam is available for some time. The net vapor from the flash vessel S41 will represent the relief load and temperature and the properties as shown in Figure 8.7.

SAFTY LOAD ONE GO_ PA FAILURE

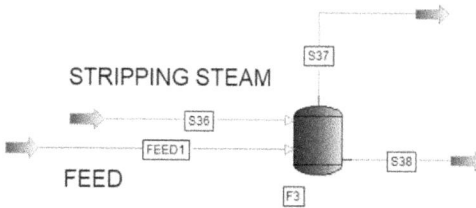

Stream Name		FEED1	S37
Stream Phase		Mixed	Vapor
Total Mass Rate	KG/HR	282000.000	105198.798
Total Molar Rate	KG-MOL/HR	1584.060	1171.844
Vapor Act. Vol. Rate	M3/HR	13471.8503	11474.1125
Temperature	C	372.000	268.700
Pressure	KG/CM2	5.000	4.500
Total Molecular Weight		178.024	89.772
Vapor Z (from k)		0.92893	0.94986
Vapor Ideal CP/CV Ratio		1.0209	1.0411

FIGURE 8.6   Load due to gas oil (GO) PA failure evaluated using a flash drum.

SAFTY LOAD TSP FAILURE

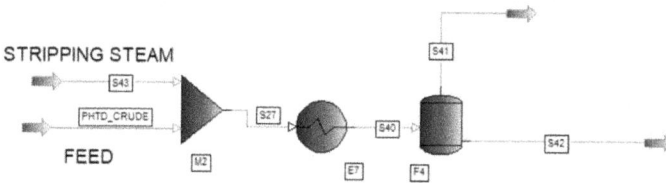

Stream Name		PHTD_CRUDE	S41
Stream Phase		Mixed	Vapor
Total Mass Rate	KG/HR	56000.000	50259.844
Total Molar Rate	KG-MOL/HR	314.565	598.345
Vapor Act. Vol. Rate	M3/HR	839.5294	6829.4211
Temperature	C	270.000	345.610
Pressure	KG/CM2	6.000	4.500
Total Molecular Weight		178.024	83.998
Vapor Z (from k)		0.92030	0.96453
Vapor Ideal CP/CV Ratio		1.0341	1.0406

FIGURE 8.7   Load site steam power (TSP) failure evaluated using a flash drum.

TABLE 8.9

Relief Loads for Different Contingencies Using Steady-State Simulation for vacuum column

Attribute	Top Reflux Failure	Gas Oil PA Failure	Site Power Failure
Relief load (MT/Hr)	120.57	105.15	50.3
Relieving temperature (°C)	284.5	268.7	345
Molecular weight	95.5	89.7	84
Moles of relieving vapor (Kg-moles)	1262.63	1171.84	598.3
Relief load in ACMH	12695.0	11474.1	6829.4

The relief load for total site power failure (50.2 MT/Hr) will be always less than that of contingency of reflux failure (120.5 MT/Hr).

The relief loads for different contingencies using the steady-state simulation method for the crude column is presented in Table 8.9 (controlling loads are highlighted).

8.4   RELIEF LOAD CALCULATION OF VACUUM COLUMN

A vacuum column is necessarily provided with a top pumparound to control the top temperature of the column. A fuel-type vacuum column is also provided, usually with two to three side draws without side strippers and typically with two additional pumparound refluxes. Side strippers with stripping steams are provided with lube-type applications.

Contingencies can be as follows:

1. Steam failure to ejectors can cause overpressure.
2. Cooling water failure to the pre-condenser and after-condensers.
3. Failure of any PA reflux may cause the column pressure to increase and need release of vapors through safety valves.
4. Site power failure.

Relief loads in three of the preceding contingencies—ejector failure, cooling water failure, highest duty PA failure—can be evaluated by configuring the flash drum as earlier.

For estimation of a fuel-type vacuum column, processing of the reduced crude oil (RCO) of the crude column considered at a rate of 110.3 MT/Hr. It has two side draws, namely, heavy diesel and vacuum gas oil (VGO). Two pumparound refluxes are provided for heat removal, namely, heavy diesel PA and VGO PA with loads 9.6 and 5.0 G-Cal/Hr, respectively. The vacuum column has a pre-condenser duty of

TABLE 8.10
Column Profile of the Fuel-Type Vacuum Tower in Section 8.4

Column Summary

Net Flow Rates in Kg/Hr

Tray	Temperature °C	Pressure Kg/CM2	Liquid	Vapor	Feed	Product	Heater Duties M*KCAL/ Hr
1	80.0	0.09	184126.1		164128.7P	1680.8V	−9.5890
2	165.8	0.09	98803.2	21678.2		164128.7P	
						9000.0L	
3	212.4	0.09	119086.8	109484.0			
4	233.9	0.10	122681.1	129767.6			
5	252.7	0.10	124800.4	133361.9			
6	270.3	0.10	336606.8	135481.1	145359.9P		−5.0000
7	283.3	0.11	27642.0	201927.7		145359.9P	
						91989.8L	
8	367.8	0.11	9995.6	130312.6			
9	401.0	0.11	4866.1	112666.2	104326.6V		
10	397.5	0.12	9787.8	3210.1	5973.6L		
11	391.3	0.12	9195.0	2158.2			
12	376.5	0.12		1565.4	1000.0V	8629.6L	

0.05 G-Cal/Hr (design) sufficient to cool the overhead to 45°C. The column profile of the vacuum tower, generated through simulation at top pressure of 0.09 Kg/cm2 (a) or 68.4 mm Hg (a), is presented in Table 8.10.

8.4.1 EJECTOR STEAM FAILURE

Ejector steam failure will lead to failure of the overhead vacuum and cause overpressure. To simulate the case a flash drum configured with feed entering at the same temperature of 415°C as maintained during normal operation, the pressure at the furnace outlet has increased to a little more than 3.5 Kg/cm2 (g) at relieving pressure of the column. As a result, the vaporization has also come down to the bare minimum (the furnace outlet stream is around 97% liquid at 3.5 Kg/cm2 and 415°C enters the flash drum). The stripping steam + coil turbulizing steam of (11+ 1) = 12 MT/Hr is considered the same as in the column simulation. The air leak to the column will stop due to pressurization above atmosphere. The production of decomposition gas (can be estimated from Liberman correlation) is assumed around 10% of hydrocarbon feed (higher is better to assume as cracking in coil will increase due to higher residence time in the coil on account of less vaporization). The pumparounds may fail as there is very little vaporization. Only the pre-condenser can remove some heat, but that is very small and thus neglected. The stripping steam (low pressure) and decomposition gas will aid in some additional vaporization in the column that will go through the flash vessel overhead. The net vapor from the flash vessel will be the relief load.

8.4.2   Cooling Water Failure

Cooling water failure to the pre-condensers and after-condensers will eventually lead to failure of the vacuum system and cause overpressure of the column as earlier. The net relief load can be considered the same as for ejector failure (vacuum system failure).

8.4.3   Pumparound Failure

Top pumparound (one pumparound with the highest duty) failure will also lead to overpressure of the column. The case can be simulated again by configuring a flash drum receiving feed at the furnace coil outlet temperature as in normal operation. The heat input to the feed is reduced owing to less vaporization due to overpressure to 3.5 Kg/cm2 (g). The VGO PA is assumed operational. So, 5 G-Cal/Hr of heat is removed from the flash drum. The net vapor from the flash drum would represent the relief load.

If both pumparounds are assumed to fail with pressurization of the column, the net vapor quantity will reduce to the vacuum system failure case and relief load will be same as the two cases discussed earlier. The relief loads and properties for the three contingencies are presented in Table 8.11 (controlling loads are highlighted).

TABLE 8.11
Relief Loads in Different Contingencies in Vacuum Column

Attribute	Ejector Steam Failure	Cooling Water Failure	Top PA Failure
Relief load (MT/Hr)	65.3	65.3	35.8
Relieving temperature (°C)	388	388	345.2
Molecular weight	85.8	85.8	53.8
Moles of relieving vapor (Kg-moles)	760.5	760.5	664.7
Relief load in ACMH	9342.4	9324.4	7694.8
Vapor ideal C_p/C_v ratio	1.0364	1.0364	1.0618

8.4.4   SITE POWER FAILURE

The relief load due to site power failure will be lower than ejector steam failure as the furnace duty will reduce to around 20% in site power failure as (only radiation from hot refractories may continue) described in the case of the crude unit. So heat input to the column will be very low. Feed will stop. All pumparounds will stop. The ejector system will discontinue and lead to overpressure. Coil purging steam or turbulizing water may continue and may displace furnace coil content. The relief load will be less than the cases described in Table 8.11.

The relief valves of vacuum columns normally discharge to the flare header. If the valve passes, this may pull vacuum in the flare header and can introduce gases from the flare header in the vacuum column and may increase the ejector system load and eventually may affect vacuum of the column. Normally, a rupture disc (at little lower set pressure than the safety valve) is provided at the upstream of the spring-loaded safety valves. The rupture disc prevents the entry of gases from the flare header to the vacuum column in case the safety valve passes. A pressure switch is provided at downstream of the rupture disc or at the upstream of the spring-loaded safety valves. In the event of rupture of the disc due to overpressure the pressure switch gives alarm. The rupture disc and spring-loaded safety valve assembly is isolated for replacement of the rupture disc. The assembly is shown in Figure 8.8.

Alternatively, a line of adequate size (without any valve) can be extended to a hot well (dipped inside the liquid level to provide a seal) to work as a safety valve. The line needs to be adequately sized to relieve the vapors to drum in case of contingency. The hot well may have a safety valve to release the during overpressure to the flare or vent stack and may be provided with a water makeup and pump-out facilities.

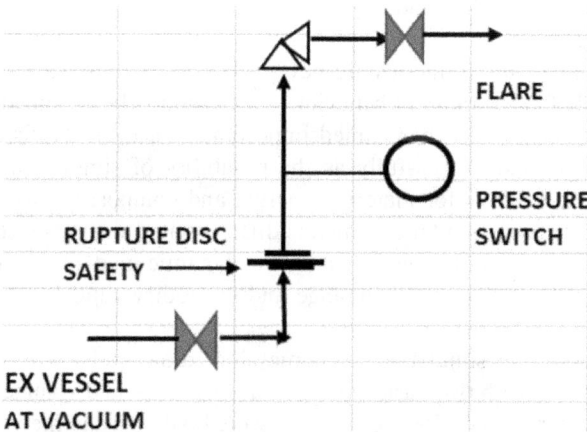

FIGURE 8.8   Rupture disc and spring-loaded safety valve assembly in a vacuum column.

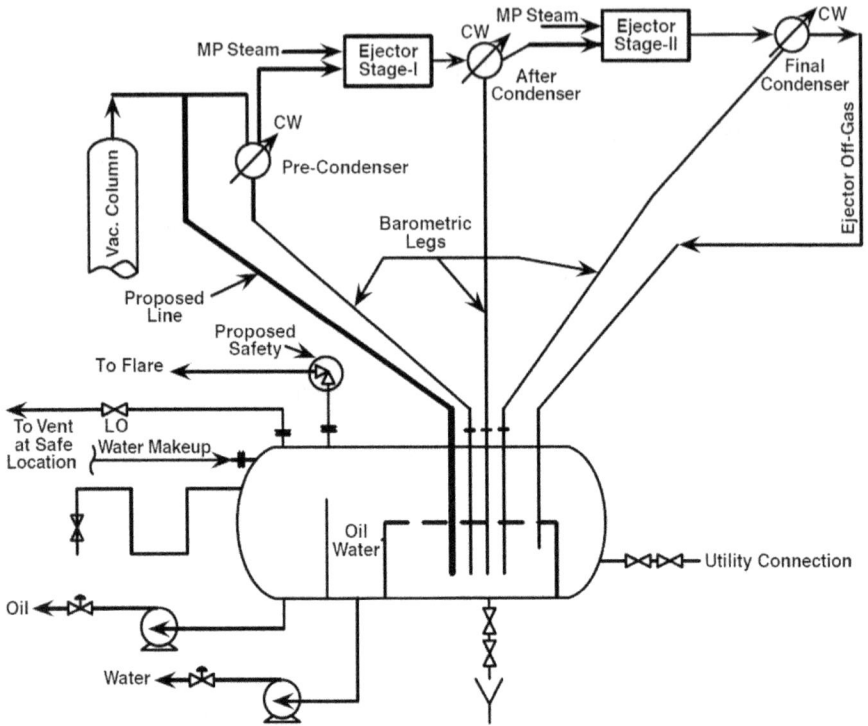

FIGURE 8.9   A simplified sketch of a hot well with no safety valve on vacuum column.

In normal operation, about 10 M of water (a higher height for water and hydrocarbon mixture) will rise in the line and during overpressure the leg will blow. The scheme is shown in Figure 8.9.

Chemical plants and refineries are never at a truly steady state and this is the case during relief. The transient behavior of a column is best studied by means of dynamic simulation, which has gained importance since the 1990s and has been used increasingly and successfully as the reliability of simulation software has increased. The equations for material, energy, and composition balances include an additional accumulation term, which is differentiated with respect to time. The inclusion of an accumulation term enables the dynamic model to rigorously calculate compositional changes at each stage and to modify vapor/liquid equilibrium over time.

Unlike steady-state simulation, dynamic simulation works within a pressure–flow (P-F) network with two basic equations as resistance and volume balance. The resistance equation defines flow between pressure hold-ups, and the volume balance equation defines material balance at pressure hold-ups.

8.5 SUMMARY OF CONTINGENCIES AND ACCUMULATION OF PRESSURE

8.5.1 OVERPRESSURE

Causes of overpressure are described and guidelines for plant design to minimize the effect are covered in API 521. Some of them are mentioned as follows:

- Closed outlet on vessels
- Inadvertent inlet valve opening
- Check valve malfunction
- Utility failure
- Instrument air or electrical power failure
- Loss of heat in series fractionation system
- Loss of instrument air or electric power
- Reflux failure in columns
- Heat exchanger tube failure
- Plant fires/external fires

8.5.2 WATER INTO HOT OIL

Water entry to hot oil is a source of potential overpressure, but no method for computing the relieving requirement is usually available. If the quantity of water and heat available is known, the relief load requirement can be calculated. But the quantity of water is seldom known. Further, the vaporization of water may be instantaneous and no valve may be able to open so quickly to relieve the pressure. Thus, no relieving device (safety valve) is provided for this contingency. Proper process design needs to be followed to eliminate water entry to hot oil. Some measures can be taken to avoid water collecting pockets, and installing proper steam condensate traps/drains and using double block and bleed on water connection to hot oil lines are recommended by experts and may be advisable.

8.5.3 ACCUMULATION OF PRESSURE IN VARIOUS CASES

Accumulation is the increase in pressure during safety discharge over the maximum allowable working pressure (MAWP) of a vessel, expressed in pressure units or usually as percentage. A safety valve starts opening at the MAWP but accumulation of pressure (pressure increase) is likely in vessels during the relieving process subject to the limits (of percentage increase in pressure) specified by standards. The maximum allowable accumulated pressure is defined as the sum of maximum allowable working pressure and the maximum allowable accumulation permitted in accordance with section VIII of the ASME Boiler and Pressure Vessel Code. Figure 8.10 conforms to the requirements of the Boiler and Pressure Vessel Code for MAWP greater

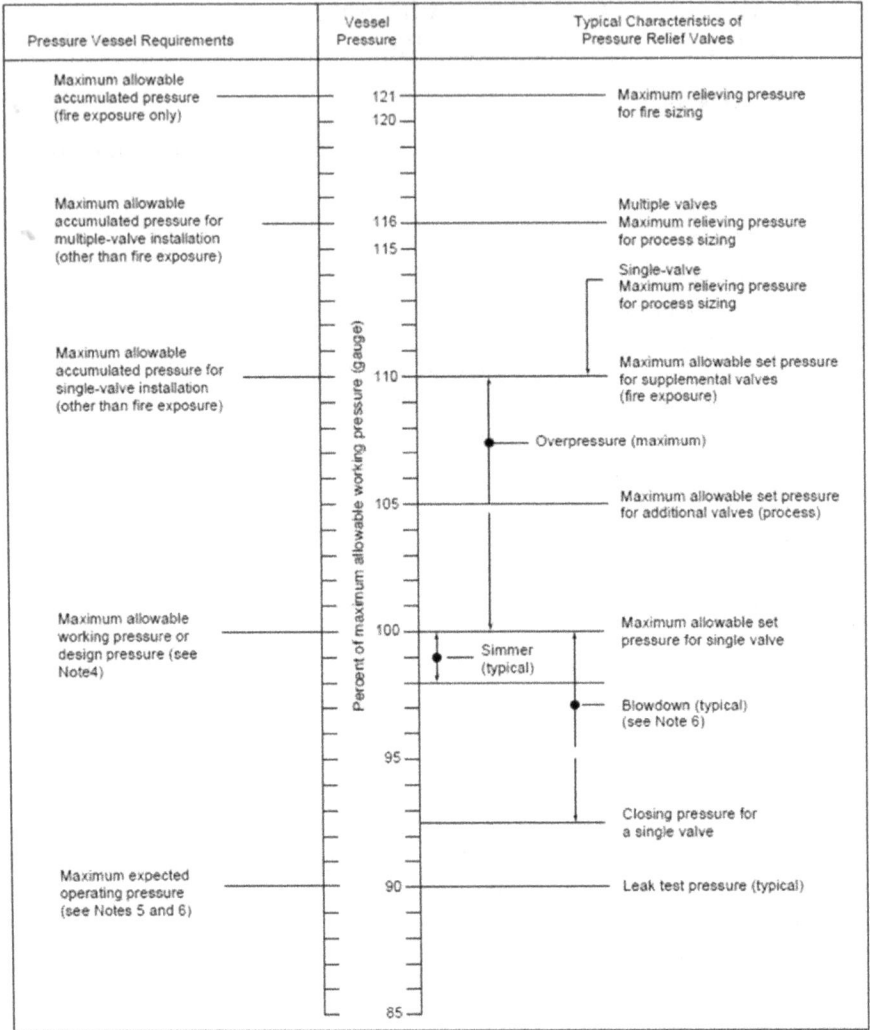

FIGURE 8.10 Pressure vessel design requirement as per section VIII of the ASME code.

than 30 psi. The MAWP may be greater and equal to the design pressure of the vessel for a coincident design temperature.

Some specific applications, necessary for ensuring safety are discussed next.

8.6 SAFETY MEASURES IN HIGH-PRESSURE AND LOW-PRESSURE INTERCONNECTIONS

A very high-pressure system preferably should not have connection with a very low-pressure system. An intermediate-pressure system may be used to ensure the safety

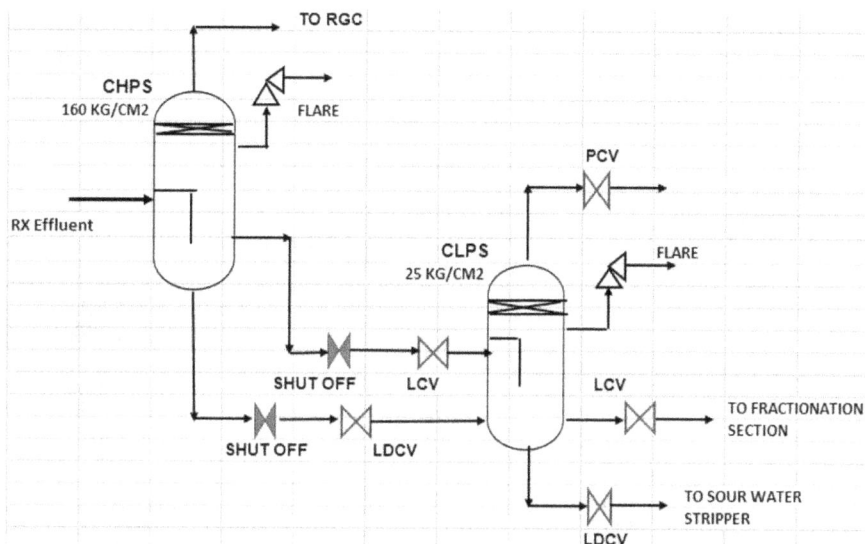

FIGURE 8.11 CHPS and CLPS in a high-pressure hydrocracker showing interconnections and safety valve locations.

of the low-pressure system. When connection with low-pressure equipment cannot be avoided, it is advisable to connect with emergency shutoff valves between them for quick isolation. Alternatively, provision may be there to close the level control valve (having tight shutoff characteristics) triggered by automatic switches. For instance, the low-level or low-low level alarm in the cold high-pressure separator (of a hydrotreater) should trigger the closure of the shutoff valve or the level control valve to avoid gassing through the liquid line and entry of a considerable quantity of gas from the high-pressure vessel to the low-pressure vessel, resulting in overpressure of the low-pressure vessel and endangering it. A typical interconnection of a high-pressure and low-pressure system is shown in Figure 8.11. There should be cushion available in the capacity of the safety valve (mounted on low pressure vessel) to take care of exigency of gassing through the liquid line. Safety valves should always be installed at upstream of the coalescer mesh in vessels, as shown in Figure 8.11.

In high-pressure system applications, the safety valve downstream spool piece and isolation valve rating is preferred as the upstream high-pressure rating valve. A bleeder at safety valve downstream, if provided should have double isolation valves and preferred to be of high pressure rating.

8.7 LOCATION OF SAFETY VALVES AND SPECIAL FEATURES

8.7.1 LOCATION OF SAFETY VALVES IN HIGH-PRESSURE SYSTEMS

A safety valve can be provided on each cold high-pressure separator (CHPS) and cold low-pressure separator (CLPS) at the upstream of the demister as per standards.

This will protect the vessel from overpressure even if the demisters are clogged/chocked. No safety valve is installed on the upstream reactor top as its popping may result in reverse flow through reactors, lifting the catalyst in the reactors (designed for flow downwards) and perhaps can damag the catalyst supports. The safety valve on the CHPS protects the entire reactor and the CHPS upstream circuit from overpressure in a high-pressure hydrotreater.

The reactor catalyst supports are designed to withstand a maximum pressure drop across the reactor in case of emergency depressurization (dumping) through the depressurization valve (dump valve) located at the top of CHPS. A reactor effluent air cooler (REAC) is designed to remove typically 30% to 40% of the designed heat duty even when the air cooler fans stop. This is to cool the hot effluent (through natural draft) from the reactor when emergency depressurization (dumping) is done with the depressurization valve at the top of the CHPS. Whenever a steam generator is provided at upstream in the reactor effluent circuit, the heat removal provision in the REAC, through natural draft, can be avoided.

In heat exchangers in a high-pressure system, high-pressure fluid may be in one side (either tube side or shell side) and much lower pressure fluid on the other side. The design pressure of the low-pressure side to be set at 10/13 times (earlier practice was 2/3 times) of the design pressure of the high-pressure side to avoid a safety valve on the low-pressure side. This would protect the low-pressure side of the heat exchanger in case of a tube leak or any internal gasket leak and pressurization of the low-pressure side. The same norm is also practiced for all heat exchangers at all pressure services as well.

8.7.2 LOCATION OF SAFETY VALVES IN COLUMNS

In most columns, safety valves are provided at the top dome or on the vapor line. In columns handling fouling services (and at relatively lower temperatures), like sour water strippers and amine regenerators, safety valves are mostly provided in the bottom. The column overpressure can be handled by safety valves located in the bottom even if upper-tray dispersers are chocked by deposition of salts (NH4HS and NH4Cl). In amine regenerators and sour water strippers, the safety valve is normally located at the bottom vapor space of the column.

8.7.3 SAFETY VALVE ASSEMBLY IN REFINERY COLUMNS

In vacuum columns, a safety valve assembly with a rupture disc at the upstream to the spring-loaded safety valve, as shown in Figure 8.8, can be followed. A rupture disc upstream of the spring-loaded safety valve is usually practiced in bitumen blowing reactors. This is to ensure that the spring-loaded safety valve is not stuck up at a closed position with solidified bitumen. A flushing oil supply with restricted orifice (to limit the flow) at the upstream of the rupture disc is usually provided to avoid chocking of the upstream section.

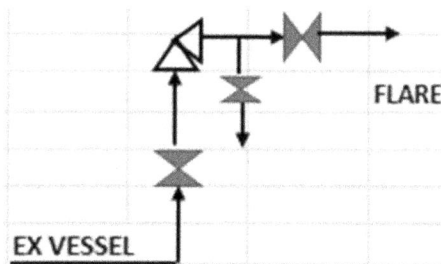

FIGURE 8.12 A typical safety valve assembly.

Inlet and outlet piping of safety valves may be designed in accordance with API RP 520 part II. A typical safety valve assembly is shown in Figure 8.12.

The isolation valves located upstream and downstream of the safety valve need to be installed such that if the block valve gates get detached they cannot drop and close the flow path upstream or downstream. This to ensure open flow path for the fluid in the vessel to discharge unobstructed during overpressure. A bleeder downstream of the safety valve can be provided for checking if the safety valve is passing. The bleeder may have two isolation valves if the system pressure is higher (preferably having the same rating as the safety downstream spool piece that usually has the same design pressure of the vessel).

Flare line isolation valves, located at individual unit battery limits may be installed horizontally or at an angle of around 30 degrees with horizontal and the valve wheel/stem may point down to avoid closure of the line in the event of dislodgement of the gate of the isolation valves. Bleeders can be provided at downstream and upstream of the isolation valves.

8.8 RELATION BETWEEN DESIGN PRESSURE AND HYDROSTATIC TEST PRESSURE

The hydrostatic test pressure of a pressure vessel is 1.3 × (MAWP or design pressure) × (Stress value at test temperature/Stress value at design temperature). However, for piping hydrostatic test pressure as 1.5 × Design pressure is used. Based on the above, the design pressure of the low-pressure side of a heat exchanger should be at least 10/13 of the design pressure of the high-pressure side to avoid a requirement of a safety valve on the low-pressure side.

BIBLIOGRAPHY

API 520 API Recommended Practice for the Design and Construction of Pressure-Relieving Systems in Refineries. Washington, DC: API Publishing Services, January, 2000.

API 521 Guide for Pressure-Relieving and Depressuring systems, API Recommended Practice 521, 4th edn, Washington, DC: API Publishing Services, March 1997.

ASME Boiler and Pressure Vessel Code.

Haribabu Chittibabu, Amudha Valli, Vineet Khanna, and Dipanjan Bhattacharya. Calculating column relief loads. *PTQ*, Q2, 2010, pp. 55–65.

J. Saha, S. Chaudhuri, and S. Groenendijk. Estimation of relief load and realistic relieving temperature for heavy-end fractionating columns. *Process Control Instrumentation*, August 2019.

ABBREVIATIONS

CHPS   cold high-pressure separator of hydrotreaters
CLPS   cold low-pressure separator of hydrotreaters
FF   feed failure
RF   reflux failure
TSP   total steam power failure

9 Challenges in Refining

Refiners are confronted with challenges like shrinking margins due to lower differentials between crude and product, stricter product specifications, stricter emission guidelines, and limited sales of heavy products. Thus, they have resorted to several means as described in this chapter to continue to remain economically viable and enhance profits.

9.1 LOWER CRUDE AND PRODUCT DIFFERENTIAL

Refining margins are shrinking due to lower differentials between crude feed and products. The differential has further declined due to higher operating costs for meeting the superior product specifications and for stricter emission control. Further, the availability of regular crude is often a problem and the supply of regular crude is declining. Heavy crudes are available at a good discount over regular and benchmark crudes, but may have higher level of contaminants.

Thus, it has become imperative to introduce cheaper crudes in the crude basket both for easy availability and profitability. Cheaper crudes can be heavy (low API gravity), have high sulfur content, have high total acid number (TAN), and have high calcium (as calcium naphthenate). Discount on cheaper crudes typically ranges from $5 to $15/barrel over benchmark crudes. This provides an incentive for refineries to include cheaper crudes to increase profitability. But there are problems associated with processing of cheaper crudes like production of higher heavy ends and the presence of contaminants (e.g., higher sulfur, naphthenic acids, and calcium; detailed in Chapter 1).

9.1.1 ACTIONS REQUIRED BY REFINERS

Refiners need to take additional measures at plants to process the cheaper crudes in isolation and in admixture. Further, there are some uncertainties regarding future discounts. Thus, some refiners are apprehensive about investing in plant modification/upgradation for processing cheaper crudes, so they start with small proportions in admixture with other crudes so that the overall nature and behavior remains by and large unaltered. Refiners also avail the opportunity of higher profit with some chemical treatments as and when required. Some refiners have completed modifications to introduce flexibility to process heavy crudes. Modifications required in different units are described in Chapter 1 for processing heavier crudes. Higher TAN is typically managed effectively by chemical treatment by experts.

The author has experience with a big refinery that increased its installed capacity by about 20% by changing to a heavier crude slate with some modifications in the primary and secondary units. Primary unit modifications involved marginal alteration in the crude preheat train and product cooling trains, addition of pumps, addition of tubes in charge heaters, and changes in column internals. The refinery had resorted to chemical treatment provided by experts to handle the fouling potential and corrosivity of a wide range of crudes.

In the secondary units like vacuum gas oil (VGO) hydrotreaters, minor modifications were carried out, including changing of catalysts with higher space velocity and some circuit augmentation with the available cushion in the balance sections. However, in the coker unit bigger modifications were carried out, e.g., the addition of a coking furnace and associated pair of drums (converted from four drums to six drums), and marginal alterations in feed preheat, product, and pumparound cooling circuits. Reduction in coking cycle time had become essential to contain the higher production of coke.

In one hydrogen generation unit, a heat transfer exchanger reactor (HTER) was added downstream of the tubular reformer, and the hydrogen generation capacity increased by 15% to 20 % to supply the additional requirement of hydrogen.

In other areas, like amine treatment and regeneration, a change of amine from diethanolamine (DEA) to methyldiethanolamine (MDEA) was introduced to lower the circulation rate, thereby achieving the required hydraulic capacity of the units and lowering the heat requirement in regenerator reboilers.

In the sulfur unit, however, a new train were added to handle higher quantities of acid gases. In the product treatment sections like sweetening some modifications were carried out. New sweetening catalyst were able to handle a higher hydraulic load and higher levels of mercaptans. Some small modifications were carried out in different circuits. Thus, the modifications would allow for processing of the easily available and cheaper crudes and at the same time increased the installed capacity of the refinery. Thus, the refiner converted challenges into opportunities.

9.2 STRICTER PRODUCT SPECIFICATIONS

9.2.1 REVISED SPECIFICATIONS

The specifications for finished products have undergone big changes with greater emphasis on reducing pollution. The major changes introduced in fuel specification are in sulfur content of motor spirit (MS) and high-speed diesel (HSD). The sulfur content in MS is reduced in steps to 10 ppm in EURO V and VI (European emission standards) along with other changes like limiting the benzene content of MS to <1% by volume, olefin content to 22% V (max), aromatic to 35% (max), and oxygenates to 5% V (max). Similarly, sulfur specification for diesel (HSD) has been revised to <10 ppm, the specified cetane number of HSD revised to >53, and the recovery specification revised to a minimum 95% by volume recovery at 350°C or 360°C. Also an additional specification for moisture content was introduced as <100 ppm wt. The sulfur content of heavy oils like furnace oil (FO) and light diesel oil (LDO) and heavy petroleum stocks (HPS) also have also been reduced.

9.2.1.1 Gasoline or Motor Spirit

The change in specifications has resulted in big changes in the constitution of MS or the gasoline pool. For meeting the low sulfur (<10 ppm) and higher research octane number (RON), together with olefin and aromatic limits, new blends of MS have been evolved. MS is being constituted by blending different naphtha streams to maintain the level of olefins, aromatics, and sulfur, and to meet the specified RON. Reformulated gasoline (MS) can be a blend of the followings constituents, in varied proportions, to meet all the properties based on requirements:

- Reformate from catalytic reforming unit/continuous catalytic reforming unit (CRU/CCRU)
- Gasoline from Fluid catalytic cracking unit (FCCU)
- Isomerate
- Alkylate
- Oxygenates like MTBE, ETBE, TAME, ethyl alcohol (EtOH)

Several new facilities have become essential and installed based on the requirements. FCCU gasoline treatment has become necessary to reduce the sulfur and benzene content of FCCU gasoline without appreciable loss of octane (due to inadvertent saturation of olefins) during the process of desulfurization and benzene saturation. Alternatively, treatment of feed to the FCCU is also carried out to lower the sulfur of FCCU gasoline and enhance conversion. But benzene saturation is still essential. Some typical schemes adopted for FCCU gasoline treatment are presented in Figure 9.1

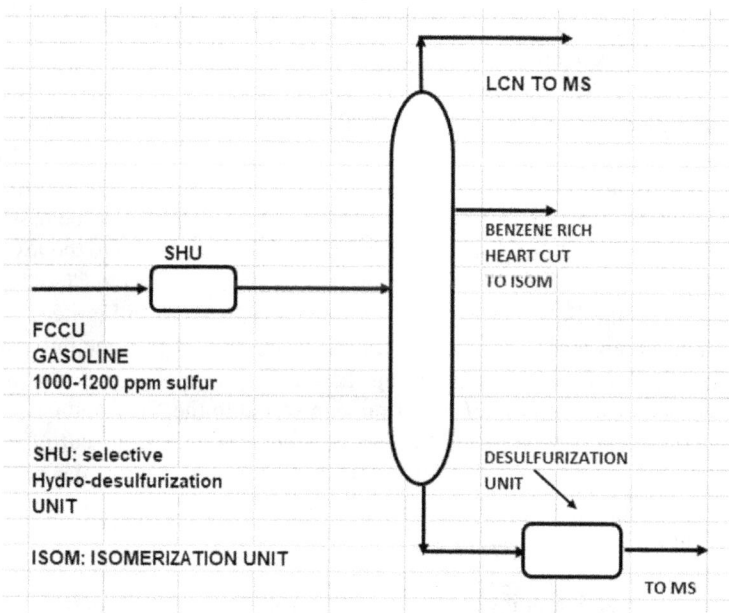

FIGURE 9.1 Schematic of FCCU gasoline treatment facility.

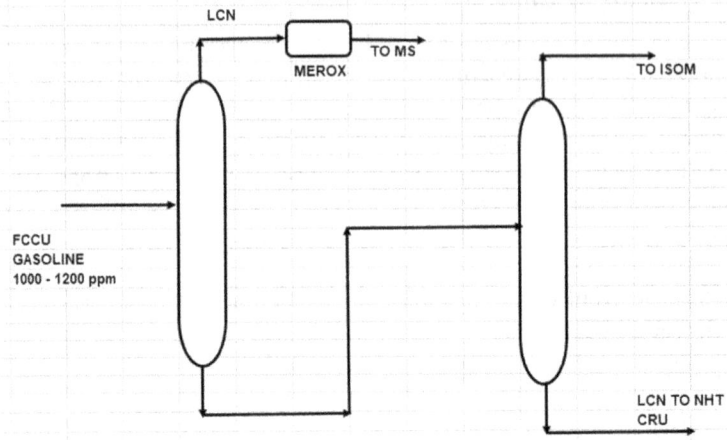

FIGURE 9.2 Alternative schematic of FCCU gasoline treatment facility.

and Figure 9.2. For FCCU using hydrotreated feed, the scheme in Figure 9.3 is an alternative. In the scheme in Figure 9.1, the light mercaptans are selectively converted to heavier mercaptans in the selective hydrogenation unit (SHU) reactor (without appreciable loss of olefins) and fed to the splitter column. The olefin-rich streams (very low in sulfur) are taken from the top and blended to MS. The sulfur-rich splitter bottom is desulfurized and blended to MS. This reduces loss of RON due to inadvertent saturation of olefins.

Often a heart cut rich in benzene is drawn from the splitter and routed to the isomerization (ISOM) unit for saturation of benzene.

In Figure 9.2, FCCU gasoline is first split into two cuts in the first splitter. The light olefin-rich top fraction, containing the sulfur species (primarily methyl and ethyl mercaptans) are treated in the extractive-type Merox unit to remove the mercaptans [initially as sodium mercaptide and finally as alkyl disulfide (RSSR)] and lower the sulfur content of the stream and blended to MS. The bottom stream of the first splitter is fed to the second splitter. The top fraction, rich in benzene, is fed to the ISOM unit and the bottom fraction is fed to the naphtha hydrotreater (NHT) and CCRU.

When feed to the FCCU is hydrotreated, then a scheme as in Figure 9.3 can be adopted. This comprises of a splitter column splitting FCCU gasoline into two fractions. The lighter top fraction containing the majority of olefins and low in sulfur can be directly blended to MS. The bottom can be fed to the NHT and CCRU as in scheme in Figure 9.2. A provision to draw of a benzene-rich heart cut is provided for routing to the ISOM feed, if necessary, to contain the benzene of the MS blend.

The following facilities are considered to produce MS with the revised specifications:

- The isomerization unit has become a necessity to convert C_5 and C_6 paraffins to isoparaffins to boost the octane number to 86–91 and saturate the benzenes and blend in MS. Often a benzene-rich heart cut from the FCCU

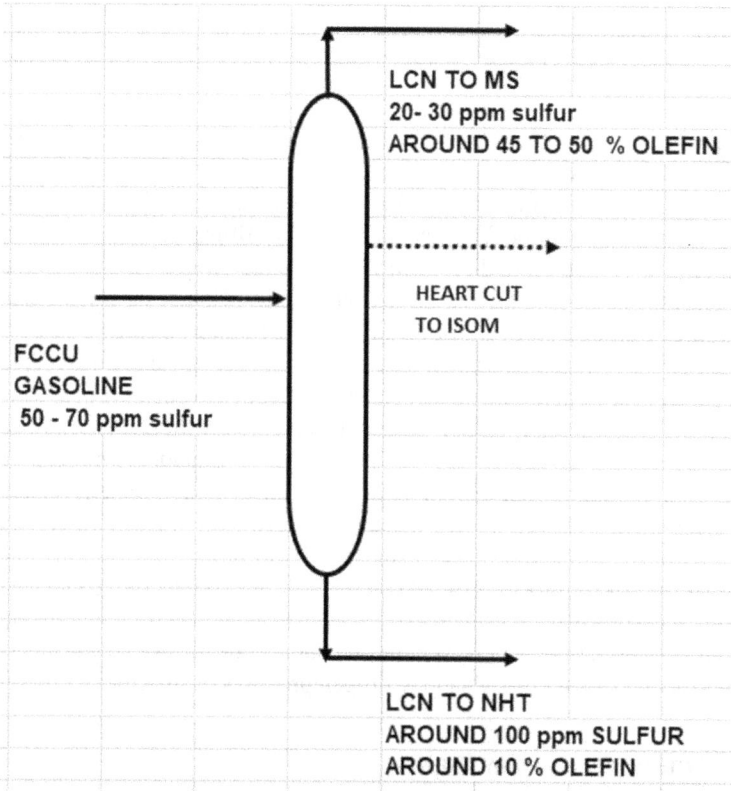

LCN TO MS
20- 30 ppm sulfur
AROUND 45 TO 50 % OLEFIN

HEART CUT
TO ISOM

FCCU
GASOLINE
50 - 70 ppm sulfur

LCN TO NHT
AROUND 100 ppm SULFUR
AROUND 10 % OLEFIN

FIGURE 9.3 Schematic of FCCU gasoline treatment facility units having feed pretreatment.

gasoline splitter is also fed to the ISOM unit, with a benzene saturation reactor upstream.

- Some refiners have installed an alkylation unit to produce high-RON alkylate by alkylation of C_4 and C_5 olefins [present in FCCU liquid petroleum gas (LPG)] with isobutane, catalyzed by either sulfuric acid or phosphoric acid. The typical RON of alkylate is around 95. A n-butane isomerization unit may be also required to produce additional isobutane (from straight run LPG) for the production of alkylate.
- Some refiners have gone for production of oxygenates like MTBE, ETBE, and TAME. MTBE and ETBE are produced by reaction of isobutylenes in FCCU LPG with methanol or ethanol, respectively. TAME is produced by reaction of methanol with isoamylenes present in FCCU gasoline. The typical RON of oxygenate is >100. This can be essential to increase the RON of gasoline without adding sulfur, olefin, and aromatics. Further, in some countries blending of oxygenate is necessary, in limited extent, to regulate CO2 emissions.
- For obtaining higher reformate RON, CCRU have become necessary. The refineries, having fixed-bed CRU, are adding continuous reactors at the end

in series along with fixed ones to achieve a higher RON. Some have added a new CCRU.

9.2.1.2 Diesel

Diesel sulfur, cetane number, and ASTM boiling temperature for 95% Vol. recovery ($T_{95\%}$) have been revised. Lowering of sulfur specifications to <10 ppm has resulted in an increase in the feed to diesel hydrotreaters as more low sulfur-bearing fractions need to be added in the feed. Further, deep desulfurization has become necessary, needing a high-pressure diesel hydrotreater unit, to lower sulfur content and to boost the cetane number to the required level. Catalyst change of low-pressure units (36–40 Kg/cm^2 at separators) is known to deliver <10 ppm sulfur.

The reduction of $T_{95\%}$ temperature to 350°C–360°C has increased the generation of feed to the vacuum distillation unit and resulted in capacity constraints in vacuum distillation units. The vacuum furnace duty and vacuum column diameter are reaching their limits. VGO production is likely to increase due to dropping of more diesel components in VGO and may need higher processing capacity in VGO hydrotreaters. To resolve the problem of higher hydraulic load of vacuum units, some refiners have gone for changes in the catalyst of diesel hydrotreaters to process diesel components, as heavy as before, and improved diesel recovery in the hydrotreater unit with changes in the catalyst in hydrotreaters, instead of loading the vacuum distillation unit and VGO hydrotreaters. This has also increased the demand for hydrogen for hydrotreating and may in turn demand augmentation of hydrogen production capacity.

9.2.2 ACTIONS INITIATED BY REFINERS

Refiners have gone for additional investment to address all the aforementioned specifications and as a result operating costs have increased, adversely affecting the profitability of refineries. The changes implemented in refineries due to changes in diesel specifications include:

- Increase in capacity of diesel hydrotreaters by way of debottlenecking, together with change of catalyst to enable more injection of hydrogen for sulfur removal and improvement in cetane number. Often isomerization or mild hydrocracking catalyst are loaded at the reactor bottom to improve the 350°C–360°C recovery ($T_{95\%}$) of treated diesel. This is to reduce the hydraulic load on the vacuum distillation unit and VGO hydrotreater or hydrocrackers (to avoid dropping of diesel components in RCO and VGO). But the incorporation of hydrocracking catalyst, at the bottom of reactors, in diesel hydrotreaters resulted in reduction of liquid yield of diesel with production of more light ends and naphtha. This has again adversely affected profitability. Thus, refiners have debottlenecked the capacity of vacuum distillation units and VGO hydrotreaters to the extent possible and also installed hydrocracking catalyst in diesel hydrotreaters to provide flexibility for marginal hydrocracking by manipulating the bed temperature to correct the $T_{95\%}$ recovery of diesel.

- Catalyst change in low pressure hydrotreaters to meet the sulfur specification of 10 ppm wt. has been found successful. Some catalyst suppliers are advocating use of Co–Mo catalysts to lower sulfur and to limit hydrogen consumption, and some are advocating nickel-based catalysts for lowering sulfur and to allow marginally higher hydrogen consumption.
- Installation of vacuum driers in diesel treaters are taken up to meet the moisture specification of finished diesel, if required.
- For supply of additional hydrogen, recovery schemes for hydrogen from hydrogen-bearing streams are pursued and small modification of existing steam reformers are carried out, as mentioned in Chapter 1 and Chapter 6.
- Sulfur unit capacity is augmented by addition of parallel trains or oxygen enrichment.

9.2.3 LUBE OIL BASE STOCK

Refineries producing lube base oil are also facing big challenges. API Group I lube base stocks have practically no demand. The quality of lube oil base stocks produced by refineries has been upgraded from Group I to quality consistent with API Group II and Group III. The Viscosity Index (VI), percentage saturates, and sulfur content specifications are revised for base oils conforming to Groups II and III (Table 9.1).

Earlier, crudes capable of producing lube oil were required for production of lubes. Units required for manufacturing included the furfural extraction unit (FEU), solvent dewaxing (SDA) unit, and hydrofinishing unit (HFU). For bright stock preparation from vacuum residue (VR), a propane deasphalting (PDA) unit is needed as well (described in Chapter 1). Lube-producing crudes are no longer required for production of lube base stocks. Unconverted stream from the hydrocracker is often used to produce Group II and Group III base oils. Now primarily catalytic routes are followed, like hydrotreating followed by catalytic dewaxing (CDW) and hydrofinishing to produce Groups II and III lube oil base stocks. Group IV lube base stocks are made of polyalphaolefins (PAO) or synthetic esters. Bright stock production still needs a propane deasphalting unit followed by catalytic processes. For lube base oil production consistent with Groups II and III grades, a catalytic dewaxing unit is installed in many refineries.

TABLE 9.1
Lube Base Oil API Groups (API 1509, Appendix E) and Related Properties

Attribute	Group I	Group II	Group III
VI	80 < VI < 120	80 < VI < 120	120 < V. I
% Saturates wt.	<90%	>90%	>90%
% S wt.	>0.03%	<0.03%	<0.03%

9.3 STRICTER EMISSION NORMS

9.3.1 EMISSION GUIDELINES

New emission norms or guidelines have been introduced by authorities for lowering emissions for gases, liquid, and solid discharges from refineries. Refiners are required to meet stricter emission norms for all types of discharges. Refineries normally are equipped with captive power plants to supply electric power and process steam. Boilers, process heaters, and other process equipment are responsible for the emission of particulates (PM), carbon monoxide (CO), nitrogen oxides (NOx), sulfur oxides (SOx), and carbon dioxide (CO2).

Volatile organic compounds (VOCs), such as benzene and toluene, are released from storage, product loading, and handling facilities; from oil–water separation systems; and as fugitive emissions from flanges, valves, seals, and drains. For gaseous emissions, the emission standards for SOx, NOx, and PM have been lowered. There are also incentives introduced to lower CO2 emissions to the atmosphere. Stricter standards for liquid effluent and solid effluent are also required to be met by refineries. Typical revised gaseous emission standards for refineries in milligrams per normal cubic meter are presented in Table 9.2.

Over and above the mentioned limits, authorities have stipulated limits for total SO2 liberation in kilograms per hour and are progressively lowering the limits. As a result, refiners are not in a position to fire liquid fuel with higher sulfur in process heaters and boilers, diverting high sulfur heavy ends to the coker or VR treaters, and inducting part natural gas as fuel to meet the SO2 emission stipulation in case of shortfall of low sulfur fuel. Heaters and boiler stack heights are increased for better dispersion of pollutants, following directives from authorities.

Petroleum refineries use relatively large volumes of water, especially for cooling systems. Oily water and sanitary wastewaters are also generated. The quantity of wastewaters generated and their characteristics depend on the process configuration. Refineries generate polluted wastewater containing biochemical oxygen demand (BOD) and chemical oxygen demand (COD) levels of approximately 150–250 milligrams per liter (mg/l) and 300–600 mg/l, respectively. Wastewater typically have

---

TABLE 9.2

Typical Gaseous Emission Standards for Refineries (Excludes NOx Emissions from Catalytic Units)

Parameter	Maximum Value (in Milligrams per Cubic Meter)
PM	50
Nitrogen oxide	460
Sulfur oxide	150 for sulfur recovery units and 500 for other units
Hydrogen sulfide	152
Nickel and vanadium (combined)	2

TABLE 9.3
Typical Treated Effluent Quality

Parameter	Maximum Values (in Milligrams per Liter)
pH	6–8
BOD	30
COD	150
TSS	30
Oil and grease	10
Chromium hexavalent	0.1
Others	0.5
Lead	0.1
Phenol	0.5
Benzene	0.05
Benzopyrene	0.05
Sulfide	1
Nitrogen (total)	10–40*

* Nitrogen limit for refineries units having hydrogenation.

phenol levels of 20–200 mg/l, oil levels of 100–300 mg/l in desalter water and up to 5000 mg/l in tank bottoms, benzene levels of 1–100 mg/l, benzo(a)pyrene levels of less than 1 to 100 mg/l, and heavy metals levels of 0.1–100 mg/l of chrome and 0.2–10 mg/l of lead and other pollutants. The typical treated liquid effluent quality desired is presented in Table 9.3. Over and above the revised qualities in Table 9.3, some standards stipulate complete reuse of treated effluent and no discharge outside (zero discharge).

Refineries also generate solid waste and sludge (ranging from 3 to 5 kg per ton of crude processed), 80% of which may be considered hazardous because of the presence of toxic organics and heavy metals. For solid waste, higher recovery of oil from tank bottom sludge are emphasized. Sludge needs to meet the stipulated quality standards to be used as landfills. Recovery of metals from catalysts have become mandatory.

9.3.2 Actions Necessary for Reduction of Air Emissions

- Use of floating roof tanks or a combination of fixed cum floating roofs for storing liquids with higher vapor pressure (to reduce emission of VOCs). Minimize hydrocarbon losses from storage tanks and product transfer areas by methods such as installation of floating roofs having double seals in tanks and vapor recovery systems in loading facilities.
- Lower fugitive emissions by proper process design and maintenance. Installation of double seals in pumps, and ensure minimum leakages through flanges, valve glands, etc.

- Minimize SOx emissions either through use of low-sulfur fuels to the extent feasible or by directing the use of high-sulfur fuels to units equipped with SOx emission controls.
- Lowering sweet gas H2S content by amine treatment of all sour gas streams and improving H2S recovery in gas amine treatment facilities.
- Use of sour water strippers for all the sour water streams. Carry out modifications in sour water strippers to further lower the stripped water H2S and NH3 content.
- Routing the acid gases from amine regenerators and sour gases to sulfur recovery units (SRU). Refiners are required to install 100% standby sulfur recovery units and put the standby in operation once an operating unit is taken out for maintenance.
- Install enhanced sulfur recovery facilities like tail gas treatment units (TGTU) or equivalents to achieve a sulfur recovery of >99.9%.
- Improve heat recovery in process units to lower use of fuel in fired heaters.
- Recover non-silica-based (i.e., metallic) catalyst and reduce particulate emissions.
- Use low NOx burners to reduce nitrogen oxide emissions.
- Treatment for SOx and PM for gases for boilers/heater stacks and FCCU regenerator stack flue gas desulfurization (FGD).
- May need to install CO2 capture facilities to limit CO2 emissions. Some refiners are known to have initiated actions for CO2 capture from gases post shift reactor and reformer flue gases. The schemes adopted are described in Chapter 5, Section 5.9. Some are known for putting up plants for production of EtOH from tail gas of hydrogen units based on new patented technology for its subsequent blending into MS. This will also reduce liberation of CO2 of tail gas from steam–methane reforming (SMR).

9.3.3 ACTIONS TO REDUCE GENERATION OF LIQUID EFFLUENT (RECYCLING AND REUSE)

- Use of cooling towers for all circulating water systems used for cooling.
- Maximize recovery of oil from oily wastewater and sludge. Minimize losses of oil to the effluent system.
- Recirculate phenolic stripped sour water as injection water to desalters and non-phenolic ones to hydrotreaters to minimize discharge.
- Improvement of effluent water quality (with addition of tertiary treatment) to enable recycling of all treated water as makeup to cooling towers and others to achieve zero discharge of effluent water from effluent treatment plants. Reverse osmosis (RO) plants are also being put up to treat and reduce suspended solids (SS) and total dissolved solids (TDS) of the water.
- Recover solvents from their spent solutions.
- Return oily sludge to coking units or crude distillation units for reprocessing.
- Install treatment of spent caustic solutions.

9.3.4 Improvement in Facilities

The key production and control practices that are identified to lead to compliance with emissions guidelines can be summarized as follows:

- Use of floating roof tanks and vapor recovery systems to reduce VOC emissions.
- Install high-sulfur recovery systems in sulfur units, wherever feasible.
- Use of low-NOx burners in heaters and boilers.
- Efficient heat recovery systems to reduce consumption of fuels.
- Burning of low-sulfur fuels.
- Segregate oily wastewater from stormwater systems and reduce effluent generation. This may include installation of oil catchers in stormwater system to arrest oil to stormwater ponds.
- Reduce oil losses during tank drainage (to remove water before product dispatch).
- Install spill prevention and control measures like improved tank gauging and level monitoring systems.
- Minimize the generation of sludge.
- Optimize frequency of tank and equipment cleaning to avoid accumulating residue at the bottom of the tanks.
- Prevent solids and oily waste from entering the drainage system.
- Establish and maintain an emergency preparedness and response plan, and carry out frequent training.
- Practice corrosion monitoring, prevention, and control in underground piping and tank bottoms.
- Efficient extraction of oil from tank bottom sludge.
- Establish leak detection and repair programs for storage tanks.
- Use non-chrome-based inhibitors (use only to the extent needed in cooling water).
- Send catalyst to outside agencies for recovery of metals.
- Further treatment of sludge is normally outsourced for bioremediation.

9.4 LIMITED SALES OF HEAVY ENDS AND COKE

Another big challenge being faced by refiners are limited demand of coke and heavy oils. The demand for high-sulfur heavy ends have reduced considerably and needs to be sold at a discount. Further, the low price of coke is also hitting the profitability of refineries. The majority of high-sulfur heavy oils are routed as feed to the coker unit to recover some distillates and produce coke. Typical coke yield from a refinery processing heavy crude is 8%–11% wt. on crude and sold at a relatively lower price, adversely affecting profitability.

9.4.1 Actions Taken by Refiners

- Some refiners are installing VR treating (LC-Fining, etc.) upstream to reduce the production of coke. The VGO generated from VR treaters along with straight-run VGO is hydrotreated/hydrocracked to produce valuable distillates. The unconverted oil stream from the residue hydrotreater is normally fed to the coker along with the balance VR, if any. But the installation of a VR hydrotreater is very capital-intensive and have considerably high operating costs.

- Recently, some refiners have added gasification units to gasify coke and heavy ends with oxygen to produce synthesis gas for production of hydrogen and other synthesis gas-based chemicals. A part can be used as fuel in furnaces. This is likely to reduce emissions to a great extent. The raw gases produced from gasifiers are extensively treated, and acid gases are routed to sulfur units (covered in detail in Chapter 5). But gasification of coke is a highly capital-intensive project. If the gasification is done with air it may produce a considerable volume of low BTU gas (LBG). The LBG may need considerable modifications in fired heaters/boilers for using the same as fuel. So, refiners are normally deploying oxygen for gasification, planning to produce hydrogen, and diversifying for production of different chemicals, their derivatives, or petrochemical intermediates from the synthesis gas produced.

9.4.2 Downstream Integration with Petrochemicals

Refiners have started integrating with downstream petrochemicals to remain viable and enhance profitability. Steam crackers are installed to produce a wide range of olefins ($C_2^=$, $C_3^=$, and $C_4^=$) with naphtha as feed, produced from refineries, or sometimes using natural gas as feedstock. Related petrochemicals plants are also built based on the olefins (described in Chapter 1). This includes the following:

- Ethylene and ethylene-derived petrochemicals like ethylene glycol (MEG) and polyethylene (polymers like LDPE, LLDPE, HDPE). Often the ethylene recovered from sponge gas from a high severity FCCU is also integrated to enhance production of ethylene-based chemicals and polymers (described in Chapter 2).
- Production of propylene-derived petrochemicals like polypropylene (PP) initially started from FCCU LPG (described in Chapter 2) and subsequently from steam crackers. Propylene produced from the FCCU and sometimes from coker is coupled with propylene of steam crackers to further enhance production of propylene and its derivatives.
- Production of aromatic-based petrochemicals like benzene, toluene, mixed xylenes, and paraxylene (and paraxylene-derived petrochemical PTA) have been initiated. The starting unit for production of aromatic is the CCRU using heavy naphtha as feed. Often aromatic-rich streams like heavy pyrolysis gasoline from steam crackers are integrated to enhance the production. Some heavy components produced from paraxylene units (having high RON) like C_9 and C_{10} aromatics are sometimes blended in MS.

- Linear alkylbenzene (LAB) plants are set up with kerosene as the basic feedstock and using benzene at a later stage in the process. LAB has good market for production of liner alkylbenzene sulfonate (LABS) used in detergent industries.
- The synthesis gas produced from the coke gasifier is used for the manufacture of hydrogen for internal use and known to be deployed for production of chemicals such as methanol.
- Some refiners are initiating actions for production of EtOH from agricultural-based feedstocks. Some refiners are setting up plants for EtOH from the tail gas of hydrogen units. The EtOH produced will be blended in MS.

Thus, at present, refineries have been converted into large chemical complexes producing fuels, lubes, petrochemical intermediates, and finished petrochemicals, and having a great deal of integration and stream sharing between the refineries and petrochemicals sections.

10 Renewable Energy

10.1 INTRODUCTION

Renewable energy is energy derived from natural resources that replenish themselves in less than a human lifetime without depleting the resources of the planet. These resources, like sunlight, wind, rain, tides, waves, biomass, and geothermal energy stored in the earth's crust, are available in one form or another nearly everywhere. They are virtually inexhaustible. And, what is even more important, they cause little climate or no environmental damage.

On the contrary, fossil fuels, such as oil, coal, and natural gas, are only available in finite quantities. As we keep extracting them, they will run out sooner or later. Although they are produced in natural processes, they do not replenish as quickly as we humans use them. The mainstream fossil fuels like coal and petroleum are nonrenewable so their reserves are depleting fast. The CO_2 produced from burning of fossil fuels results in greenhouse effect leading to global warming. Alternative energy sources have a much lower carbon footprint than natural gas, coal, and other fossil fuels.

10.2 TYPES OF RENEWABLE ENERGY

1) Solar energy. Sunlight is one of the most abundant and freely available energy resources.
2) Wind energy. Wind is a plentiful source of clean energy.
3) Hydro energy.
4) Tidal energy.
5) Geothermal energy.
6) Biomass energy.
7) Hydrogen.

Renewable energy is generally more efficient than non-renewable energy. The energy we get from the wind, the sun, and hydroturbines can be reused without relying on exhaustible or finite sources. The most efficient renewable forms are hydrothermal, tidal, wind, and solar.

Solar energy has been proven to be the most efficient and effective among renewable energy sources for home and commercial use. Solar and wind energy is harnessed to generate electricity. Wind energy makes use of air flow to move massive wind turbines. There is a vast potential for generation of electricity from solar and wind energy, and is growing rapidly.

Hydroelectric power is currently the cheapest renewable energy source. Hydroelectric power is the cheapest because the infrastructure has been in place for a long time, and it consistently produces electricity.

DOI: 10.1201/9781003268246-10

Photosynthesis

In the process of photosynthesis, plants convert radiant energy from the sun into chemical energy in the form of glucose—or sugar.

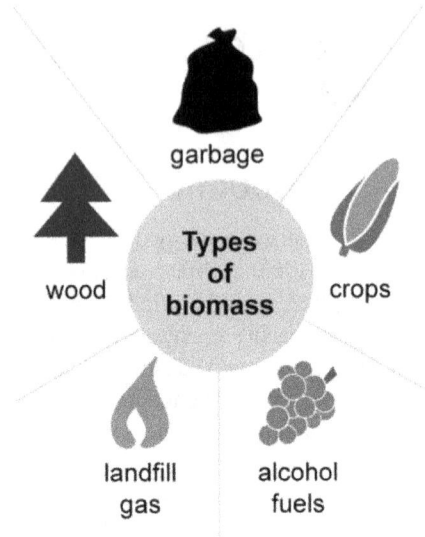

$$\underset{\text{(water)}}{6\,H_2O} + \underset{\substack{\text{(carbon} \\ \text{dioxide)}}}{6\,CO_2} + \text{(sunlight)} \atop \text{radiant energy} \rightarrow \underset{\text{(glucose)}}{C_6H_{12}O_6} + \underset{\text{(oxygen)}}{6\,O_2}$$

garbage

wood

Types of biomass

crops

landfill gas

alcohol fuels

FIGURE 10.1   The types of biomass.

Geothermal energy is harnessed to produce electricity. The United States is the largest producer of geothermal energy in the world and hosts the largest geothermal field.

Biomass energy is renewable and obtained from plants and animals. Biomass provides the biggest share among deployed renewable energy sources with a contribution of 50%, followed by hydroelectricity at 26%, and wind power at 18%. The types of biomass are shown in Figure 10.1.

10.3   TYPES OF BIOMASS

Biomass is renewable organic material that comes from plants and animals. Biomass continues to be an important fuel in many countries, especially for cooking and heating in developing countries. The use of biomass fuels for transportation and for electricity generation is increasing in many developed countries as a means of avoiding carbon dioxide emissions from use of fossil fuels. In 2020, biomass provided nearly 5 quadrillion British thermal units (BTU) and about 5% of total primary energy use in the United States.

Biomass contains stored chemical energy from the sun. Plants produce biomass through photosynthesis. Biomass can be burned directly for heat or converted to renewable liquid and gaseous fuels through various processes. Biomass sources for energy include:

- Wood and wood processing wastes—firewood, wood pellets, and wood chips; lumber and furniture mill sawdust and waste; and black liquor from pulp and paper mills
- Agricultural crops and waste materials—corn, soybeans, sugar cane, switchgrass, woody plants, algae, and crop and food processing residues

- Biogenic materials in municipal solid waste (MSW)—paper, cotton, and wool products; and food, yard, and wood wastes
- Animal manure and human sewage

The vision is to develop technology for biorefineries that will convert biomass into a range of valuable fuels, chemicals, materials, and products like oil refineries and petrochemical plants.

One of the major advantages of biomass energy is that it produces a smaller amount of harmful greenhouse gases than fossil fuel. Biomass is considered as a renewable energy source *because supply does not run out*. As opposed to fossil fuels that takes millions of years to form, biomass grows and regrows relatively quickly through the photosynthetic process.

Biomass fuels are organic materials produced in a renewable manner. Two categories of biomass fuels, namely woody fuels and animal wastes, comprise the vast majority of available biomass fuels. Municipal solid waste (MSW) is also a source of biomass fuel. Biomass fuels have low energy densities compared to fossil fuels (need higher volume to generate same amount of energy as fossil fuel).

Wood wastes of all types make excellent biomass fuels and can be used in a wide variety of biomass technologies. Combustion of woody fuels to generate steam or electricity is a proven technology and is the most common biomass-to-energy process. Different types of woody fuels can typically be mixed together as a common fuel.

There are at least six subgroups of woody fuels. The difference between these subgroups mainly have to do with availability and cost.

1. Forestry residues—forest woody debris and slash from logging and forest management activities.
2. Mill residues—byproducts such as sawdust, wood chips from different mills e.g. plywood manufacturing, and other wood processing facilities.
3. Agricultural residues—byproducts of agricultural activities including crop wastes, straws, residues after pruning of trees and vines, and rejected agricultural products.
4. Urban wood and yard wastes—residential organics collected by municipal programs or recycling centers and construction wood wastes.
5. Dedicated biomass crops—trees, oilseed from plants like Jatropha, joboba, and other crops grown as dedicated feedstocks for a biomass project.
6. Chemical recovery fuels (black liquor)—woody residues recovered out of the chemicals used to separate fiber for the pulp and paper industry. By and large, the chemical recovery facilities are owned by pulp and paper facilities and are an integral part of the facility operation.

10.4 CONVERSION OF BIOMASS TO ENERGY AND BYPRODUCTS

Bioenergy technologies are primarily intended to produce the following:

1. Biofuels. Biofuels are transportation fuels, such as bioethanol and biodiesel, created by converting biomass into liquid fuels to meet transportation needs.

2. Bioproducts. In addition to electricity and fuels, biomass can also be converted into chemicals for making polymers and other products that typically are made from petroleum.

10.4.1 CONVERSION PROCESSES

Biomass is converted to energy through various processes, including:

- Direct combustion (burning) to produce heat
- Thermochemical conversion to produce solid, gaseous, and liquid fuels
- Chemical conversion by use of solvent to produce liquid fuels
- Biological conversion to produce liquid and gaseous fuels

Direct combustion is the most common method for converting biomass to useful energy. All biomass can be burned directly for heating buildings and water, supply of industrial process heat, and for generating electricity in steam turbines.

Thermochemical conversion of biomass includes pyrolysis and gasification and are described next:

- Pyrolysis entails heating organic materials to (400°C–500°C) in the near-complete absence of free oxygen. Biomass pyrolysis produces fuels such as charcoal, bio-oil, renewable diesel, methane, and hydrogen.

  Hydrotreating is used to further process bio-oil (produced by fast pyrolysis) with hydrogen under elevated temperatures and pressures in the presence of a catalyst to produce renewable diesel, renewable gasoline, and renewable jet fuel.
- Gasification entails heating organic materials to 800°C–900°C with injections of controlled amounts of free oxygen and/or steam into the vessel to produce a gas rich in carbon monoxide and hydrogen called synthesis gas or syngas. Syngas can be used as a fuel for diesel engines, for heating, and for generating electricity in gas turbines. It can also be treated to separate the hydrogen from the gas, and the hydrogen can be burned or used in fuel cells. The syngas can be further processed to produce liquid fuels using the Fischer–Tropsch (FT) process (production of hydrocarbons from CO and H2).

10.5 FORMS OF BIOFUELS AND BENEFITS OF BIOFUELS

10.5.1 BIOGAS

Renewable natural gas—also called biogas or biomethane—is produced in anaerobic digesters at sewage treatment plants and at dairy and livestock operations. It can be produced from solid waste landfills. The biogas plant consists of a concrete tank (10–15 feet deep) in which biowaste are collected and a slurry of dung is fed. A floating cover is placed over the slurry, which keeps on rising as the gas is produced in the tank due to the microbial activity. The biogas thus produced can be used for cooking and lighting.

Properly treated renewable natural gas has the same uses as fossil fuel natural gas. Biogas is an energy-rich gas, composed mostly of methane (CH_4) and carbon dioxide (CO_2). The methane content of raw (untreated) biogas may vary from 40%–60%, with CO_2 making up most of the remainder along with small amounts of water vapor and other gases. Biogas can be burned directly as a fuel or treated to remove the CO_2 and other gases for use just like natural gas.

Anaerobic decomposition of biomass is normally used to produce biogas. Anaerobic bacteria occur naturally in soils, in water bodies such as swamps and lakes. Biogas can be collected from municipal solid waste landfills and livestock manure holding ponds. Biogas can also be produced under controlled conditions in special tanks called anaerobic digesters. The material remaining after anaerobic digestion is complete is called digestate, which is rich in nutrients and can be used as a fertilizer.

Thermochemical conversion of biomass to biogas can be achieved through gasification. Nearly all of the biogas now consumed in the United States is produced from anaerobic decomposition and used for electricity generation.

10.5.2 GASOHOL (A BLEND OF GASOLINE AND ETHANOL)

Biological conversion includes fermentation to convert biomass into ethanol and anaerobic digestion to produce renewable natural gas. Ethanol is used as a vehicle fuel.

Ethanol contains approximately 34% less energy per unit volume than gasoline, and therefore releases 34% less energy compared to burning of pure gasoline. However, since ethanol has a higher octane rating, the engine can be made more efficient by raising its compression ratio.

For E10 (10% ethanol and 90% gasoline), the effect on fuel economy is small (~3%) when compared to conventional gasoline, and even smaller (1%–2%) when compared to oxygenated and reformulated blends. For E85 (85% ethanol), the effect on the fuel economy becomes significant. E85 produces lower mileage than gasoline. Based on US Environmental Protection Agency (EPA) tests for all 2006 E85 models, the average fuel economy for E85 vehicles was 25.56% lower than gasoline derived from petroleum. E85 is a high-performance fuel, with an octane rating of about 94–96, and should be compared to premium gasoline.

Brazilian flex fuel vehicles can operate with ethanol mixtures up to E100, which is hydrous ethanol (with up to 4% water) having lower vapor pressure. Thus, Brazilian flex vehicles are built with a small secondary gasoline reservoir located near the engine for a cold start. An improved flex engine version was launched later that is known to eliminate the need for the secondary gas storage tank for starts in cold climates.

10.5.3 BIODIESEL

A chemical conversion process known as transesterification is used for converting vegetable oils, animal fats, and greases into fatty acid methyl esters (FAME), which are used to produce biodiesel.

Blends of biodiesel and conventional hydrocarbon-based diesel are most commonly distributed for use in the retail diesel fuel marketplace. Much of the world uses a system known as the "B" factor to state the amount of biodiesel in any fuel mix:

- 100% biodiesel is referred to as B100.
- 20% biodiesel, 80% petro-diesel is labeled B20.
- 5% biodiesel, 95% petro-diesel is labeled B5.
- 2% biodiesel, 98% petro-diesel is labeled B2.

Blends of 20% biodiesel and lower can be used in diesel equipment with no or only minor modifications, although certain manufacturers do not extend warranty coverage if equipment is damaged by these blends. The B6 to B20 blends are covered by the ASTM D7467 specification. Biodiesel can also be used in its pure form (B100), but may require certain engine modifications to avoid performance problems.

Biodiesel can be used in pure form (B100) or may be blended with petroleum diesel at any concentration in most injection pump diesel engines.

Biodiesel has different solvent properties from petro-diesel and is likely to degrade natural rubber gaskets and hoses in vehicles (particularly, for older ones). Although, the components are known to have already been replaced with fluroelastomer (FKM), which is nonreactive to biodiesel. It is recommended by experts to change the fuel filters in engines and heaters shortly after first switching to a biodiesel blend to avoid filter clogging.

Raw material for biodiesel production can be used cooking oils and oils from nonedible plant seeds like flax seeds (containing 30–45% oil) and safflower seed (containing 30–40% oil). Other plants can be Tung, Cotton, Avocado, Jojoba, and Jatropha. Microalgae have high potential for production of biodiesel as they can have higher oil content. There are several methods for carrying out this transesterification reaction, including the common batch process, heterogeneous catalyst, supercritical processes, ultrasonic methods, and even microwave methods.

Transesterified biodiesel comprises a mix of mono-alkyl esters of long-chain fatty acids. The most commonly used alcohol is methanol, the cheapest available alcohol to produce methyl esters referred to as fatty acid methyl ester (FAME). Ethanol can be also used to produce an ethyl ester (commonly referred to as fatty acid ethyl ester or FAEE) biodiesel, and higher alcohols such as isopropanol and butanol have also been used. Using alcohols of higher molecular weights improve the cold flow properties of the resulting ester. Any free fatty acids (FFAs) in the base oil are known to be either converted to soap and removed from the process, or they are esterified (yielding more biodiesel) using an acidic catalyst. After this processing, unlike straight vegetable oil, biodiesel has combustion properties very similar to those of petroleum diesel and can replace it directly or can be further improved by hydrotreatment.

10.5.4 BENEFITS OF BIOMASS FUEL

Use of biomass fuels can provide an array of benefits.

1. *Reduction of greenhouse gas emissions.* The use of biomass fuels has the potential to greatly reduce greenhouse gas emissions. Biomass releases carbon dioxide that is largely balanced by the carbon dioxide captured for its own growth. However, studies have found that clearing forests to grow biomass results in a carbon penalty that takes decades to recoup, so it is best if biomass is grown on previously cleared land, such as underutilized farmland.

2. *Support to forest product industry.* Biomass energy can support agricultural and forest-product industries by encouraging farming of biomass crops. For biomass fuels, the most common feedstocks used today are corn grain (for ethanol) and soybeans and jatropha (for biodiesel). In the near future, agricultural residues such as corn stover (the stalks, leaves, and husks of the plant) and wheat straw may also be used. Long-term plans include growing and using dedicated energy crops, such as fast-growing trees and grasses, and algae. These feedstocks can grow on land that does not support food crops.

10.6 BIOREFINERY

A biorefinery is a refinery that converts biomass to energy and other byproducts (such as chemicals). The International Energy Agency (Bioenergy Task 42) defined biorefining as "the sustainable processing of biomass into a spectrum of bio-based products (food, feed, chemicals, materials) and bioenergy (biofuels, power and/or heat)". As refineries, biorefineries can produce many intermediate chemicals by initial processing like pressing, filtration and fractionation of the raw material (biomass) that can be further converted into value-added products. The use of biomass as feedstock can provide a benefit by reducing the impacts on the environment. In addition, biorefineries are intended to achieve the following goals:

1. Produce fuels and chemical building blocks (*building block* is a term in chemistry that is used to describe a virtual molecular fragment or a real chemical compound the molecules of which possess a reactive functional group) and can produce many new chemicals
2. Valorization of waste (agricultural, urban, and industrial waste)
3. Achieve the ultimate goal of reducing greenhouse gas emissions

Biorefineries can be classified based on four main features: platforms, products, feedstocks, and processes.

Platforms refers to key intermediates between raw material and final products. Some important intermediates are

- Biogas from anaerobic digestion
- Syngas from gasification

- C6 sugars from hydrolysis of sucrose, starch, cellulose, and hemicellulose
- C5 sugars (e.g., xylose, arabinose: $C5H_{10}O_5$), from hydrolysis of hemicellulose and food and feed side streams
- Liquid from pyrolysis (pyrolysis oil)

Products from biorefineries can be grouped in two main categories:

- Energy-driven biorefinery systems—The main product is an energy carrier as biofuels, power, and heat.
- Material-driven biorefinery systems—The main product is a bio-based product.

Feedstocks may be dedicated feedstocks (sugar crops, starch crops, lignocellulosic crops, oil-based crops, grasses, marine biomass); and residues (oil-based residues, lignocellulosic residues, organic residues and others).

Conversion processes to transform biomass into a final product are described earlier in Section 10.4.1.

Some examples of biorefinery classifications include:

- C6 sugar platform biorefinery for bioethanol and animal feed from starch crops.
- Syngas platform biorefinery for FT diesel and phenols from straw.
- C6 and C5 sugar and syngas platform biorefinery for bioethanol, FT diesel, and furfural from saw mill residues.

Bioethanol plants and sugar mills are well-established processes where the biorefinery concept can be implemented since sugarcane bagasse is a feasible feedstock to produce fuels and chemicals like lignocellulosic bioethanol (second generation, 2G). TEA (tri-ethanol amine) production using mild liquefaction of bagasse is known to be a feasible option. The production of different chemicals like xylitol, citric acid and glutamic acid, and lactic acid from sugarcane lignocellulose (bagasse and harvesting residues) are feasible and can be economically viable.

The biodiesel production industry also has the potential to integrate biorefinery systems to convert residual biomasses and wastes into biofuel, heat, electricity and bio-based green products. Glycerol is the main coproduct in biodiesel production and can be transformed into valuable products.

The pulp and paper industry is considered as the first industrialized biorefinery system. In this process, several other coproducts, like tall oil, rosin, and vanillin, can be produced. This industry is the highest consumer of biomass and is capable of processing agricultural waste: rice straw, and corn stover as feedstock and has established logistics for the production and supply of biomass as feed.

Lignin is a natural polymer cogenerated in the plant and is generally used as boiler fuel to generate heat or steam to cover the energy demand in the process. Since lignin accounts for 10–30 wt.% of the available lignocellulosic biomass, the viability and economics of biorefineries depend on the cost-effective processes to transform lignin into value-added fuels and chemicals.

The waste biomass can be an attractive source for conversion to valuable products, and several biorefinery routes have been proposed to upgrade waste streams in valuable products. The production of biogas from waste banana peel is possible including the production of other coproducts like ethanol, xylitol, syngas, and electricity. This process can also provide high profitability for higher scales of production.

Life cycle assessment (LCA) is a methodology to evaluate the environmental load of a process. LCA can be used to investigate the potential benefits of biorefinery systems.

10.7 HYDROGEN FUEL

10.7.1 INTRODUCTION

Hydrogen is the fuel of the future. Hydrogen is an energy carrier and not a source of energy. It can be produced from a wide variety of energy sources. Traditionally, hydrogen has been predominantly produced from fossil sources. Hydrogen as an energy carrier can be used in internal combustion engines or fuel cells producing virtually no greenhouse gas emissions when burnt with oxygen and produces only water vapor as exhaust. Hydrogen can be produced and converted into energy through a variety of ways. Steam methane reforming (SMR) is currently the major route to hydrogen production. All the major routes for hydrogen gas production like steam reforming, gasification, and electrolysis of water involve the use of fuel and electricity. Thus, the production of hydrogen cannot be emissions-free unless some renewable source can be deployed for its production.

Hydrogen production at present, from different sources, is estimated as follows:

- Natural gas 48%
- Oil 30%
- Coal 18%
- Electrolysis 4%

10.7.2 RENEWABLE HYDROGEN PRODUCTION PATHWAYS

Hydrogen production and storage is currently undergoing extensive research. A solar-hydrogen system can provide the means of a totally emissions-free method of producing hydrogen. Among all routes for production of hydrogen, the solar–hydrogen system can produce hydrogen involving no emissions and with reasonably high efficiency of 65%.

The most established technology options for producing hydrogen from renewable energy sources are water electrolysis and steam reforming of biomethane/biogas.

Biomass gasification and pyrolysis, thermochemical water splitting, photocatalysis, supercritical water gasification of biomass, and anaerobic digestion processes are being developed further as economically viable methods for the production of hydrogen.

An *electrolyzer* is capable of producing hydrogen and can be operated with electricity derived from renewable sources. An electrolyzer is a device that splits water into hydrogen and oxygen using electricity. Electrolyzers offer a flexible load that can regulate hydrogen production (and hence regulate electricity consumption) and provide low-cost balancing services to the power system while producing hydrogen for industrial uses, or injection into the gas grid. The built-in storage capacity of downstream sectors (gas infrastructure) can be used as a buffer to adjust hydrogen production (and hence electricity consumption) depending on the requirement of power.

Alkaline (ALK) electrolyzers have been used by industry for nearly a century. ALK electrolyzer technology is almost fully developed. It has been used by industry particularly in the chemicals industry for chlorine manufacture.

Mainly two types of *electrolyzers* deploying solid electrolytes are in use: Polymer electrolyte membrane (PEM) electrolyzer cell and Solid oxide electrolyzer cell (SOEC). Proton exchange membrane or polymer electrolyte membrane electrolyzers are commercially available today and are rapidly gaining market.

Solid oxide electrolyzers are known to have the potential to improve energy efficiency but are undergoing further development and, unlike alkaline and proton exchange membrane electrolyzers, are known to work at high temperatures.

10.7.3 RELATIVE MERITS OF THE TECHNOLOGIES

PEM electrolyzer technology is known as rapidly emerging and entering commercial deployment. Further, the CAPEX for PEM has dropped significantly in recent years. The lifetime of an ALK electrolyzer is currently twice as long as that of PEM electrolyzers.

State-of-the-art PEM electrolyzers can operate more flexibly and reactively than current ALK technology. PEM technology is known to offer a wider operating range and has a shorter response time.

PEM electrolyzers can produce hydrogen at a higher pressure than ALK electrolyzers. As a result, the need for downstream compression to reach the desired end-use pressure may be lower and can be relevant in applications where high pressure is important, such as for mobility applications and injection to gas grid.

10.7.4 HYDROGEN STORAGE AND TRANSPORT

Hydrogen storage and transportation is a critical issue involving intense research. The problem is the low density of hydrogen gas. Three possible solutions have been proposed by experts:

- Hydrogen can be injected into the gas grid to reduce natural gas consumption. Injection could be an additional revenue source for electrolyzer operators beyond their hydrogen sales to mobility or industrial markets. In the short term, this could be of significant help to reach the volumes necessary to effect cost reductions through economies of scale (EOS) and improve the competitiveness of hydrogen from renewable power in the long term.

- The other potential hydrogen delivery systems may include compressed tube trailers and liquid storage tank trucks. One major disadvantage of each system is the high capital costs. Hydrogen and electricity, as energy carriers, are complementary in the energy transition.

- Power generated from renewable sources (like wind, solar) can be deployed to produce hydrogen by use of electrolyzers and the hydrogen can be transported and stored on a large scale in gas grids during seasonal swings in demand for electricity. This can be also relevant when the capacity of the electricity grid is insufficient to absorb the fullest potential of electricity generation from the renewable sources. Thus, power can be generated up to the full potential offshore with wind as the source, and hydrogen could be produced at the location and can be transported to the shore via natural gas pipelines instead of laying costly submarine cables for transmission of electricity to onshore.

Hydrogen carriers, such as liquid organic hydrogen carriers (LOHCs), could be more suitable for long-distance transport than gaseous or liquid hydrogen. To date, however, pipelines remain the most economic route to transporting hydrogen in large volumes, hence "greening the gas grid" allows the economies of scale necessary to reduce the cost of hydrogen production.

The use of metal hydrides is known to be the most promising storage material. The advantages are high volume efficiencies, easy recovery, and advanced safety. The common metal hydrides in current research are known to be $PdH_{0.6}$, $ZrV_2H_{5.5}$, $LaNi_5H_6$, and Mg_2NiH_4. They can have reasonably high hydrogen content (0.56% to 3.59% wt.) They can be stored at low to moderate pressures and can release hydrogen at low to moderate temperatures.

10.7.5 Use of Hydrogen and Fuel Cells in Transport Sector

Hydrogen can be used as the primary fuel in an internal combustion engine or in a fuel cell. A hydrogen internal combustion engine is similar to that of a gasoline engine, where hydrogen combines with oxygen in the air and produces expanding hot gases that directly move the parts of an engine.

Fuel cells can also be used for the generation of electricity for drives that often uses hydrogen as fuel and oxygen of air. They are mainly of two types: PEM fuel cells and solid oxide fuel cells (SOFC). A single fuel cell can produce electricity continuously till fuel is available but can generate very low voltage. Thus, the fuel cells are connected in series to produce adequate voltage required for the desired application. Both types are used for drives and vehicles.

A PEM fuel cell uses a polymer as electrolyte and allows only hydrogen ions to pass to the cathode where the ions react with oxygen of air to produce water.

A SOFC is a highly efficient electrochemical device that can use hydrogen, other hydrocarbons and even carbon monoxide as fuel to produce power. It uses oxygen ion conducting ceramics as electrolyte. SOFCs are promising candidates for future energy systems due to their much higher efficiencies compared to heat engines and other forms of fuel cells. The fuel cell electric vehicles (FCEVs) can have similar driving performance to conventional vehicles (driving range, fueling time). FCEVs are complementary to battery electric vehicles (BEVs) and can replace them. They can be used in heavy-duty segments (long-range or high utilization rate vehicles, e.g., trucks, trains, buses, taxis, ferry boats, cruise ships, aviation, forklifts) where batteries are currently limited.

BIBLIOGRAPHY

Alternative Fuels Data Center. *Fuel Blends*. Energy Efficiency & Renewable Energy (EERE), SW, Washington, DC: US Department of Energy.

Bioethanol production from renewable raw materials and its separation and purification: A review. *FTB, Food Technology and Biotechnology*, 56, no. 3, 2018, pp. 289–311.

Husam Al-Mashhadani and Sandun Fernando. Properties, performance, and applications of biofuel blends: A review. *AIMS Energy*, 5, no. 4, 2017, pp. 735–767.

Hydrogen and Fuel Cell Technologies. *Fuel Cells*. EERE, SW, Washington, DC: US Department of Energy.

Hydrogen and Fuel Cell Technologies. *Hydrogen Fuel Basics*. EERE, SW, Washington, DC: US Department of Energy.

International Renewable Energy Agency (IRENA). *Hydrogen from Renewable Power: Technology Outlook for the Energy Transition*. International Renewable Energy Agency (IRENA), Masdar City, September 2018.

L. G. Roberts and T. J. Patterson. Bio fuels. In *Encyclopedia of Toxicology*. 3rd edition, Amsterdam: Elsevier, 2014.

ABBREVIATIONS

ALK	alkaline
BEV	battery electric vehicle
CAPEX	capital expenditure
CCS	carbon capture and storage
CCU	carbon capture and utilization
CSP	concentrated solar power
EPA	US Environmental Protection Agency
FCEV	fuel cell electric vehicle
FT	Fisher–Tropsch
GHG	greenhouse gas
HRS	hydrogen refueling station
LDV	light-duty vehicle
LOHC	liquid organic hydrogen carrier
OPEX	operating expenditure
PEM	polymer electrolyte membrane
PV	photovoltaic
SMR	steam–methane reforming
SOEC	solid oxide electrolyzer cell
VRE	variable renewable energy

Index

A

ABD, *see* Apparent bulk density
Acid gas
  advantages of blending amines, 237
  amine-based solvents, 231–233
  CCS, 231–233
  chemical reactions, 235
  chemical solvents, 220, 221
  components, 217
  conventional processes, 218
  cryogenic fractionation, 219
  DEPG, 222–224
  direct conversion, 219
  downstream process, 217–218
  ethanol amines, 220
  flow configuration, 228
  hybrid/composite solvents, 219
  membrane process, 219
  methanol processes, 222–223
  mixed solvents, 221
  NMP, 224–225
  non-amine-based solvents, 234–235
  non-amine solvents, 219
  PC, 225–226
  physical adsorption, 219
  physical-chemical solvents, 228–231
  physical properties, 226–228
  physical solvents, 218, 221, 222
  piperazine, 238
  solvent blends, 237–238
  solvent regeneration, 228
  sterically hindered amines, 233–234
  synthesis gas treatment, 221
  treatment, 3, 4
Air cooler, 13, 42, 53, 136, 144, 197, 229, 247, 270
Aircraft turbine fuel (ATF), 180
Alkaline (ALK) electrolyzers, 346
Amine-based chemical absorption process,
    231–233
Amine units
  arresting contamination
    iron sulphide, 251
    knockout drum, 250, 251
    lean amine temperature, 251
    water wash facility, 250
  higher lean amine flow, 258
  industry, 244–246
    condenser duties, 244–247
    strength on reboiler, 244–247

  lowering contaminations
    carbon/charcoal bed filters, 252
    coalescer, 252
    hydrocarbon skimming, 252
    kidney units, 253
    particulate filters, 252
    purge water, 253
    skimming locations, 252
  problems
    corrosion, 247–249
    degradation, 246
    foaming, 246–247
    fouling, 248–249
    poor sweet product quality, 249–250
    reflux drum, 249
  refinery operating, 256
  sources of contamination
    feed gas, 249
    hydrocarbons, 249
    oxygen ingress, 250
    water, 249–250
  sour gas, 255
  sweet gas, 255, 257
  vapor–liquid composition, 256, 257
Anaerobic digesters, 339
Anti-jump baffles, 10, 12
Apparent bulk density (ABD), 96–97, 120
APS, *see* Average particle size
Aromatic extraction units, 3–4
Aromatics, 147
Aromatic saturation (HDA), 180, 184, 207
Asphaltenes, 147
ATB, *see* Atmospheric tower bottoms
ATF, *see* Aircraft turbine fuel
Atmospheric tower bottoms (ATB), 41
Atmospheric vacuum unit (AVU), 32–37
Average particle size (APS), 103
AVU, *see* Atmospheric vacuum unit

B

Banana effect syndrome, 165–166
Basic sludge and water (BS&W), 54
Battery electric vehicles (BEVs), 346
BBU, *see* Bitumen blowing unit
BDS, *see* Blowdown scrubber
Bed density, 97, 113, 167
Benfield process, 234, 235
BEVs, *see* Battery electric vehicles
Biodiesel, 340

Bioethanol plants, 342
Biogas, 339, 343
Biomass
   advantage, 337
   availability and cost, 337
   benefits, 341
   conversion processes, 338
   fuels, 337
   sources, 336–337
Biomethane, 338, 343
Biorefineries
   classifications, 342–343
   conversion processes, 342
   feedstocks, 342
   goals, 341
   platforms, 341–242
   products, 342
Bitumen and Visbreaker unit (VBU), 3
Bitumen blowing unit (BBU), 3
Blowdown scrubber (BDS), 133
Bonded/agglomerated shot coke, 166
Brazilian flex fuel, 339
British thermal units (BTU), 336
BTU, *see* British thermal units
Bubbling velocity, 97

C

Calcium, 65–66
Captive power plant (CPP), 6, 328
Carbocation, 122
Carbon capture, 259–262
Carbon capture and storage (CCS), 231–233
Carbon/charcoal bed filters, 252
Carbon dioxide removal (CDR), 217
CAT, *see* Catalyst average temperature
CATACARB process, 234
Catalyst
   coking/fouling, 182
   contaminants, in hydrotreater feed
      arsenic, 185
      calcium, 185
      iron, 186
      nickel, 186
      phosphorous, 186
      silicon, 185
      sodium, 185
      vanadium, 186
   deactivation, 182
   mechanical failure, 183
   poisons
      inhibiting agents, 184–185
      metals, 185
      nitrogen compounds, 183–184
   sintering, 183
   sulfiding, 181–182

Catalyst average temperature (CAT), 184
Catalytic cracking, 123
Catalytic dewaxing units (CDW), 4
Catalytic reformer unit (CRU), 3, 282
Caustic injection, 55–58
CBA, *see* Cold bed adsorption
CCD, *see* Coke condensate drum
CCR, *see* Conradson carbon residue
CCRU, *see* Continuous catalytic reforming unit
CCS, *see* Carbon capture and storage
CDR, *see* Carbon dioxide removal
CDW, *see* Catalytic dewaxing units
Chemical solvents, 220
CHPS, *see* Cold high-pressure separator
Clarified oil (CLO), 75
CLO, *see* Clarified oil
Cobalt–molybdenum catalyst, 178
Coil outlet temperature (COT), 36, 49, 70, 142,
      150–153, 169, 312
Coke condensate drum (CCD), 136
Coke drum
   banana effect syndrome, 165–166
   coking cycle, 159
   cycle time, 158
   drum deformities, 159
   fast quench issues, 159–160
   junction stress, 160–163
   salient features, 158–159
Coker
   BDS/drum, 137
   case studies, 170–172
   decoking operation, 155–157
      coking cycles, 147
      cooling, 144
      drilling/cutting, 144
      drum heating, 146
      feed switch, 146–147
      flow and pressure estimation, 145–146
      heading, 146
      pressure testing, 146
      steaming/wax tailing, 143–144
      unheading, 144
      water draining, 144
   effect of feedstock, 147–148
   feed device, 139–140
   fouling problems, 169
   furnace fouling
      coil outlet temperature, 150
      double-fired, 149
      metallurgy of tubes, 150
      turbulizing/velocity steam, 148
   gasoline, 27
   generalized flow scheme, 135
   main sections, 134
   operating variables
      COT, 142

drum pressure, 143
recycle rate, 142–143
process scheme, 133–139
reasons for rapid fouling, 151
burners, 153
catalyst, 152
feed interruptions, 152
feed preheating, 152–153
feed quality, 151–152
low mass velocity, 151
skin temperatures, 153
run length
firebox, 155
locations in transfer line, 154
online spalling, 155
pigging/mechanical coke removal,
154–155
problematic transfer line, 154
steam-air decoking, 154
transfer line configuration, 154
storage and handling safety, 169–170
valves, 138
wash zone, 140–142
yields of products, 134
Cold bed adsorption (CBA), 287
Cold high-pressure separator (CHPS), 181, 317
Cold low-pressure separator (CLPS), 199, 317
Column flooding mechanisms, 9
Compressed air system, 6
Compressor, 13
Condenser duty (QC), 7, 69, 172, 299, 304, 308, 312
Conradson carbon residue (CCR), 75, 89, 133
Continuous catalytic reforming unit (CCRU),
5, 323
Corrosion
amine units, 247–248
FCCU
ammonia chloride deposition, 126–127
inhibitors, 129
NH4Cl deposition, 127–128
polysulfide, 129
salt deposition, 128
water wash, 128
heavy crudes
acid, 52–55
caustic injection, 55–58
caustic reacting, 56, 59
filmers, 62
tramp amines, 59–62
high-TAN crudes, 63–64
NA, 64–65
COT, *see* Coil outlet temperature
CPP, *see* Captive power plant
CRU, *see* Catalytic reformer unit
Crude distillation unit (CDU), 34, 40, 68, 70, 184,
306–310

Crudes, 28
Crude–vacuum unit, 38–39
Cryogenic fractionation, 219
Cyclones
areas of failures, 105
catalyst loss, 103–105
coke, 106
design limits, 102–103
dipleg diameter, 115
dipleg sealing, 112
dipleg termination, 115–117
erosion, 109
failures, 107
FCC standpipe
aeration gas, 122
aeration practice, 120–121
catalyst behavior, 118–119
catalyst flux, 120
guidelines, 119–120
pressure profiles, 117–118
RCSP, 122
SCSV, 122
fine generation, 106
flooded diplegs, 104
formation of holes, 108
nomenclature, 102
poor catalyst circulation, 106
poor product yields, 107
primary, 102–103
problems and causes, 105
reactor arrangements, 110
regenerators, 113–115
remedies, 107–109
riser separation systems, 109–113
secondary, 103
trickle valve, 101
unsealed dip legs, 104
vortex stabilizer, 109

D

DAO, *see* Deasphalted oil; Dewaxed oil
DDSV, *see* Double-disc slide valve
DE, *see* Dethanizer
DEA, *see* Diethanolamine
Deasphalted oil (DAO), 4
Deflagration, 169
Delta coke
blower and generation, 90
definition, 88
types, 88–89
Demethanizer (DM), 76
DEPG, *see* Dimethyl ether of polyethylene glycol
Dethanizer (DE), 76
Dewaxed oil (DAO), 75
DGA, *see* Diglycolamine

DHDS, *see* Diesel hydrodesulfurization
Diesel, 326
Diesel hydrodesulfurization (DHDS), 138
Diesel hydrotreater, 3, 176, 192–193
    color, 194
    corrosion, 204–206
    fluorescence species, 207
    problems, 193–194
Diethanolamine (DEA), 232–233
Digestate, 339
Diglycolamine (DGA), 76, 232
Diisopropanolamine (DIPA), 221, 228, 232
Dimethyl disulfide (DMDS), 76, 181
Dimethyl ether of polyethylene glycol (DEPG),
        222–224
DIPA, *see* Diisopropanolamine
Direct combustion, 338
Direct conversion process, 219
Direct fired heaters, 287
Distillation, 6
    column internals, 8
    dimensions, 8–9
    simulation, 7
DM, *see* Demethanizer
DMDS, *see* Dimethyl disulfide
Double-disc slide valve (DDSV), 76, 81, 89
Downcomer backup flooding, 9
Downcomer choke flooding, 9
Dry flue gas analysis, 82
Dry-out ratio, 47
Dry reflux drums, 41
Dumping, 11

E

ECAT, *see* Equilibrium catalyst
EFB, *see* Empty fruit bunches
Electric overhead traveling (EOT), 137
Electrolyzers, 345
Electro static precipitator (ESP), 104
Embedded shot coke, 166
End of run (EOR), 150, 182, 206
EOR, *see* End of run
EOT, *see* Electric overhead traveling
Equalization line open, 20
Equilibrium catalyst (ECAT), 93, 97, 103, 120,
        124
ESP, *see* Electro static precipitator
Ethanol, 339
Ethanol amines, 219, 220
Extra-heavy crude oils, 28–31

F

FAME, *see* Fatty acid methyl esters
Fatty acid methyl esters (FAME), 340

FCCU, *see* Fluid catalytic cracking unit
FCEV, *see* Fuel cell electric vehicles
Feed gas, 250
Fenske–Underwood and Erbar–Maddox
        (FUEM), 6
Fenske–Underwood–Gilliland (FUG), 6
FEU, *see* Furfural extraction unit
FFAs, *see* Free fatty acids
FGD, *see* Flue gas treatment
FGDU, *see* Flue gas desulfurization unit
Filmers, 62
Fine generation, 106
Fischer–Tropsch (FT) process, 271, 274, 338
Flare line isolation valves, 319
Flue gas desulfurization unit (FGDU), 104
Flue gas treatment (FGD), 76
Fluid catalytic cracking unit (FCCU), 75–77
    additives, 124–125
    air grids, 98
    configurations, 95–96
    corrosion, 125–126
        ammonia chloride deposition, 126–127
        inhibitors, 129
        NH4Cl deposition, 127–128
        polysulfide, 129
        salt deposition, 128
        water wash, 128
    cyclones (*see* Cyclones)
    dry flue gas analysis, 82
    equipment/components, 79, 93–94
    external cyclones, 99
    flapper valves, 100
    flow scheme, 78
    flue gas components, 84
    fluidization, 93
    integration of main sections, 78
    MFT, 95
    operating data, 86
    operating variables
        catalyst makeup rate, 93
        feed preheat temperature, 92
        fresh feed rate, 92
        gasoline end point, 93
        recycle rate, 91–92
        riser temperature, 91
    petrochemical, 5
    physical properties
        ABD, 96–97
        bed density, 97
        bubbling velocity, 97
        fluidization velocity, 97
        particle density, 100
        PV, 100
        SD, 100
        stabilizer, 25, 27
        superficial velocity, 101

reaction mechanism, 122–124
reactor arrangements, 94–95
reactor cyclones, 96
reactor regenerator section, 80
recovery of ethylene, 77
R-R heat balance (*see* Reactor–regenerator
    heat balance)
R-R section, 94
sections, 77
simulation for design, 129–130
two-stage regenerator, 99
wet flue gas analysis, 83
Fluidization velocity, 97
Foaming, 247
Free fatty acids (FFAs), 340
Froth entrainment flooding, 9
Fuel cell electric vehicles (FCEV), 346
Fuel-oriented refinery, 2
FUEM, *see* Fenske–Underwood and
    Erbar–Maddox
FUG, *see* Fenske- Underwood-Gilliland
Furfural extraction unit (FEU), 3, 4
Furnace fouling
    coil outlet temperature, 150
    double-fired, 149
    metallurgy of tubes, 150
    turbulizing/velocity steam, 148

G

Gallons per minute (GPM), 9, 10
Gas concentration unit (GCU), 91, 93
Gas/gas exchangers, 287
Gasification, 218, 332, 338
Gasohol, 339
Gasoline, 3, 27, 93, 124, 323–326
GCU, *see* Gas concentration unit
Glycerol, 342
GPM, *see* Gallons per minute

H

HCGO, *see* Heavy coker gas oil
HDA, *see* Aromatic saturation
Heat transfer exchanger reactor (HTER), 52,
    282, 322
Heavy coker gas oil (HCGO), 75, 133, 182, 195
Heavy crudes, 28
    coker units, 50–51
    corrosion
        acid, 52–55
        caustic injection, 55–58
        caustic reacting, 56, 59
        filmers, 62
        tramp amines, 59–62
    crude units, 40–42

FCCU, 51
    material balance, 35, 37
    processing issues, 32
    properties, 32
    requirement, 32
    specifications, 31
    sulfur unit, 52
    vacuum section, 36
    vacuum units, 42–45
        barometric legs, 47–48
        de-entrainment devices, 46–47
        dry-out ratio, 47
        minimum wetting value, 47
        upgrades in metallurgy, 50
        vacuum furnace transfer line pressure
            drop, 48–49
        VR quench, 49
        wash zone efficiency, 45–46
Heavy oil hydrotreaters, 194–195
    effluent circuit, 195–198
    frequency of backwash, 195
    higher differential pressure drop, 195
Heavy oil processing, 42
Heavy oil upgradation units, 1
HFU, *see* Hydrofinishing unit
HGU, *see* Hydrogen generation unit
High-pressure separator (HPS), 181, 184, 187, 199
High-TAN crudes, 62–63
    bottom coking, 72
    corrosion treatment, 63–64
    distillation units, 66–70
    methodologies, 63
    NA corrosion, 64–65
    vacuum unit, 71–72
High temperature hydrogen attack (HTHA), 199
Hi-Pure process, 236
Hollow fiber membrane, 241
Hot flue gas, 75, 81, 89–91, 282
Hot gas bypass, 287
Hot reflux drum, 41–42
Hot vapor bypass method, 18–20
HPS, *see* High-pressure separator
HTHA, *see* High temperature hydrogen attack
Hybrid/composite solvents, 219
Hydrocarbons, 251
Hydrocarbon skimming, 252
Hydrocrackers
    capacity augmentation, 209–210
    configurations, 199
    heat exchangers, 201
    high-pressure drop, 202–204
    metallurgy, 199–202
    single recycle compressor, 200
    three reactors in series, 201
    ULSD, 206–207
    unconverted recycle, 200

Hydrocracking, 180–181
Hydrodemetallization (HDM), 180
Hydrodenitrogenation (HDN), 178–180
Hydrodesulfurization (HDS), 175–178, 206
Hydroelectric power, 335
Hydrofinishing unit (HFU), 4
Hydrogen fuel, 343–345
    production pathways, 343–344
    storage and transportation, 344–345
    technologies, 344
Hydrogen generation unit (HGU), 3, 27, 265–282
Hydroprocessing
    catalyst (*see* Catalyst)
    process parameters
        H_2/oil ratio, 188–189
        pressure, 187–188
        recycle gas, 188–189
        space velocity, 189, 190
        temperature, 188
    reactions, 175
Hydrostatic test pressure, 319
Hydrothermal deactivation, 100
Hydrotreating, 338
Hydrotreatment, 175; *see also* Hydroprocessing

I

IBP, *see* Initial boiling point
ILs, *see* Ionic liquids
Initial boiling point (IBP), 41
Initial condensation point (ICP), 41, 52
Intermediate-pressure system, 316–317
Ionic liquids (ILs), 239
ISOM, *see* Isomerization unit
Isomerization unit (ISOM), 3, 189, 324

L

LAB, *see* Linear alkyl benzene
LCA, *see* Life cycle assessment
LCGO, *see* Light coker gas oil
LCO, *see* Light cycle oil
Life cycle assessment (LCA), 343
Light coker gas oil (LCGO), 138, 182
Light cycle oil (LCO), 75, 207
Lignin, 343
Linear alkyl benzene (LAB), 5, 333
Line burners, 287
Liner alkylbenzene sulfonate (LABS), 333
Liquefied petroleum gas (LPG), 75, 249, 283
Loose shot coke, 166
LPG, *see* Liquefied petroleum gas
Lube base oil, 4, 327

M

Main air blower (MAB), 75
Main column bottom pump around (MCB PA), 129

Main combustion chamber (MCC), 283, 285, 292
Main fractionators (MF), 125
Makeup gas compressor (MUG), 211
Makeup hydrogen compressor (MUG), 175
MAWP, *see* Maximum allowable working
    pressure
Maximum allowable working pressure (MAWP),
    295, 315
MCB PA, *see* Main column bottom pump around
MCC, *see* Main combustion chamber
McCabe–Thiele method, 6
MCR, *see* Micro-carbon residuum
MDEA, *see* Methyldiethanolamine
MEA, *see* Monoethanolamine
Membrane process, 219
    hollow fiber membrane, 241
    one-stage membrane process, 241–242
    pretreatment, 242
    spiral wound membrane, 240–241
    three-stage membrane process, 243
    two-stage membrane process, 242
    types, 240
Mercaptans, 235
Methanol (MEOH), 222–223
Methyldiethanolamine (MDEA), 233
MF, *see* Main fractionators
MFT, *see* Multifunctional two-riser
MHC, *see* Mild hydrocrackers
Micro-carbon residuum (MCR), 166
Mild hydrocrackers (MHC), 30
Mixed solvents, 221
Monoethanolamine (MEA), 221, 232
MUG, *see* Makeup hydrogen compressor
Multifunctional two-riser (MFT), 95

N

Naphtha hydrotreater (NHT), 189–190
    capacity augmentation, 208–209
    increase pressure drop, 190–192
    NH4CL deposition, 192
    stabilizer system, 210–211
Naphtha vaporizer, 27
Naphthenic acid (NA) corrosion, 64–65
NBP, *see* Normal boiling point
Needle coke, 168
Negative pressure separator, 111
Nickel passivators, 89, 124
N-Methyl-2-pyrrolidone, 224–225
Non-amine-based solvents, 234–235
Non-amine solvents, 219
Non-condensable vents, 21
Normal boiling point (NBP), 129

O

Once-through reboilers, 22
One-stage membrane process, 241–242

Onstream factor, 150
Overaeration, 121
Oxygen enrichment, 52
Oxygen-lean regeneration, 88

P

Paraffinic feeds, 147, 148
Particle size distribution (PSD), 103
Particulate filters, 252
PDA, *see* Propane deasphalting unit
PDS, *see* Predesulfurisation section
PEM, *see* Polymer electrolyte membrane
PFV, *see* Pre-flash vessel
Phosphate esters, 63, 186
Physical adsorption process, 219
Physical solvents, 218, 221, 222
Piperazine, 238
Plant safety, 295–319
Plastic Strain Index (PSI), 164
PNA, *see* Polynuclear aromatic molecules
Polymer electrolyte membrane (PEM), 344
Polynuclear aromatic (PNA) molecules, 207
Polysulfide, 129
Pore volume (PV), 100
Positive pressure separator, 111
Potassium bisulfide (KHS), 235
Potassium carbonate, 234
Power recovery train (PRT), 90
Predesulfurisation section (PDS), 266
Pre-flash vessel (PFV), 33
Pressure control
    air cooler, 13
    balance around column, 17–18
    compressor, 13
    critical flux, 25
    hot vapor bypass, 18–20
    LPG components, 21
    original processing scheme, 22
    submergence variation, 14–17
    vapor line restrictions, 20 21
Pressure relieving device, 295
Pressure swing adsorption (PSA), 212, 219–220,
    282
Primary alkanol amines, 232
Product quality improvement units, 3
Propane deasphalting unit (PDA), 3, 4, 327
Propylene carbonate (PC), 225–226
Proton exchange membrane, 344
PRT, *see* Power recovery train
PSA, *see* Pressure swing adsorption
PSD, *see* Particle size distribution
Pyrolysis, 338

R

RAFD, *see* Regenerator amine flash drums
Rare earth, 123

RCO, *see* Reduced crude oil
RCSP, *see* Regenerated catalyst standpipe
RCSV, *see* Regenerated catalyst slide value
REAC, *see* Reactor effluent air cooler
Reactor effluent air cooler (REAC), 175, 205, 318
Reactor inlet temperature (RIT), 181, 188, 189,
    191, 207
Reactor outlet temperature (ROT), 80, 82, 94, 188
Reactor–regenerator heat balance
    catalyst cooler, 84–88
    heat balance, 81
    regenerator heat removal, 82–84
    R-R section, 80–81
Reactor termination device (RTD), 94
Reboiler duty (Q_R), 7, 28, 68, 210, 232, 256, 257,
    261, 298, 303, 304
Reboilers, 22, 25–28
Recirculation-type reboilers, 23–24
Rectisol process, 222–223
Recycle gas compressor (RGC), 175, 181, 211
Recycle gas (RG) purity, 211–213
Reduced crude oil (RCO), 75, 133, 310
Refinery
    air emissions, 329–330
    changes implemented, 326–327
    coke and heavy oil gasification, 4
    definition, 1
    demand
        actions taken, 331–332
        petrochemicals, 332–333
    diesel, 326
    downstream integration, 5
    emission guidelines, 328–329
    fractions, 4
    fuel-oriented, 1
    gasoline, 323–326
    improvement in facilities, 331
    liquid effluent, 330
    lower crude, 321–322
    lube base oil, 327
    modifications, 321–322
    petrochemical integration, 5
    petrochemical units, 5
    process units, 1–3
    product differential, 321–322
    production of aromatics, 5
    revised specifications, 322
    specifications, 324–326
Reflux drum arrangement, 41
Reformulated gasoline, 3
Regenerated catalyst slide value (RCSV), 75, 80,
    91, 94
Regenerated catalyst standpipe (RCSP), 75, 80,
    119, 122
Regenerator amine flash drums (RAFD), 245,
    253–255
Regenerators, 113–115
Reid vapor pressure (RVP), 25

Relief loads, 295–296
accumulation of pressure, 315–316
overpressure, 315
water entry to hot oil, 315
computation, 298–299
crude distillation unit
pumparound failure, 306, 308
reflux failure, 306
site power failure, 308–310
vapor–liquid profile, 307–308
side strippers, 310
steady-state simulation method, 297–298, 302
feed failure, 304–305
reflux failure, 303
site power failure, 303–304
UBH method, 296–297
loss of feed, 300–301
loss of reflux, 299–300
SWPF, 301
vacuum column, 310–312
cooling water failure, 312
ejector steam failure, 311–312
fuel-type vacuum tower, 311
pumparound failure, 312
site power failure, 313–314
Relief valve, 295
Renewable energy, 335
Research octane number (RON), 3, 323
Resins, 147
RGC, *see* Recycle gas compressor
Riser terminating devices (RTD), 110
RIT, *see* Reactor inlet temperature
RON, *see* Research octane number
Room-temperature ionic liquids (RTILs), 239
ROT, *see* Reactor outlet temperature
RTD, *see* Riser terminating devices
RTILs, *see* Room-temperature ionic liquids
Rupture disc, 295
RVP, *see* Reid vapor pressure
Ryan/Holmes Process, 219

S

Safety relief valve, 295
Safety valves, 295
columns, 318
high-pressure systems, 318
refinery columns, 318–319
Saturates, 147
SBR, *see* Styrene butadiene rubbers
SCO, *see* Synthetic crude oil
SCOT, *see* Shell Claus off-gas treatment
SCSP, *see* Spent catalyst standpipe
SCSV, *see* Spent catalyst slide valve
SDA, *see* Solvent dewaxing unit
SDU, *see* Solvent dewaxing unit

Secondary alkanol amines, 232–233
Selective hydrogenation unit (SHU), 324
Shell Claus off-gas treatment (SCOT), 288
Shot coke, 166
Single-reflux drums, 41
Site-wide power failure (SWPF), 301
Skeletal density (SD), 100
Skimming, 252
Solar energy, 335
Solid oxide electrolyzers, 344
Solids flux, 115
Solvent dewaxing unit (SDU), 4, 327
Solvent regeneration, 235
SOR, *see* Start of run
Souders–Brown equation, 9
Sour gas, 255
Sour water strippers, 3, 27–28
Spent catalyst slide valve (SCSV), 75, 81, 94
Spent catalyst standpipe (SCSP), 75, 122
Spiral wound membrane, 240–241
Splash baffle, 10, 12
Split flow variant, 236
Sponge coke, 168
Spray entrainment flooding, 9
SRU, *see* Sulfur recovery unit
Stable expansion ratio, 119
Start of run (SOR), 207
Steady-state simulation method, 297–298
Steaming/wax tailing, 143–144
Steam reformer unit
block diagram, 267
capacity augmentation, 282
case studies, 272–282
cooled synthesis gas, 270
deactivation of catalysts
HDS catalyst, 272–273
pre-reformer catalyst, 273
reformer catalyst, 273–274
shift catalyst, 274–275
ZNO catalyst, 273
horizontal convection section, 268
HT and LT shift reactors, 269
hydrogen production, 266
hydrogen recovery, 282
pre-reforming, 266, 268
problems
chloride slippage, 272
overheating of tubes, 272
quality of steam, 272
steam-to-carbon ratio, 271
sulfur slippage, 270–271
tubular reformer, 271
PSA, 269
sections, 265
shift reaction, 270
Steam reheaters, 287

Sterically hindered amines, 233–234
Styrene butadiene rubbers (SBR), 5
Sulfinol-D process, 221
Sulfur recovery unit (SRU), 250
Sulfur recovery units (SRU)
    capacity augmentation, 292
    Claus catalytic stage, 286
    Claus sections, 284
    Claus thermal stage, 285–286
    enhance, 288–291
    flow scheme, 285
    problems, 291–292
    reaction mechanism, 283–285
    reheat methods, 287–288
    sections, 284
Sulfur units, 3
Superficial velocity, 101
Suspended solids (SS), 272
Sweet gas, 255
SWPF, *see* Site-wide power failure
Syngas, 338
Synthetic crude oil (SCO), 30

T

Tail gas cleanup unit (TGCU), 288
Tail gas treatment (TGT), 283, 288
Tail gas treatment unit (TGTU), 259, 288
TAN, *see* Total acid number
TDS, *see* Total dissolved solids
TEA, *see* Triethanolamine
Tertiary amines, 233
Tertiary stage separator (TSS), 91
TGTU, *see* Tail gas treatment unit
Thermal degradation, 109–110
Thermochemical conversion, 338
Thermosiphon reboiler, 25
Three-stage membrane process, 243
Total acid number (TAN), 31, 321
Total dissolved solids (TDS), 272
Tramp amines, 59–62
Transesterification, 340
Tray support ring (TSR), 11
Triethanolamine (TEA), 233
TSS, *see* Tertiary stage separator
Two-pass tray, 10
Two reflux drums, 41–42
Two-stage membrane process, 242

U

UCO, *see* Upgraded crude oil
ULSD, *see* Ultra-low sulfur specification of diesel

Ultra-low sulfur specification of diesel (ULSD),
    206–207
Unbaffled arrangement, 23
Under deposit corrosion, 53
UOP process, 234, 236
Upgraded crude oil (UCO), 29–31
Upstream integration, 5

V

Vacuum distillation unit (VDU), 31, 39, 42, 52,
    326
Vacuum furnace transfer line pressure drop,
    48–49
Vacuum gas oil (VGO), 30, 75, 182, 186, 194, 322
Vacuum residue (VR), 4, 55, 72, 133, 198–199, 327
Vanadium traps, 124–125
Vapor collapse, 16
Vapor hood, 10
VBU, *see* Bitumen and Visbreaker unit
VCM, *see* Volatile combustible matter
VDU, *see* Vacuum distillation unit
Vent stack (API 521), 295
Very high-pressure system, 316–317
Very low-pressure system, 316–317
VGO, *see* Vacuum gas oil
Volatile combustible matter (VCM), 142, 150
Vortex separation system (VSS), 110, 111
VR, *see* Vacuum residue
VSS, *see* Vortex separation system

W

WABT, *see* Weighted average bed temperature
Wash oil circulation, 43
Waterfall pool effect, 28
Weighted average bed temperature (WABT), 51,
    184, 188
Wet flue gas analysis, 83
Wet gas compressor (WGC), 143, 214
Wet reflux drum, 41
WGC, *see* Wet gas compressor
Wind energy, 335
Wood wastes, 337

Y

Y-type zeolites, 123

Z

Zeolites, 123–124
ZSM-5, 125

For Product Safety Concerns and Information please contact our EU
representative GPSR@taylorandfrancis.com
Taylor & Francis Verlag GmbH, Kaufingerstraße 24, 80331 München, Germany